T0133222

Verification of Branch and Bound Algorithms applied to Water Distribution Network Design

---

Vom Promotionsausschuss der
Technischen Universität Hamburg-Harburg
zur Erlangung des akademischen Grades
Doktorin der Naturwissenschaften
genehmigte Dissertation

von
Angelika Christina Hailer
aus Hamburg, 2006

Bibliografische Information Der Deutschen Bibliothek

Die Deutsche Bibliothek verzeichnet diese Publikation in der Deutschen Nationalbibliografie; detaillierte bibliografische Daten sind im Internet über http://dnb.ddb.de abrufbar.

©Copyright Logos Verlag Berlin 2006

Alle Rechte vorbehalten.

ISBN 3-8325-1316-7

Logos Verlag Berlin
Comeniushof, Gubener Str. 47,
10243 Berlin
Tel.: +49 030 42 85 10 90
Fax: +49 030 42 85 10 92
INTERNET: http://www.logos-verlag.de

1. Gutachter: Prof. Dr. rer. nat. Siegfried M. Rump
2. Gutachter: Prof. Dr.-Ing. Knut Wichmann
Tag der mündlichen Prüfung: 6.7.2006

To Manfred

Acknowledgments

First of all I would like to express my sincere gratitude to my academic advisor Professor Dr. Siegfried M. Rump for the freedom he gave me to accomplish this thesis in an excellent environment during my work at the Institute for Reliable Computing at Hamburg University of Technology.

In particular, I would like to thank Professor Dr.-Ing. Knut Wichmann for his willingness to act as co-advisor, for introducing me to the engineering point of view while asking important questions concerning the practical application of the calculations, and I would like to thank PD Dr. Christian Jansson, who initiated this thesis with the idea to verify the algorithm of Sherali, Subramanian and Loganathan, for valuable comments.

Finally, special thanks go to my family, to all colleagues and friends who encouraged me by their interest and personal commitment to complete this thesis.

Contents

List of Figures

List of Tables

Notation

Theorems and lemmata will be proven explicitly insofar as they have a major impact on this thesis, in other cases an appropriate reference will be provided. In the interest of readability the abbreviation "iff" is used instead of the term "if and only if".

In general, vectors are regarded as column vectors. The term $x = (x_1 , x_2)^\top \in I\!R^{m+n}$ denotes the compound vector of the vectors $x_1 \in I\!R^m$ and $x_2 \in I\!R^n$. To be correct, one had to write $x = (x_1^\top , x_2^\top)^\top$ but for better readability the "inner" transposition symbols are omitted throughout the text. Comparisons for vectors are to be understood by the natural componentwise partial order, which implies for $x \in I\!R^n$ that $x \geq 0$ iff $x_i \geq 0$ for all $i \in \{1, \ldots , n\}$.

The following notation is chosen for intervals: If a real vector is expressed in lower case, for example x , then according to the context, the degenerate interval $[x, x]$ or a generic interval is denoted with squared brackets $[x]$, analogously $[A]$ denotes an interval matrix. An inclusion isotonic interval enclosure of the real valued function $f : I\!R^n \to I\!R$ is denoted by the superscript "I", i.e. $f(x) \in f(x)^I$.

The gradient of a differentiable function $f : I\!R^n \to I\!R$ is considered to be a row vector. For a differentiable function $f = (f_1, \ldots , f_m)^\top : I\!R^n \to I\!R^m$, for $i \in \{1, \ldots , m\}$ the i-th row of the Jacobian matrix J_f is the gradient of f_i .

If not otherwise specified, numerical results are rounded to nearest, except that computational times lower than one second are always rounded upwards. Intervals as result of verified calculations are naturally rounded outwards.

In the bibliography, the page numbers where a reference is cited are provided at the end of each item. The index includes references to the extracts of source code part of the appendix, but does not contain occurences in parameter lists of functions.

Additional to standard notation the following symbols are used:

$I\!N$	Set of natural numbers
$I\!R$	Set of real numbers
$I\!R_+$	Set of non-negative real numbers, i.e. $\{x \in I\!R \mid x \geq 0\}$
$I\!I\!R$	Set of real intervals

$[x^L, x^U]$	Interval with bounds x^L and x^U, analogously denoted by $[\underline{x}, \overline{x}]$ or $[x]$
$A^\top$	Transpose of matrix A
$0^{n\times m}$	n-by-m-matrix of zeros
$\operatorname{diag}\langle v\rangle$	Diagonal matrix, where the vector v is on the main diagonal and the other elements are zero
$[x]^+$	Positive part of the interval vector $[x]$, i.e. $[x]^+ = \{x \in [x] \mid x \geq 0\}$
$[x]^-$	Negative part of $[x]$, i.e. $[x]^- = \{-x \mid x \in [x] \text{ and } -x \geq 0\}$
δ_{ij}	Kronecker delta, $\delta_{ij} = \begin{cases} 1, & \text{if } i = j \\ 0, & \text{if } i \neq j \end{cases}$
$\operatorname{sgn}(x)$	Signumfunction, i.e. $\operatorname{sgn}(x) = \begin{cases} x/\lvert x\rvert & \text{if } x \neq 0 \\ 0 & \text{else} \end{cases}$
$\ln(x)$	Natural logarithm of x, i.e. logarithm to base e
$\log(x)$	Logarithm of x to base 10
$J_f(x)$	Jacobian matrix of $f : I\!R^n \to I\!R^m$, $J_f(x) = \left(\frac{\partial f_i}{\partial x_j}(x)\right)_{\substack{i=1,\dots,m \\ j=1,\dots,n}}$
$\rho(A)$	Spectral radius of a real square matrix $A \in I\!R^{n\times n}$, i.e. $\rho(A) = \max\limits_{i=1,\dots,n} \lvert\lambda_i\rvert$, where λ_i, $i = 1, \dots, n$ are the eigenvalues of A
$\overset{\circ}{M}$	Interior of a set M
$\partial(M)$	Boundary of a set M
$\lceil x\rceil$	Smallest integer greater than or equal to x
$x \underline{\cup} y$	Convex hull of x and y, for $x, y \in I\!R^n$
w.l.o.g.	Without loss of generality
$\square$	End of proof

The optimization problems used are abbreviated as follows:

NOP	Nonlinear optimization problem (Section 2.2.1, page 39)
LNOP	Linear lower bounding problem for NOP (Section 2.3.1, page 51)
DNOP	NOP with Darcy-Weisbach friction loss calculation (Section 3.3, page 86)

The following abbreviations are used for the head-loss formulae:

HW	Hazen-Williams (Equation 2.2.1)
DW	Darcy-Weisbach (Equation 3.1.1)
AHW	adjusted Hazen-Williams (Section 3.6, Equations 3.6.1 and 3.6.2)

Zusammenfassung

"The purpose of computing is insight, not numbers."[1] [H62]

Das Design von Verteilungsnetzen ist ein wesentlicher Bestandteil des Planungsprozesses von Wasserversorgungssystemen. Modelle, die diesen Designprozess abbilden, führen auf das Problem, Rohrdurchmesser für ein gegebenes Netzlayout auszuwählen. Unter der Annahme, daß der Verbrauch für einen bestimmten Zeitpunkt geschätzt ist und die Kosten für unterschiedliche Durchmesser bekannt sind, kann dieses stationäre Problem als globales nichtlineares Optimierungsproblem formuliert werden.

Dieses Problem hat eine ähnliche Struktur wie das bekannte Transportproblem. Die Vorgabe von Grenzen für den Druck sowie die Berechnung des Druckverlustes aufgrund von Reibung in den Rohren führt zu nichtlinearen Nebenbedingungen.

Sowohl heuristische als auch deterministische Algorithmen, die auf Basis von Gleitpunkt-Arithmetik implementiert sind, wurden in den letzten 30 Jahren zur Lösung dieses Problems entwickelt. Die deterministischen Algorithmen garantieren theoretisch, daß das globale Optimum gefunden wird, allerdings liegt bisher keine Fehlerabschätzung für diese Berechnungen vor. In einem Branch und Bound-Algorithmus ist die Entscheidung in jedem Zweig von numerischen Resultaten abhängig. Daher können selbst kleine Fehler weitreichende Konsequenzen haben.

Wendet man das Motto "The purpose of computing is insight, not numbers"[1] von Hammings Buch [H62] auf diese Situation an, wird ein Ingenieur Einsichten aus numerischen Rechnungen gewinnen. Solange er sich auf die berechneten Resultate nicht verlassen kann, sollte er sich der numerischen Fehler bewußt sein, mit denen die Resultate behaftet sind. Sobald ihm zuverlässige Resultate oder zumindest Resultate zusammen mit einer Abschätzung für den maximal möglichen Fehler vorliegen, kann er Einsichten gewinnen, ohne sich Gedanken um die zugrundeliegenden Algorithmen zu machen.

[1] "Zweck numerischen Rechnens sind Erkenntnisse, nicht Zahlen."

Im Rahmen dieser Dissertation ist das Softwarepaket WaTerInt entwickelt worden, das den ersten verifizierten Algorithmus für das Wasserversorgungsnetz-Optimierungsproblem enthält. Mit Hilfe von Intervallarithmetik ist die Fehleranalyse Teil der Berechnung und als Ergebnis werden untere und obere Schranken für die optimale Lösung ermittelt.

Die in WaTerInt implementierten Algorithmen beruhen auf einer Kombination des von Sherali, Subramanian und Loganathan entwickelten Branch- und Bound-Algorithmus mit verifizierten unteren Schranken für lineare Optimierungsprobleme, die vor kurzem von Neumaier, Shcherbina [N&S04] und Jansson [J04] entwickelt wurden. Um die Rechenzeiten für Erweiterungsnetzwerke zu verringern, wird eine weitere neue Constraint Propagation vorgestellt.

Desweiteren wird die hydraulische Berechnung des Druckverlustes näher betrachtet. Das Netzwerkmodell, das für die Optimierung des Designs von Verteilungsnetzen verwendet wird, wurde zuerst für die Konstruktion des dritten Wassertunnels des Rohrnetzes von New York City entwickelt und beruht daher auf der Hazen-Williams Formel. Diese Formel wird insbesondere in Nordamerika verwendet, während in Europa der Gebrauch der theoretisch korrekteren Darcy-Weisbach Formel üblich ist. Das Optimierungsmodell wird dahingehend umformuliert, daß es die Darcy-Weisbach und die Colebrook-White Formeln enthält.

Die implizite Definition der Darcy-Weisbach Reibungszahl in der Colebrook-White Gleichung erhöht die nichtlineare Struktur des Optimierungsproblems. Diese Struktur wird im Detail untersucht und gleichzeitig wird die Adjazenzeigenschaft, die von Fujiwara und Dey für das Hazen-Williams Problem gezeigt wurde, für dieses neue Problem bewiesen. Ein Algorithmus zur Bestimmung der global minimalen Kosten wird analog zu dem von Sherali, Subramanian und Loganathan vorgeschlagenen entwickelt und wiederum wird eine verifizierte Version bereitgestellt.

Um die Vorteile des Darcy-Weisbach Optimierungsproblems mit den deutlich geringeren Rechenzeiten des Hazen-Williams Problems zu verbinden, wird eine Anpassung der Hazen-Williams Formel vorgeschlagen und letztendlich die berechneten Resultate für unterschiedliche hydraulische Parameter verglichen.

Kurz zusammengefaßt, werden vor kurzem entwickelte Verifikationstechniken mit Branch- und Bound-Algorithmen kombiniert, um zu untersuchen, wie praktikabel es ist, eine Abschätzung des maximalen Fehlers zusammen mit der Berechnung einer optimalen Lösung eines nichtlinearen Optimierungsproblems zu bestimmen. Diese Verifikationstechniken werden auf das Design von Wasserverteilungsnetzen angewandt, sowohl auf den bekannten Branch- und Bound- Algorithmus von Sherali, Subramanian und Loganathan als auch auf den Algorithmus für das realitätsnähere Netzwerkmodell, das die Darcy-Weisbach Formel beinhaltet.

Als Ergebnis kann festgehalten werden, daß es möglich ist, mit derzeit bekannten Techniken der Intervallarithmetik die Berechnung der optimalen Lösung des nichtlinearen Wasserverteilungsnetz-Optimierungsproblems zu verifizieren.

Bei den unverifizierten Berechnungen auf Basis von Gleitpunktarithmetik treten numerische Artefakte auf. Es ist beispielsweise möglich, das ein leerer zulässiger Bereich identifiziert wird, obwohl eine optimale Lösung existiert, was anhand des "two-loop expansion" Netzwerks auf einem HP Superdome Server gezeigt wird. Des weiteren existieren Beispiele, bei denen die untere Schranke größer als die obere ist, sowohl innerhalb derselben Rechnung, als auch bei Rechnungen auf unterschiedlichen Rechnern oder bei Verwendung verschiedener linearer Löser.

Bei verifizierten Rechnungen werden diese Artefakte sämtlich vermieden und die so erhaltenen Resultate sind stets zuverlässig. Gleichwohl beträgt der zusätzlich benötigte Rechnenaufwand bei der momentanen Implementierung für die Verifikation durchschnittlich etwa das fünfzehnfache des Aufwandes für reine Gleitpunktrechnung. Dieser Faktor wäre vermutlich deutlich geringer, wenn der derzeit benötigte Aufwand aufgrund des Interprationsoverhead von MATLAB, insbesondere bei Objektorientierung, verhindert werden könnte.

Im Wesentlichen werden hierbei zwei Resultate der Intervallanalysis verwendet. Nach außen gerundete Intervallarithmetik ermöglicht einerseits, verifizierte Schranken für die Lösung eines nichtlinearen Gleichungssystems zu bestimmen, sowie Existenz und Eindeutigkeit der Lösung innerhalb dieser Schranken zu garantieren. Andererseits können optimale Lösungen, die von linearen Gleitpunktlösern berechnet sind, im nachhinein verifiziert werden. Diese a posteriori Verifikation der unteren Schranken eines linearen Optimierungsproblems ist zentraler Bestandteil der Verifikation einer globalen optimalen Lösung eines nichtlinearen Problems mit Hilfe eines Branch- und Bound-Algorithmus.

Ein großer Vorteil der Intervallanalysis besteht in der Möglichkeit, Fehlerschranken zusammen mit der Lösung des numerischen Problems zu erhalten. Hierfür ist es notwendig, mehrere Prozeduren anzupassen, beziehungsweise zu erweitern. Um eine verifizierte obere Schranke zu bestimmen muß garantiert werden, daß der Lösungsvektor zulässig ist. Dieser Teil des Verifikationsprozesses hängt wesentlich von der Güte der approximativen Lösung und somit von dem linearen Löser ab, der während des Branch und Bound Algorithmus verwendet wird.

Gleichzeitig erfordert diese Verifikation der Zulässigkeit derzeit eine gezielte Auseinandersetzung mit der Stuktur des Problems und bleibt somit weiterhin eine größere Herausforderung. Der im Rahmen dieser Dissertation entwickelte Algorithmus verwendet spezielle Eigenschaften der Funktionen, die die Nebenbedingungen beschreiben. Weitergehende Arbeiten könnten sich somit mit allgemeinen Methoden zur Verifizierung der Zulässigkeit, insbesondere mit der Einschließung von Lösungen eines nichtlinearen Systems auf dem Rand eines Intervalls, beschäftigen.

Die Stuktur der Nebenbedingung wurde verwendet, um die zugrundeliegenden Gleitpunktalgorithmen zu verbessern. Wesentlich ist die Constraint Propagation aufgrund der Druckgrenzen für Erweiterungsnetze, mit der die Rechenzeit in etwa um ein drittel bis zur Hälfte reduziert werden kann.

Eine weitere Verbesserung der Gleitpunktalgorithmen mit Hinblick auf die verifizierte Version ist möglich, beispielsweise könnten mit Gleitpunktalgorithmen bestimmte Lösungen leicht angepaßt werden, so daß sie die Adjazenzeigenschaft erfüllen. Dies zeigt, daß durch die Anwendung von Intervalltechniken auf einen existierenden Gleitpunktalgorithmus Teile identifiziert werden können, die sich numerisch verbessern lassen.

Zusätzlich zu dem bis jetzt angesprochenem Problem anhand der Hazen-Williams Formel wurde der Algorithmus um die Möglichkeit, das Darcy-Weisbach Optimierungsproblem zu lösen, erweitert. Hierfür ist eine detaillierte Betrachtung der Nichtlinearität der Nebenbedingungen notwendig, um bestimmte Monotonie- und Konvexitätseigenschaften weiterhin garantieren zu können und die Adjazenzeigenschaft wurde auch für dieses neue Problem bewiesen. Die Rechenzeit ist jedoch etwa vierzig mal höher als für das Hazen-Williams Problem.

Es ist möglich, die Koeffizienten der Hazen-Williams Formel dahingehend anzupassen, daß eine genauere Approximation der Darcy-Weisbach Formel erreicht wird. Die hierbei erzielten Resultate zeigen eine höhere Genauigkeit, wobei die Rechenzeiten in etwa mit denen des ursprünglich verwendeten Hazen-Williams Problems übereinstimmen.

Schließlich werden verifizierte Resultate zum Vergeich der Lösungen bei unterschiedlichen hydraulischen Parametern verwendet. Der Einfluß des Rauhigkeitswertes kann nicht vernachlässigt werden, und bei Betrachtung der gesamten Lebensdauer eines Rohrnetzes sind die Energiekosten ein wesentlicher Teil der Gesamtkosten, es sei denn das Wasser wird hauptsächlich durch Schwerkraft geliefert.

Die zugrundeliegenden Optimierungsprobleme stellen insofern eine Herausforderung dar, als die Nebenbedingungen höchst nichtlinear sind. Das größte der gerechneten Testbeispiele hat 1.478 Variablen. Mit der entwickelten Software WaTerInt werden etwa 10 Stunden auf einem HP Superdome Server für die Bestimmung rigoroser Schranken für das Optimum des Hazen-Williams Problems benötigt.

Für das New York Standardproblem, das 554 Entscheidungsvariablen, 58 lineare und 33 nichtlineare Gleichheitsnebenbedingungen enthält, werden auf einem Notebook etwa 4 Stunden für die Bestimmung einer optimalen Lösung mit Hazen-Williams und ungefähr 54 Stunden für das Darcy-Weisbach Problem benötigt.

In Anlehnung an Schaake und Lai [S&L69] könnte das vorangestellte Zitat von Hamming auf mathematische Programmierung angewendet werden, indem sie als einen entscheidenden Vorzug ihres Modells ein verbessertes Verständnis des Designs von Wasserverteilungssystemen aufzeigen, das sich aus diesen Einsichten ergibt.

Mit WaTerInt wird ein Softwarepaket zur Verfügung gestellt, das ausgehend von einer modularen Struktur leicht erweitert werden kann. Auf der einen Seite können somit Ingenieure im Designprozess von Wasserversorgungssystemen unterstützt werden und auf der anderen kann es als Basis für weitere Forschung im Bereich der Verifikation von Optimierungsalgorithmen in Verbindung mit INTLAB dienen.

Introduction and Summary

"The purpose of computing is insight, not numbers." [H62]

The design of distribution networks is an integral factor in the planning process of water supply systems. Models concerning this design process contain the problem of selecting pipe diameters for a given layout of the network. Assuming the demand to be estimated for a certain point of time and the costs for pipes of different diameters to be known, this stationary problem can be formulated as a nonlinear global optimization problem.

This problem has a similar structure to the well-known transportation problem. Restrictions to the pressure head and the calculations of head-loss due to friction in the pipes imply nonlinear constraints.

Heuristic as well as deterministic algorithms have been developed during the last 30 years to solve this problem. Their implementation is based on floating point arithmetic. Theoretically, the deterministic algorithms guarantee to obtain the global best solution, but so far no estimate for the arithmetic error has been presented. In a branch and bound algorithm, the decision for every branch depends on numerical results. Therefore, even small errors may have far-reaching consequences.

Transforming the motto of Hamming's book "The purpose of computing is insight, not numbers" to this situation, an engineer is going to use insight he gets from calculations for the design of water networks. As long as he cannot rely on the results of his calculations, he should be aware about numerical errors that might impact his results. When he gets reliable results, or at least results together with an estimate for the maximum possible error, he can get insight without having to care about the underlying algorithms.

In the context of this thesis, the software package WaTerInt has been developed, which contains the first verified algorithm of the water distribution design optimization problem. Based on interval arithmetic, the error analysis is part of the calculations and as result lower and upper bounds for the optimal solution are provided.

The algorithms implemented in WaTerInt are based on the branch and bound algorithm developed by Sherali, Subramanian and Loganathan in combination with verified lower bounds for linear programming, which have recently been introduced by Neumaier, Shcherbina [N&S04] and Jansson [J04]. To reduce the computational times for expansion networks, a new additional constraint propagation technique is introduced.

Further on, the hydraulic calculation of friction loss is investigated. The network optimization model used in water distribution network design was first developed for the construction of the third water tunnel of the New York City network. This model is based on the Hazen-Williams formula, which is customarily used in North America, whereas in Europe the theoretically more correct Darcy-Weisbach formula is usually applied. For this, the optimization model is reformulated containing Darcy-Weisbach and Colebrook-White equations.

The implicit definition of the Darcy-Weisbach friction factor in the Colebrook-White equation increases the nonlinear structure of the optimization problem. This structure is investigated in detail and the adjacency property, which was shown by Fujiwara and Dey for the Hazen-William problem, is also proved for this new problem. An algorithm for finding the global minimum costs is developed analogous to the one suggested by Sherali, Subramanian and Loganathan and again a verified version is provided.

To combine the advantages of the Darcy-Weisbach optimization problem with the significantly lower computational times of the Hazen-Williams problem, an adjustment of the Hazen-Williams formula is introduced. Finally, computational results for different hydraulic parameters are compared.

In summary, recently developed verification techniques are combined with a branch and bound algorithm to investigate the practicability of obtaining error estimates along with the calculation of the optimal solution of a nonlinear optimization problem. They are applied to water network distribution design, both to the known branch and bound algorithm of Sherali, Subramanian and Loganathan and to the one for the more realistic network model based on the Darcy-Weisbach equation.

This thesis is organized as follows: *Chapter 1* contains basic definitions in the field of interval analysis and optimization theory as far as needed for the algorithms developed. In particular, bounds for linear programming are used as one of the main factors in the verification process and are part of Lemma 1.2.3.

In *Chapter 2*, the water distribution network optimization model is introduced. The engineering context is explained and suggestions for expansions of the model are discussed. An overview of known algorithms for this problem is provided, where the focus is on the branch and bound algorithm developed by Sherali, Subramanian and Loganathan. This algorithm is explained in detail and a proof is provided for one of the main procedures, the maximum spanning tree based reduction (MSTR). The new constraint propagation technique based on head bounds is introduced for expansion networks. Properties of the optimization problem are discussed and an equivalent formulation is developed to demonstrate, on the basis of the New York Network, that in general one cannot expect to find a solution in the relative interior with respect to the head bounds.

To improve the optimization problem, in *Chapter 3* the calculation of head-loss with the Hazen-Williams formula is substituted with the theoretically more correct Darcy-Weisbach friction loss equation. The differences between the formulae are opposed and the nonlinearity of the Colebrook-White equation is investigated in detail. The algorithm developed for the Darcy-Weisbach problem is described and finally the adjusted Hazen-Williams problem is introduced.

Techniques for verifying the solutions obtained are presented in *Chapter 4* and two aspects concerning the verification of nonlinear optimization are covered: First the changes to be made to a floating-point algorithm are explained and second the additional time needed for verification of a branch and bound algorithm is investigated on the optimization problems presented.

Chapter 5 contains details of the software package WaTerInt, developed as computational basis for the results presented. Standard test cases are one small two-loop network and the distribution networks of New York and Hanoi as well as one small expansion network derived from the solution of the two-loop network. These test cases are introduced with a description of their layout and the computational results are discussed.

Finally, *Appendix A* contains the input data of the test networks and in *Appendix B*, extracts of the source code of WaTerInt containing the main procedures are provided. Details for the evaluation of computational time are enclosed in Appendix C. Appendix D contains optimal solutions of the test networks calculated with WaTerInt for the standard optimization problem, the problem with Darcy-Weisbach friction loss formula and the one with adjusted Hazen-Williams formula, as well as comparisons of the results. Illustrations of the difference of the head-loss formulae can be found in *Appendix E* and details needed for the proof of the adjacency property for the Darcy-Weisbach optimization problem are part of *Appendix F*.

Chapter 1

Interval Analysis and Global Optimization

"Numerical errors are rare, rare enough not to care about them all the time, but yet not rare enough to ignore them." – Vel Kahan [R99]

This chapter covers definitions and theorems used throughout this thesis. Basic concepts of interval arithmetic and optimization theory are introduced, particularly with regard to their application in the algorithms developed for water distribution network design. An essential aspect are the bounds for linear optimization problems, as these form the basis for the verification of the branch and bound algorithm.

Inclusions of the zeros of one-dimensional continuous monotone functions are needed while determining verified relaxations. For this, a procedure following the bisection developed by Vrahatis is introduced.

The procedure `verifynlss` *of the software package INTLAB [R03] forms a central part of the verification of feasibility in Chapter 4. Therefore, a detailed description of the theoretical background of this procedure is provided.*

1.1 Interval Analysis

1.1.1 Introduction and Definitions

After earlier publications e.g. by Dwyer and Sunaga, interval arithmetic has gained widespread attention in 1962 ([A&M00], [H02], [H&W04]) with the thesis of Moore, where he summarizes the objective as follows:

> "A complete a priori error analysis for an extensive numerical computation can become a formidable task if it is to answer in advance and in every detail all questions of the accuracy of approximations to be made during the computation and their effect on the accuracy of the final result. [...]
>
> In order to take full advantage of the great speed of the automatic stored-program digital computer it is obviously desirable to mechanize as much as possible to the error analysis required for a computation so that it can be carried out by the machine itself. [M62]"

This fact, to obtain bounds for the error along with the calculation when using intervals instead of simple floating point numbers, provides the possibility to get error bounds for the optimal solution of the water distribution network introduced in Chapter 2.

The following definitions are closely related to the book of Hansen and Walster [H&W04] as well as to papers on interval analysis, especially to the survey article of Alefeld and Mayer [A&M00], the thesis of O. Knüppel [K95], a number of publications of S. M. Rump ([R83], [R99], [R01]) and introductions to INTLAB ([R98], [H02]), a MATLAB toolbox for self-validating algorithms.

In the following, let n, $m \in \mathbb{N}$. It is common to use only closed intervals in the field of interval analysis, while elsewhere the term interval includes open and half-open sets as well.

1.1.1 Definition. An *n-dimensional interval* or an *n-dimensional interval quantity* $[x]$ is a closed bounded set of the form

$$[x] = [\underline{x}, \overline{x}] = \{x = (x_1, \dots, x_n)^\top \in \mathbb{R}^n \mid x_j \in [\underline{x}_j, \overline{x}_j] \text{ for } j = 1, \dots, n\},$$

where $\underline{x} = (\underline{x}_1, \dots, \underline{x}_n)^\top$, $\overline{x} = (\overline{x}_1, \dots, \overline{x}_n)^\top$ with $\underline{x}_j$, $\overline{x}_j \in \mathbb{R}$ and $\underline{x}_j \leq \overline{x}_j$ for $j = 1, \dots, n$. It is called degenerate if $\underline{x}_j = \overline{x}_j$ for at least one $j \in \{1, \dots, n\}$.

The system of all closed and possibly degenerate n-dimensional intervals $[x] \subset \mathbb{R}^n$ is denoted $\mathbb{IR}^n$.

A one-dimensional interval is mostly just referred to as interval. $[x] \in \mathbb{IR}^n$ is also called *interval vector* and accordingly a matrix $[A] \in \mathbb{IR}^{n \times m}$, where the elements are one-dimensional intervals, is called *interval matrix*. Obviously, $\mathbb{IR}^n$ and $\mathbb{IR}^{n \times 1}$ coincide.

Real vectors $x \in \mathbb{R}^n$ are naturally embedded into the space of interval vectors $\mathbb{IR}^n$ by identifying x with $[x, x]$. According to the context, the real notation x is kept instead of $[x, x]$ and bounds of an interval are denoted with the superscripts L and U as well, i.e. $[x] = [x^L, x^U]$.

1.1.2 Definition.

(i) Let $\underline{A} = \left(\underline{A}_{ij}\right)_{\substack{i=1,\dots,n\\ j=1,\dots,m}}$ and $\overline{A} = \left(\overline{A}_{ij}\right)_{\substack{i=1,\dots,n\\ j=1,\dots,m}} \in I\!R^{n\times m}$ be the bounds of the interval matrix $[A] \in I\!I\!R^{n\times m}$.

Then, the *center* or *midpoint* of an interval matrix $[A] \in I\!I\!R^{n\times m}$ is the matrix $m([A]) \in I\!R^{n\times m}$ with elements

$$A_{ij} = \frac{1}{2}(\underline{A}_{ij} + \overline{A}_{ij}) \quad \text{for all } i = 1, \dots, n \text{ and } j = 1, \dots, m\,.$$

(ii) The diameter $d([A]) \in I\!R^{n\times m}$ of an interval matrix $[A] \in I\!I\!R^{n\times m}$ is defined componentwise as

$$d([A]) = \overline{A} - \underline{A}\,.$$

1.1.3 Definition. Let $[x]\,,\ [y] \in I\!I\!R$. Then any order relation $\bowtie\, \in \{<, \leq, >, \geq\}$ is defined by

$$[x] \bowtie [y]\,, \quad \text{iff} \quad x \bowtie y \quad \text{for all } x \in [x] \text{ and for all } y \in [y]\,.$$

As two intervals need not be comparable, these relations define a partial order. These order relations are defined componentwise for interval vectors and matrices.

1.1.4 Definition. The *interval hull* $H(M)$ of a real set $M \in I\!R^n$ is the tightest interval containing M, i.e.

$$H(M) := \bigcap \{\, [x] \in I\!I\!R^n \mid M \subseteq [x]\}\,.$$

1.1.5 Definition. Let $[x]\,, [y] \in I\!I\!R$ be interval quantities. For a binary operation $\circ \in \{+, -, \cdot\,, /\}$ the *interval operation* $[x] \circ [y]$ is the interval hull of all possible real results, i.e.

$$[x] \circ [y] := \bigcap \{\, [z] \in I\!I\!R \mid x \circ y \in [z] \quad \text{for all } x \in [x]\,, y \in [y]\,\}\,,$$

where in case of division it is assumed that $0 \notin [y]$. For vectors $[x] \in I\!I\!R^n$ and matrices $[A] \in I\!I\!R^{n\times m}$ these operations are defined in analogy to the usual evaluation in $I\!R^n$ and $I\!R^{n\times m}$, respectively.

1.1.6 Remark.

(i) For $[x]\,,\ [y] \in I\!\!R$ and for any binary operation $\circ \in \{+, -, \cdot\,, /\}$ the interval operation is equivalent to the *power set operation* $[x] \circ [y]$ defined by

$$[x] \circ [y] := \{x \circ y \mid \text{ for all } x \in [x]\,, y \in [y]\,\}\,,$$

where in case of division again it has to be assumed that $0 \notin [y]$.

(ii) In the case of matrix-vector multiplication (i) generally does not hold true, i.e. for $[A] \in I\!\!R^{m \times n}$ and $[x] \in I\!\!R^n$ the set

$$\{y \mid y = Ax \text{ for some } A \in [A] \text{ and some } x \in [x]\}$$

is often not an interval and therefore a proper subset of $[A]\,[x]$.

The fundamental property of interval arithmetic, inclusion isotonicity, follows directly from Definition 1.1.5:

1.1.7 Theorem. All operations between interval quantities satisfy the property of *inclusion isotonicity,* i.e. for all binary operations $\circ \in \{+, -, \cdot\,, /\}$ and for arbitrary but fixed intervals $[w]$, $[x]$, $[y]$, $[z] \in I\!\!R$ with $[w] \subseteq [y]$ and $[x] \subseteq [z]$ it holds true that

$$[w] \circ [x] \subseteq [y] \circ [z]\,.$$

1.1.8 Lemma. Let $[x] = [\underline{x}, \overline{x}]$ and $[y] = [\underline{y}, \overline{y}] \in I\!\!R$ be interval quantities. Then the basic binary interval operations can be calculated with

$$\begin{aligned}
[x] + [y] &= [\underline{x} + \underline{y}\,,\ \overline{x} + \overline{y}] \\
[x] - [y] &= [\underline{x} - \overline{y}\,,\ \overline{x} - \underline{y}] \\
[x] \cdot [y] &= [\min\{\underline{x}\cdot\underline{y}\,,\ \underline{x}\cdot\overline{y}\,,\ \overline{x}\cdot\underline{y}\,,\ \overline{x}\cdot\overline{y}\}\,,\ \max\{\underline{x}\cdot\underline{y}\,,\ \underline{x}\cdot\overline{y}\,,\ \overline{x}\cdot\underline{y}\,,\ \overline{x}\cdot\overline{y}\}] \\
[x] \,/\, [y] &= [\underline{x}\,,\ \overline{x}] \cdot [\frac{1}{\overline{y}}\,,\ \frac{1}{\underline{y}}]\,, \qquad \text{if} \quad 0 \notin [y]\,.
\end{aligned}$$

1.1.9 Remarks.

(i) Let $[x] \in I\!\!R$ be an interval. Then Lemma 1.1.8 implies

$$[x] - [x] = \{x_1 - x_2 \mid x_1, x_2 \in [x]\}$$

instead of the probably expected result $\{x - x \mid x \in [x]\} = 0$. This effect causes an increased diameter of the interval obtained, and according to its origin is called *dependence problem.* More illustratively, it is said that intervals "have no memory" [R01].

(ii) If every interval variable occurs only once in a given expression, then the dependence problem does not arise and exact interval evaluation yields the exact range of the expression.

1.1.10 Lemma.

(i) For addition and multiplication, the associative and commutative laws hold true, but the distributive law does not hold in general and has to be substituted with the *subdistributivity law:* For $[x]\,,\ [y]\,,\ [z] \in I\!I\!R$ it is

$$[x]\,([y] + [z]) \subseteq [x]\,[y] + [x]\,[z]\ .$$

(ii) The structures $(I\!I\!R\,,\ +)$ and $(I\!I\!R\,,\ \cdot\ )$ are commutative semigroups with neutral elements 0 and 1, respectively. As a nondegenerate interval $[x] \in I\!I\!R$ has no inverse with respect to addition or multiplication, $(I\!I\!R\,,\ +\,,\ \cdot\ )$ is not a ring.

Proof. For a proof see [A&M00].

The definition of positive and negative part according to Collatz and Wetterling ([C&W71], p. 42) is expanded to intervals:

1.1.11 Definition. Let $[x] \in I\!I\!R^n$ be an interval quantity. Then *the positive and the negative part of* $[x]$ are defined as

$$\begin{aligned} [x]^+ &:= \{x \in [x] \mid x \geq 0\} \quad \text{and} \\ [x]^- &:= \{-x \mid x \in [x] \text{ and } -x \geq 0\}\ . \end{aligned}$$

1.1.12 Lemma. For all $[x] \in I\!I\!R^n$, Definition 1.1.11 implies

$$[x] = [x]^+ - [x]^-\ .$$

1.1.13 Definition. Let $f : I\!R^n \to I\!R$ be a real function.

(i) Then the interval function $f^E : I\!I\!R^n \to I\!I\!R$ is said to be an *interval extension* of f iff

$$f(x) = f(x)^E \quad \text{for all } x \in I\!R^n$$

and

$$\{f(x) \mid x \in [x]\} \subseteq f([x])^E \text{ for all } [x] \in I\!I\!R^n\ .$$

(ii) The interval function $f^I : I\!I\!R^n \to I\!I\!R$ is said to be an *interval enclosure* of f iff

$$\{f(x) \mid x \in [x]\} \subseteq f([x])^I \text{ for all } [x] \in I\!I\!R^n\ .$$

1.1.14 Definition. If $f : I\!R^n \to I\!R^m$ is a continuously differentiable function, then analogously an *interval enclosure of the Jacobian of* f is denoted by $J_{f^I} \in I\!I\!R^{m\times n}$ with

$$\{J_f(x) \mid x \in [x]\} \subseteq J_{f^I}([x]) \quad \text{for all } [x] \in I\!I\!R^n .$$

1.1.15 Definition. An interval function $f^I : I\!I\!R^n \to I\!I\!R^m$ is said to be *inclusion isotonic* if $[x] \subseteq [y]$ implies

$$f([x])^I \subseteq f([y])^I .$$

1.1.16 Remarks. Let $f : I\!R^n \to I\!R$ be a real function.

(i) Let f be a rational function. Then the interval function $f^E : I\!I\!R^n \to I\!I\!R$, where every real operation of f is replaced by the corresponding interval operation, is called an arithmetic evaluation of f. If f^E is evaluated only in one single form of expression, then Theorem 1.1.7 implies f^E to be an inclusion isotonic interval extension of f.

(ii) The interval extension f^E obtained in (i) is not necessarily identical to the interval hull of all possible results, $\bigcap\{[z] \in I\!I\!R \mid f(x) \in [z] \text{ for all } x \in [x]\}$, since dependence can cause an inflation of intervals.

(iii) As explained in detail by Hansen and Walster ([H&W04], p. 28ff), for an irrational continuous function f, an inclusion isotonic interval enclosure f^I can be determined by approximating f with a rational function and adding an appropriate error bound, i.e. by determining an inclusion isotonic interval bound.

The next theorem taken from Hansen and Walster [H&W04] is an immediate consequence of the inclusion isotonicity and provides conditions for Definition 1.1.13.

1.1.17 Fundamental Theorem of Interval Analysis. Suppose $[x] \in I\!I\!R^n$. Let $f : I\!R^n \to I\!R$ be a real function and $f^E : I\!I\!R^n \to I\!I\!R$ be an infinite precision inclusion isotonic interval function with $f(x) = f(x)^E$ for all $x \in I\!R^n$.

Then $f([x])^E$ contains the range of values of $f(x)$ for all $x \in [x]$, i.e. f^E is an interval extension of f.

In the rest of this thesis, let f^I denote an inclusion isotonic interval enclosure of a real function f.

1.1.2 Machine Computations

Hamming [H62] explains the basic application of interval arithmetic, called range arithmetic in his book:

> "Perhaps the simplest and most useful approach to the roundoff problem is *range arithmetic.* In this method each number is, in fact, represented by two numbers, the maximum and the minimum values that it might have. In a sense, each number is replaced by a range in which the correct answer must lie — hence the name. When two numbers are combined, the new range is computed in the appropriate fashion from the given ranges (using proper rounding). Thus at every stage there are safe bounds within which the true answer must lie."

Hence, with outward rounding, the true value of a real operation can be included in an interval with floating point numbers as boundary points.

1.1.18 Definition.

(i) The set of all floating point numbers available on a computer is called $\mathbb{F}$. The cardinality of $\mathbb{F}$ is finite and $\mathbb{F} \subset \mathbb{R}$.

(ii) For any binary operation $\circ \in \{+, -, \cdot\, , /\}$ it is often the case that

$$c = a \circ b \notin \mathbb{F}$$

for two floating point numbers $a\,, b \in \mathbb{F}$.

In this case the computer returns $\widetilde{c} \in \mathbb{F}$ with

$$\widetilde{c} = \Box c\,,$$

where the operator $\Box : \mathbb{R} \rightarrow \mathbb{F}$ is called rounding.

(iii) The IEEE 754-Standard for floating point arithmetic defines four rounding modes, round to nearest, round down, round up and round towards zero. These rounding modes allow to switch to the required rounding when implementing interval operations, to guarantee outward rounding of the boundary points of an interval.

If outward rounding is used, interval arithmetic is inclusion isotonic even when rounding occurs. Analogously to Corollary 7.8 of [R83, p. 82], the Inclusion Theorems of Sections 1.1.4 and 1.1.5 are valid when using outwardly rounded interval arithmetic.

Therefore, the superscript I can be used for such an interval enclosure as well: Then the interval enclosure $f(x)^I$ denotes an interval, resulting from the fact that a real function f of a real value x is computed using outwardly rounded interval arithmetic to bound rounding errors.

1.1.3 One-dimensional Bisection

During the construction of the relaxations of Sherali, Subramanian and Loganathan of the water distribution network problem, a verified inclusion of the zero of a strictly monotone increasing continuous function $f : \mathbb{R} \to \mathbb{R}$ in the interval $[\underline{x}\,,\,\overline{x}]$ is needed.

Only one verified bound is needed at once and the focus for determining this bound is more on calculation time than on accuracy. Therefore, a verified "approximation" of the one-dimensional bisection method is used, according to the formulation of Vrahatis (see e. g. [H99, p. 23]):

1.1.19 Lemma. Suppose $f(\underline{x}) < 0$ and let h be defined by $h = \overline{x} - \underline{x}$. For the verified lower bound let t_0 be initialized with $t_0 = \underline{x}$.

If the iteration

$$t_{n+1} = \begin{cases} t_n + \frac{h}{2^n} & \text{if } f(t_n + \frac{h}{2^n})^I < 0 \\ t_n & \text{else} \end{cases}$$

is applied for $n = 0, 1, \ldots, N$, $N \in \mathbb{N}$,

then it is guaranteed that $f(t_n) < 0$ for all t_n, $n \in \{1, \ldots, N\}$.

Proof: The property $f(t_n) < 0$ for all $n \in \mathbb{N}$ is a direct consequence of the definition of the iteration. □

1.1.20 Remarks.

(i) Suppose $x^* \in [\underline{x}\,,\,\overline{x}]$ with $f(x^*) = 0$. Then with exact computations and $f(t_n + \frac{h}{2^n})$ instead of the interval inclusion $f(t_n + \frac{h}{2^n})^I$ a maximal distance of

$$|t_n - x^*| \leq \frac{h}{2^n}$$

is obtained. No accuracy can be guaranteed a priori when using the interval inclusion f^I of f.

(ii) In WaTerInt $N = \lceil \log_2(\frac{h}{\varepsilon}) \rceil$ iterations are executed with a constant value (BEPS) of $\varepsilon = 0.01$.

(iii) Analogously for the verified upper bound with $t_0 = \overline{x}$, $f(\overline{x}) > 0$ and

$$t_{n+1} = \begin{cases} t_n - \frac{h}{2^n} & \text{if } f(t_n - \frac{h}{2^n})^I > 0 \\ t_n & \text{else,} \end{cases}$$

it is guaranteed that $f(t_n) > 0$ for all t_n, $n \in \{1, \ldots, N\}$.

1.1.4 Nonlinear Interval-Systems

To verify feasibility of a solution for the water distribution optimization problem, an inclusion of the zero of a continuously differentiable nonlinear function $f : \mathbb{R}^n \to \mathbb{R}^n$ has to be determined. This can be done for example using the INTLAB procedure `verifynlss`, which can guarantee existence and uniqueness as shown by S. M. Rump [R83]. The steps of this procedure are summarized in Table 1.1.

First, the Brouwer fixed-point theorem and results of Perron-Frobenius theory, which are needed to understand the INTLAB procedure `verifynlss`, are cited. Theorem 1.1.27 provides the proof of existence and uniqueness of an inclusion of a zero and is based on Theorem 1.1.26, which can be proved with Theorem 1.1.25, developed in 1942 by Collatz.

1.1.21 Brouwer fixed-point theorem. Let $X \subset \mathbb{R}^n$ be non-empty, convex and compact and let $f : \mathbb{R}^n \to \mathbb{R}^n$ be continuous with

$$f(X) \subseteq X \,.$$

Then there exists at least one $\hat{x} \in X$ with

$$f(\hat{x}) = \hat{x} \,.$$

Proof. An intuitive proof is based on the negative retract principle, "that there is no continuous map from a closed ball in the n-space onto its boundary which leaves the boundary pointwise fixed" [Z86, p. 48]. This intuitive fact can be proved with the aid of the topological degree, see Zeidler [Z86]. □

1.1.22 Perron's theorem. Let $A \in \mathbb{R}^{n\times n}$ and suppose $A > 0$.

Then

(i) $\rho(A) > 0$,

(ii) $\rho(A)$ is an eigenvalue of A, and

(iii) there is a vector $x \in \mathbb{R}^n$ with $x > 0$ such that $A\,x = \rho(A)\,x$.

Proof. For a proof see [HJ85, Theorem 8.2.11, p. 500].

For a nonnegative matrix A it can be shown that items (ii) and (iii) hold true analogously:

1.1.23 Lemma. Suppose $A \in \mathbb{R}^{n\times n}$ and $A \geq 0$.

Then $\rho(A)$ is an eigenvalue of A and there is a nonnegative vector $x \in \mathbb{R}^n$, $x \geq 0$, $x \neq 0$ with $A\,x = \rho(A)\,x$.

Proof. For a proof see [HJ85, Theorem 8.3.1, p. 503].

1.1.24 Lemma. Let $A \in \mathbb{R}^{n\times n}$.

Then

$$\rho(A) \leq \rho(|A|) .$$

Proof. For a proof see [HJ85, Theorem 8.1.18, p. 491].

1.1.25 Theorem of Collatz. Let $A \in \mathbb{R}^{n\times n}$, $A \geq 0$ and $x \in \mathbb{R}^n$ with $x > 0$.

Then

$$\min_{1\leq i\leq n} \frac{1}{x_i} \sum_{j=1}^{n} a_{ij}x_j \;\leq\; \rho(A) \;\leq\; \max_{1\leq i\leq n} \frac{1}{x_i} \sum_{j=1}^{n} a_{ij}x_j \; .$$

Proof. According to Lemma 1.1.23 there exists $y \geq 0$, $y \neq 0$ with $A^\top y = \rho(A)y$. Let $\mu_i := \frac{(Ax)_i}{x_i}$ denote the i'th component of the product $\frac{1}{x_i}Ax$.

Then

$$\sum_{i=1}^{n}(\mu_i - \rho(A))\, x_i\, y_i = y^\top Ax - x^\top A^\top y = 0 \; .$$

Since $y \neq 0$ and for all $i = 1, \dots, n$ it is $x_i > 0$ and $y_i \geq 0$, there exists at least one i with $x_i\, y_i > 0$.

Let then $S := (\mu_i - \rho(A))\, x_i\, y_i + \sum_{\substack{j=1\\ j\neq i}}^{n} (\mu_j - \rho(A))\, x_j\, y_j$.

Assume that $\mu_j > \rho(A)$ for all $j = 1, \dots, n$. This implies the first term of S to be positive and the second to be nonnegative, so $S > 0$, a contradiction to $S = 0$ as stated above. With an analogous argument the assumption that $\mu_j < \rho(A)$ for all $j = 1, \dots, n$ implies a contradiction. Thus there exist $i, j \in \{1, \dots, n\}$ with $\mu_i \leq \rho(A) \leq \mu_j$. □

1.1.26 Theorem [R83]. Suppose $[z]$, $[x] \in \mathbb{IR}^n$, $[C] \in \mathbb{IR}^{n\times n}$ and $[z] + [C]\,[x] \subseteq \overset{\circ}{[x]}$. Then for every $C \in [C]$ it holds that

$$\rho(C) < 1 .$$

Proof. Let $z \in [z]$, $C \in [C]$ be arbitrary but fixed. Then

$$z + C\,[x] \subseteq \overset{\circ}{[x]} \; .$$

Using midpoint-radius notation, the interval $[x]$ can be written as $[x] = m([x]) \pm r([x])$. Hence

$$z + C\,[x] = z + C\,m([x]) \pm |C|\,r([x]) \;\subseteq \overset{\circ}{[x]}\;.$$

Therefore

$$\begin{aligned} m([x]) - r([x]) \;&<\; z + C\,m([x]) - |C|\,r([x]) \\ &\leq\; z + C\,m([x]) + |C|\,r([x]) \;<\; m([x]) + r([x])\;, \end{aligned}$$

which implies $|C|\,r([x]) < r([x])$. As $z + C\,[x] \subseteq \overset{\circ}{[x]}$ one obtains $\overset{\circ}{[x]} \neq \emptyset$, which implies $r([x]) > 0$. As $|C| \geq 0$, the Theorem of Collatz 1.1.25 implies

$$\rho(|C|) \leq \max_{i \in 1,\dots,n} \frac{(|C|\,r([x]))_i}{(r([x]))_i} < 1\;.$$

Then with Lemma 1.1.24 it follows $\rho(C) \leq \rho(|C|) < 1$, which proves the theorem. □

The proof of existence and uniqueness of a zero of the function f with the INTLAB procedure `verifynlss` is provided by Rump [R83, Theorem 7.4, p. 82]. The following theorem contains a slightly modified version, which corresponds to version 4.1.2 of INTLAB [R03]. Excluding the proof of uniqueness of the zero, it is a special case of [R96, Theorem 1, p. 79]. As the procedure `verifynlss` forms a central part of the verification procedure developed in Chapter 4, the proof, which is analogous to [R83], is provided in detail.

1.1.27 Theorem (cf. [R83]). Let $f : I\!R^n \to I\!R^n$ be a continuously differentiable function, let J_f be the Jacobi matrix of f and let $\tilde{x} \in I\!R^n$ and $R \in I\!R^{n \times n}$.

Suppose for some $[x] \in I\!I\!R^n$ with $0 \in [x]$ and for some $[y] \in I\!I\!R^n$ with $[y] \subseteq [x]$

$$-Rf(\tilde{x}) + \Big(I - R\,J_{fI}(\tilde{x} + [x])\Big)\,[y] \;\subset \overset{\circ}{[y]}\;.$$

Then the matrix R and each $B \in I\!R^{n\times n}$ with $B \in J_{fI}(\tilde{x} + [x])$ is non-singular and there exists one and only one $\hat{x} \in \tilde{x} + [y]$ with

$$f(\hat{x}) = 0\;.$$

Proof. It follows from Theorem 1.1.26 that for all $B \in J_{fI}(\tilde{x} + [x])$ it holds true that

$$\rho(I - RB) < 1\;,$$

which implies R and B to be non-singular.

INTLAB procedure `verifynlss`: Interval iteration

Determine $R \approx J_f^{-1}(\tilde{x})$ and $[z] = -Rf(\tilde{x})^I$

set $[x] = [z]$

for $k = 0, \dots, k_{\max}$

  $[y] =$ interval hull $(0, [x] \circ \varepsilon)$ and $[y_{old}] = [y]$

  $[C] = I - R\, J_{f^I}(\tilde{x} + [y])$

  for $i = 1, 2$ (additional iteration)

    $[x] = [z] + C\,[y]$

    if $[x] \subseteq \overset{\circ}{[y]}$

      stop: $[x] = \tilde{x} + ([x] \cap [y_{old}])$ contains one and only one zero of f

    else

      $[y] = [x] \cap [y_{old}]$

    fi

  rof

rof

Table 1.1: Determination of an inclusion for the zero of a continously differentiable function $f : I\!R^n \to I\!R^n$ according to Theorem 1.1.27.

Let $g : I\!R^n \to I\!R^n$ be defined by $g(x) = x - Rf(\tilde{x} + x)$ and let $y \in [y]$. According to the mean value theorem there exist $\xi_1, \ldots, \xi_n \in \tilde{x} \,\underline{\cup}\, \tilde{x} + y$ with

$$g(y) = y - R\left(f(\tilde{x}) + \begin{pmatrix} \frac{\partial f_1}{\partial y}(\xi_1) \\ \frac{\partial f_2}{\partial y}(\xi_2) \\ \vdots \\ \frac{\partial f_n}{\partial y}(\xi_n) \end{pmatrix} (\tilde{x} + y - \tilde{x})\right).$$

Since $[y] \subseteq [x]$ and $0 \in [x]$ it follows $\xi_1, \ldots, \xi_n \in \tilde{x} + [x]$. Thus, according to the assumption

$$g(y) = -Rf(\tilde{x}) + (I - R \begin{pmatrix} \frac{\partial f_1}{\partial y}(\xi_1) \\ \frac{\partial f_2}{\partial y}(\xi_2) \\ \vdots \\ \frac{\partial f_n}{\partial y}(\xi_n) \end{pmatrix}) y \subset \overset{\circ}{[y]} .$$

By the Brouwer fixed-point Theorem 1.1.21 one $\hat{y} \in [y]$ exists with $g(\hat{y}) = \hat{y}$. As R is non-singular, $f(\tilde{x} + \hat{y}) = 0$.

It remains to be shown that $\hat{y}$ is unique. Assume $\hat{z} \in [y]$ with $f(\tilde{x} + \hat{z}) = 0$. Then by the mean value theorem there exist $\xi_1, \ldots, \xi_n \in (\tilde{x} + \hat{z}) \,\underline{\cup}\, (\tilde{x} + \hat{y})$ and $B := \left(\frac{\partial f_1}{\partial y}(\xi_1), \frac{\partial f_2}{\partial y}(\xi_2), \ldots, \frac{\partial f_n}{\partial y}(\xi_n)\right)^\top$ with

$$f(\tilde{x} + \hat{z}) = f(\tilde{x} + \hat{y}) + B\left((\tilde{x} + \hat{z}) - (\tilde{x} + \hat{y})\right) .$$

Then $B \in J_{f'}(\tilde{x} + [x])$ and therefore is non-singular. So the assumption $f(\tilde{x} + \hat{z}) = f(\tilde{x} + \hat{y}) = 0$ results in $\hat{z} = \hat{y}$, which proves the theorem. □

1.1.28 Remarks.

(i) The expression $-Rf(\tilde{x}) + \left(I - R\, J_{f'}(\tilde{x} + [x])\right)[y] \subset \overset{\circ}{[y]}$ instead of

$$-Rf(\tilde{x}) + \left(I - R\, J_{f'}(\tilde{x} + [x])\right)[x] \subset \overset{\circ}{[x]}$$

is needed to guarantee existence and uniqueness of a zero within the additional iteration in Table 1.1.

(ii) In the case of successful termination, the additional intersection $[x] \cap [y_{old}]$ is no restriction, as $[y] \subseteq [y_{old}]$ and therefore the condition

$$[x] \subseteq \overset{\circ}{[y]} \quad \text{implies} \quad [x] \cap [y_{old}] = [x] .$$

(iii) In the case $[x] \cap [y_{old}] = \emptyset$ no error occurs, as INTLAB returns `NaN` as interval bounds. `NaN` is also used to indicate that no inclusion could be found.

(iv) In Theorem 1 of [R96, p. 79] existence of a Jacobian matrix is not required. Instead it is assumed that $S(\tilde{x}, [x]) \in I\!I\!R^{n\times n}$ such that for all $x \in [x]$ there exists $M \in S(\tilde{x}, [x])$ with

$$f(\tilde{x} + x) - f(\tilde{x}) = Mx .$$

Therefore this theorem is valid for slopes as well. If an inclusion for the Jacobian $J_f(\tilde{x} + [x])$ exists and $0 \in [x]$, it follows directly from the mean value theorem that such an M exists.

The operator used in Theorem 1.1.27 is known as the Krawczyk operator. Neumaier summarizes some general properties [N01]:

1.1.29 Definition. Let $f : I\!R^n \to I\!R^n$ be continuously differentiable, and $R \in I\!R^{n\times n}$. Then the *Krawczyk operator* $K(\tilde{x}, [x]) : I\!R^n \times I\!I\!R^n \to I\!I\!R^n$ is defined as

$$K(\tilde{x}, [x]) := \tilde{x} - Rf(\tilde{x}) + \Big(I - RJ_{f'}([x])\Big)\,([x] - \tilde{x}) .$$

1.1.30 Lemma. Suppose $\tilde{x} \in [x]$. The Krawczyk operator $K(\tilde{x}, [x])$ has the following properties:

(i) If $x^* \in [x]$ with $f(x^*) = 0$, then $x^* \in [x] \cap K(\tilde{x}, [x])$.

(ii) If $[x] \cap K(\tilde{x}, [x]) = \emptyset$ then there exists no $x^* \in [x]$ with $f(x^*) = 0$.

(iii) If $K(\tilde{x}, [x]) \overset{\circ}{\subseteq} [x]$ then there exists one and only one $x^* \in [x]$ with $f(x^*) = 0$.

Proof. (i) is a direct consequence of the mean value theorem and the construction of $K(\tilde{x}, [x])$, (ii) follows directly from (i).

(iii) Let $[z] := -\tilde{x} + [x]$. As $\tilde{x} \in [x]$, it follows $0 \in [z]$. Applying Theorem 1.1.27 proves the assertion. □

So, as summarized in [R98], one main point is to verify that a certain interval is mapped into its interior. If this is not the case for an initial test interval, an interval iteration is started. For this iteration the term epsilon-inflation was introduced by S. M. Rump for $[x] \in I\!I\!R$ as

$$[x] \circ \varepsilon := \begin{cases} [x] + d([x]) \cdot [-\varepsilon, \varepsilon] & \text{for } d([x]) \neq 0 \\ [x] + [-\eta, \eta] & \text{otherwise.} \end{cases}$$

This enlarges the intervals prior to the next iteration to increase chances of self-mapping.

As discussed in [R98] and implemented in INTLAB, the algorithms presented in this chapter are based on a simplified version:

1.1.31 Definition. Let $[x] \in I\!I\!R^n$ and $\varepsilon, \eta \in I\!R$, $\varepsilon > 0$ and suppose η is the smallest positive normalized floating point number on the computer used. Then the ε *-inflation* is defined as

$$[x] \circ \varepsilon := [x] + d([x]) \cdot [-\varepsilon, \varepsilon] + [-\eta, \eta] .$$

In INTLAB, a value of $\varepsilon = 0.05$ is used for the relative factor.

1.1.32 Remarks.

(i) For the verification of the upper bound of the water distribution optimization problem introduced in Chapter 2, a bounded underdetermined nonlinear system has to be solved:

Let $f : I\!R^m \to I\!R^n$ and $m < n$. Let $\tilde{x} \in I\!R^n$ be an approximate solution of the underdetermined bounded nonlinear system

$$f(x) = 0 , \quad \underline{x} \leq x \leq \overline{x}$$

with $\underline{x} \leq \tilde{x} \leq \overline{x}$.

Then an inclusion $[x]$ "close to" $\tilde{x}$ with $\underline{x} \leq [x] \leq \overline{x}$ has to be determined so that at least one $x^* \in [x]$ exists with $f(x^*) = 0$.

(ii) In WaTerInt this problem is solved with the procedure `verify_udnlss`. In addition to the approximate solution, an index-vector can be passed to this procedure, which contains selected indices where the values of the approximate solution should be fixed:

Per default the indices for the square sub-function are chosen with the aid of the standard LU-decomposition part of MATLAB, which for a matrix X returns a unit lower triangular matrix L, an upper triangular matrix U, and a permutation matrix P so that $PX = LU$. Using the center of the Jacobian $m(J_f([x]))$, the order implied by P, i.e. $P \cdot (1, 2, \ldots, n)^\top$, is used for ordering the variables of the $n \times n$-submatrix (cf. [J04]). All indices not to be used according to the a priori determination are then deleted from this list.

To accelerate convergence, at the beginning of this procedure all elements $[x_i]$ of the interval vector, which are not fixed a priori, are enlarged about an absolute factor as $[x_i] = [x_i] + 10^{-12} \cdot [-1, 1]$. Finally the INTLAB algorithm `verifynlss` is used to determine an inclusion for the square nonlinear subfunction.

1.1.5 Linear Interval-Systems

1.1.33 Definition. Let $[A] \in I\!I\!R^{m\times n}$, $x \in I\!R^n$ and $[b] \in I\!I\!R^m$. Then the system

$$[A]\,x = [b]$$

is called a *system of linear interval equations* or *linear interval system.*

The solution is defined to be the set $\Sigma([A],[b])$ with

$$\Sigma([A],[b]) := \{x \in I\!R^n \mid Ax = b \quad \text{for some } A \in [A] \text{ and for some } b \in [b]\,\}\ .$$

1.1.34 Remarks.

(i) The solution set is connected and piecewise convex with up to 2^n pieces, i.e. it is generally not an n-dimensional interval. When all $A \in [A]$ are non-singular, the set $\Sigma(A,b)$ is bounded.

(ii) Let $[A] \in I\!I\!R^{n\times n}$, $[b] \in I\!I\!R^n$, $x \in I\!R^n$ and $R \in I\!R^{n\times n}$. Then changing the linear system $[A]x = [b]$ to

$$R[A]x = R[b]$$

is called *preconditioning.* The matrix R is called *preconditioning matrix.* This process implies for the solution sets

$$\Sigma([A],[b]) \subseteq \Sigma(R[A],R[b])\ .$$

(iii) The determination of the hull $H(\Sigma([A],[b]))$ is known to be an NP-hard problem. It is common practice to compute outer bounds for the hull $H(\Sigma([A],[b]))$ instead, i.e. an n-dimensional interval containing $\Sigma([A],[b])$.

1.1.35 Theorem (cf. [R83]). Let $R \in I\!R^{n\times n}$, $\tilde{x} \in I\!R^n$, $[A] \in I\!I\!R^{n\times n}$ and $[b] \in I\!I\!R^n$. For the interval vector $[x] \in I\!I\!R^n$ let the inclusion

$$R([b] - [A]\,\tilde{x}) + (I - R\,[A])\,[x] \ \subseteq\ \overset{\circ}{[x]}$$

hold true.

Then R and all $A \in [A]$ are non-singular and

$$\Sigma([A],[b]) \subseteq \tilde{x} + [x]\ .$$

Proof. Let $A \in [A]$ and $b \in [b]$ be arbitrary but fixed. Suppose $f : I\!R^n \to I\!R^n$ is defined by $f(x) = Ax - b$. Then $J_f(x) = A$ independent of x, and the assumption results in

$$-Rf(\tilde{x}) + (I - RJ_{f'})\,[x] \subseteq \overset{\circ}{[x]} \; .$$

By Theorem 1.1.27 R and A are non-singular and there exists one and only one $\hat{x} \in \tilde{x} + [x]$ with $A\hat{x} = b$. As $A \in [A]$ and $b \in [b]$ are chosen arbitrarily, the assertion follows. □

1.1.36 Remarks.

(i) Analogous to Remarks 1.1.32, a bounded underdetermined linear system has to be solved for the water distribution problem:

Let $[A] \in I\!I\!R^{m \times n}$, $[b] \in I\!I\!R^m$ and $m < n$. For arbitrary but fixed $A \in [A]$ and $b \in [b]$ let $\tilde{x} \in I\!R^n$ be an approximate solution of the underdetermined bounded linear system

$$Ax = b\,, \quad \underline{x} \le x \le \overline{x}$$

with $\underline{x} \le \tilde{x} \le \overline{x}$.

Then an inclusion $[x]$ "close to" $\tilde{x}$ with $\underline{x} \le [x] \le \overline{x}$ has to be determined so that for all $A \in [A]$ and $b \in [b]$ at least one $x^* \in [x]$ exists with $Ax^* = b$.

(ii) Analogously to the procedure for the nonlinear function, this is part of the WaTerInt procedure `verify_udlss`, where the indices are again selected using the LU-decomposition of A and a vector containing indices to be fixed is considered.

1.1.6 Numerical Differentiation

An inclusion of the Jacobian matrix has to be calculated for the interval Newton method used in WaTerInt to verify the upper bound of the optimal solution.

There exist three commonly used methods to calculate the Jacobian at a fixed point: approximation by difference quotient, automatic differentiation and symbolic differentiation. With difference quotients, only an approximation of the derivative of a function can be determined. This would not be sufficient during the verification of water distribution network design, as an inclusion of the Jacobian is needed.

Automatic differentiation is based on the mechanical application of the chain rule to obtain derivatives of a function at a fixed point. In connection with the data type "`gradient`" and the concept of operator overloading, the forward mode of automatic differentiation is implemented as part of INTLAB [R03]. The main property is the possibility to calculate the derivative of any arithmetic expression at a certain point without knowing a symbolic expression for the derivative of f. Kearfott [Ke96] summarizes this parallel to interval arithmetic as "the forward mode of differentiation is analogous to interval arithmetic in the sense that an arithmetic is defined on an extended set of objects, intervals in the case of interval arithmetic and function/derivative pairs in the case of automatic differentiation."

Using symbolic differentiation according to the rules of differentiation, a symbolic expression for the derivative is determined, which afterwards can be evaluated at certain points. With respect to interval calculations, the main advantage is the possibility to further transform this expression to reduce effects of dependence as described in Lemma 1.1.9. As emphasized by Hansen and Walster "a symbolic expression for a derivative can be written to reduce the effect of dependence" [H&W04, p. 135].

The nonlinear constraints of the water distribution optimization problem contain the expression $\operatorname{sgn}(x)\,|x|^{c_d}$, where c_d is a constant which has a value out of $[1.5, 2]$. Derivatives needed for these constraints cannot be calculated with INTLAB, because of a lack of an interval definition for the sign-function.

But for the functions needed, the derivatives can easily be determined manually. This manual determination of the Jacobian can be regarded as the simplest form of symbolic differentiation and is used in WaTerInt. Hence, additional computational time for the interpretation overhead, especially when using object orientation for gradients, is avoided. Consequently, the INTLAB procedure `verifynlss` cannot be used, so it is reimplemented as part of the module `vnop.m`, see Appendix B.5.

1.2 Optimization Theory and Verification

"Nature (and man) loves to optimize, and the world is far from linear." [G&L02]

1.2.1 Basic Definitions

The following definition of an optimization problem is closely related to [N03] and adjusted to the problems considered in this thesis:

1.2.1 Definition. Let $\underline{x} \in (\mathbb{R} \cup \{-\infty\})^n$ and $\overline{x} \in (\mathbb{R} \cup \{\infty\})^n$ be such that $\underline{x} \leq \overline{x}$.

Let $[b] \in \mathbb{IR}^m$ be a possibly degenerate interval and let $f : [\underline{x}, \overline{x}] \to \mathbb{R}$ and $F : [\underline{x}, \overline{x}] \to \mathbb{R}^m$ be continuous functions.

Then a *continuous global optimization problem* is specified in the form

$$\begin{cases} f(x) \to \min \\ \\ F(x) \in [b] \\ x \in [\underline{x}, \overline{x}] \\ x \in \mathbb{R}^n \,, \end{cases}$$

where $C = \{x \in [\underline{x}, \overline{x}] \mid F(x) \in [b]\}$ is the *feasible domain*, a point $x \in C$ is a *feasible point* and a point $\widehat{x} \in C$ with

$$f(\widehat{x}) = \min_{x \in C} f(x)$$

is called *optimal solution.*

A feasible point $\widehat{x}_\varepsilon \in C$ with

$$f(\widehat{x}_\varepsilon) \leq \min_{x \in C} f(x) + \varepsilon$$

is called ε-optimal solution.

The problem is called *linear,* iff f and F are linear functions.

A number of numerical algorithms exist to solve optimization problems. Neumaier provides an impression for the computation time needed, see Table 1.2.1. These estimates are based on floating point calculations. Additional effort and time is necessary to obtain rigorous results using interval analysis.

Time for solving a linear program	LP
for solving a convex quadratic program	$QP = 5 \times LP$
for solving a convex nonlinear program	$SQP = 30 \times QP$
for finding a global minimizer of a nonconvex nonlinear program	$GLP_f \geq 100 \times SQP$
for verifying that it is a global minimizer	$GLP_v \geq 1000 \times SQP$

Table 1.2: Neumaier suggests these rules for getting a rough idea of times and difficulties for solving problems of comparable size and sparsity structure. According to Neumaier these are "completely unreliable but catchy", they are adapted from [N03]. Verifying in this context does not mean reliable in the sense of interval analysis, but verified in the sense of not heuristically determined.

Most algorithms for determining a global optimal solution of a nonlinear, nonconvex problem incorporate branch and bound techniques. Such a branch and bound method is based on the principle of splitting a problem recursively into subproblems (branches) and eliminating those that cannot contain an optimal solution according to bounds calculated.

As emphasized by Neumaier [N03], branching techniques are basic to almost all complete global optimization algorithms, i.e. methods where the global minimum is reached "with certainty, assuming exact computations and indefinitely long run time."

For the verified solution of optimization problems concerning floating point arithmetic, Neumaier [N03] summarizes the reliability of optimization with the following four aspects:

(i) *Rounding in the problem definition:* The set $\mathbb{R} \setminus \mathbb{F}$ contains real numbers that cannot exactly be represented as floating point numbers. Therefore an error occurs when input data of the optimization problem contain these numbers.

(ii) *Rounding in the solution process:* Often algorithms valid in exact arithmetic are implemented using floating point arithmetic without taking into account rounding errors and accumulation of these during the computations.

(iii) *Certification of upper bound:* An upper bound always implies the feasibility of a certain point. Often, this is satisfied only within defined tolerances. Instead, it is theoretically correct to use verification techniques to guarantee existence of a feasible point within certain bounds.

(iv) *Certificates of infeasibility:* To exclude regions in the branch and bound algorithms due to infeasibility, it has to be rigorously ensured that no admissible points can be part of these regions.

In the verification algorithm for water distribution developed in Chapter 4, these points are addressed. Some of the input data can be provided as intervals, capturing the problem of rounding in the problem definition.

As shown in the next section, it is possible to verify a posteriori the solution of a linear optimization problem obtained with a floating point solver. This fact forms a central part in the verified calculation of water distribution.

Sections 1.1.4 and 1.1.5 provide algorithms for proving admissibility, and certificates of infeasibility are discussed in Section 1.2.3.

1.2.2 Bounds for Linear Optimization Problems

Recently, rigorous error bounds have been developed (see [N&S04], [J02], [J04] and [J04c]) independently in Vienna and Hamburg. Whereas the focus of Neumaier and Shcherbina [N&S04] was on linear optimization with finite bounds, Jansson investigated the more general cases of unbounded linear and of convex problems. Theorem 6.1 of [J02] contains this bound for interval linear optimization problems where the bounds may be infinite.

1.2.2 Definition. A *linear optimization problem* is an optimization problem of the following form:

$$\begin{cases} c^\top x \to \min \\ \\ Ax \leq a \\ Bx = b \\ x \in I\!R^n \\ \underline{x} \leq x \leq \overline{x} \end{cases}$$

This formulation implies the corresponding dual problem

$$\begin{cases} -a^\top y + b^\top z + \underline{x}^\top u - \overline{x}^\top v \to \max \\ \\ -A^\top y + B^\top z + u - v = c \qquad\qquad (D) \\ (y, z, u, v)^\top \in I\!R^{m+p+2n} \\ y \geq 0\,,\ u \geq 0\,,\ v \geq 0\,. \end{cases}$$

Commonly in optimization theory, formulations are used that contain only equalities or inequalities. As these transformations can have a negative impact on the condition of the problem, in the following the linear problem is regarded in this original form.

The lower bounding problem of the standard nonlinear optimization problem for water distribution network design (NOP) is bounded, the cost vector c and the matrix A describing the inequalities can be represented by a pure floating point vector or matrix, respectively. The following lemma contains this special case of Theorem 6.1 [J02] developed by Jansson:

1.2.3 Lemma. Let $c \in \mathbb{IR}^n$, $A \in \mathbb{R}^{m\times n}$, $a \in \mathbb{R}^m$, $[B] \in \mathbb{IR}^{p\times n}$ and $[b] \in \mathbb{IR}^p$. Then let a family of linear programming problems $[P]$ with finite bounds $\underline{x}$, $\overline{x} \in \mathbb{R}^n$, $\underline{x} < \overline{x}$, be in the form

$$\begin{cases} c^\top x \to \min \\ \\ A\,x \le a \\ [B]\,x = [b] \\ x \in \mathbb{R}^n \\ \underline{x} \le x \le \overline{x}\,. \end{cases} \qquad [P]$$

Additionally let vectors $y \in \mathbb{R}^m$ and $z \in \mathbb{R}^p$ be given with $y \le 0$. Let the interval vector $[d]$ and for fixed $B \in [B]$ the vector d be defined as

$$[d] := c - A^\top y - [B]^\top z \quad \text{and} \quad d := c - A^\top y - B^\top z\,.$$

Then a verified lower bound $\underline{f}^*$ of the objective value for all linear programs of $[P]$ is

$$\underline{f}^* = \min\left\{a^\top y + [b]^\top z + \underline{x}^\top\,[d]^+ - \overline{x}^\top\,[d]^-\right\}\,. \tag{1.2.1}$$

Proof. Let $P \in [P]$ be arbitrary but fixed. Then for all primal feasible x of the linear program P it is

$$\begin{aligned} a^\top y + b^\top z + \underline{x}^\top d^+ - \overline{x}^\top d^- &\le x^\top(A^\top y + B^\top z) + \underline{x}^\top d^+ - \overline{x}^\top d^- \\ &\le x^\top c + (\underline{x} - x)^\top d^+ + (x - \overline{x})^\top d^- \\ &\le x^\top c\,, \end{aligned}$$

which implies $\underline{f}^*$ being a lower bound of the family $[P]$. □

The main advantage of this bound is the possibility to certify the results of linear and convex optimization problems by pre- and postprocessing without having to modify the solvers.

1.2.3 Verification of Infeasibility of a Linear Problem

The relaxations used in the branch and bound algorithm for the water distribution network in the following chapters result in linear lower bounding problems (LNOP).

Neumaier and Shcherbina introduce a certificate of infeasibility, which they discuss in [N&S04] for linear problems of the form

$$\begin{cases} c^\top x \to \min \\ \\ \underline{b} \le Ax \le \overline{b} \\ \underline{x} \le x \le \overline{x} \\ \underline{b}, \overline{b}, \underline{x}, \overline{x} \in I\!R^n \, . \end{cases}$$

Lemma 1.2.5 is analogous to this certificate and corresponds to the notation used in WaTerInt. The basis for verifying non-solvability of linear problems is provided by the Lemma of Farkas:

1.2.4 Corollary to the Lemma of Farkas. Let $A \in I\!R^{m\times n}$ and $b \in I\!R^m$ be arbitrary but fixed. Then exactly one of the following alternatives holds true:

(i) The linear system $Ax \le b$ has a solution $x \in I\!R^n$.

(ii) $A^\top y = 0$, $b^\top y < 0$ has a nonnegative solution $y \in I\!R^m_+$.

Proof. For a proof see for example [J99, p. 37]. □

1.2.5 Lemma. Assume $c \in I\!R^n_+$, $\underline{x}$, $\overline{x} \in I\!R^n$ and let a family of linear lower bounding problems $[P]$ be of the form

$$\begin{cases} c^\top x \to \min \\ \\ Ax \le a \\ [B]\, x = [b] \\ \underline{x} \le x \le \overline{x} \\ x \in I\!R^n \, . \end{cases} \qquad [P]$$

Then all problems (P) of $[P]$ are infeasible, iff for all $B \in [B]$ and $b \in [b]$ the system

$$A^\top y + B^\top z - u + v = 0 \, , \quad a^\top y + b^\top z - \underline{x}^\top u + \overline{x}^\top v < 0$$

has a solution $y \in I\!R^m_+$, $z \in I\!R^p$ and u , $v \in I\!R^n_+$.

Proof. The objective function for all $(P) \in [P]$ is bounded below, so an optimal solution for all (P) exists, iff the set of admissible points

$$\{x \in I\!R^n \mid \begin{pmatrix} A \\ B \\ -B \\ -I \\ I \end{pmatrix} x \leq \begin{pmatrix} a \\ b \\ -b \\ -\underline{x} \\ \overline{x} \end{pmatrix}\}$$

is nonempty for all $b \in [b]$, $B \in [B]$. Corollary 1.2.4 implies the assertion. □

So using floating point calculations and interval arithmetic with outward rounding, the verification of infeasibility of all $(P) \in [P]$ can be obtained by determining an approximate solution of the linear problem

$$\begin{cases} a^\top y + b^\top z - \underline{x}^\top u + \overline{x}^\top v \to \min \\ \\ A^\top y + B^\top z - u + v = 0 \\ y \geq 0, \; u \geq 0, \; v \geq 0 \quad , \end{cases} \qquad (P_{aux})$$

where b is the center of $[b]$ and B the center of $[B]$.

It has to be ensured that $y \geq 0$, then the values obtained for y and z are regarded as degenerate intervals. With the definitions of

$$[d] := A^\top [y] + [B]^\top [z] \,, \quad [u] := [d]^+ \quad \text{and} \quad [v] := [d]^-$$

the infeasibility is proven, if $a^\top [y] + [b]^\top [z] - \underline{x}^\top [u] + \overline{x}^\top [v] < 0$. This algorithm is summarized in Table 1.3.

Verification of Infeasibility

Determine approximate solution $(y, z, u, v)^\top$ of (P_{aux}) with linear solver
Set negative elements of y to zero
Determine $[u] = [d]^+$ and $[v] = [d]^-$, where $[d] = A^\top [y] + [B]^\top [z]$
if $a^\top [y] + [b]^\top [z] - \underline{x}^\top [u] + \overline{x}^\top [v] < 0$
    Stop: all $(P) \in [P]$ are infeasible
else
    no proof of infeasibility can be provided
fi

Table 1.3: Algorithm for verification of infeasibility of a linear optimization problem.

Chapter 2

Water Distribution Network Design

"Although optimization may increasingly serve as another tool for design engineers, it is not likely to replace good engineering judgment." [W&a03]

The optimization model for determining the diameters of pipes for a given network layout of a water distribution system is introduced in this chapter. The engineering background of the application in water distribution design is explained and possible improvements of the model are discussed. Among these are the addition of energy costs for the standard test networks and the choice of the head-loss formula which is investigated in detail in Chapter 3.

An overview of the algorithms developed for the standard model during the last 30 years is provided with the focus on the branch and bound algorithm developed by Sherali, Subramanian and Loganathan. Where this algorithm contains alternatives, the one implemented as part of WaTerInt is described in detail. A new constraint propagation technique is introduced for expansion networks.

The single loading problem – as exhibited by all standard test networks – is discussed and the adjacency property [F&D87] is explained. An equivalent formulation is developed to heuristically demonstrate difficulties that occur during the verification process in Chapter 4.

2.1 Modeling Water Distribution Systems

2.1.1 Engineering Context

"Water-distribution-system design has remained intriguing because of its complexity and utility. Standard hydraulics textbooks treat only the problem of finding flows in a given network layout for specified pipe diameters. The real design aspects of layout selection and choosing suitable diameters for the pipes however, are quite cumbersome" ([L&a95]).

In this citation, Loganathan, Greene and Ahn summarize a problem which has received a great deal of attention in the United States: the design of cost effective water distribution systems. Pipes get old and are burdened with increasing demand. According to [S&a01], investments for the water distribution account for the largest proportion in municipal maintenance budgets.

This explains the practical relevance of the optimization problem discussed in this chapter: the least-cost optimization for a given network layout with pipe diameters acting as primary decision variables.

For practical relevance, different stages of the planning process demanding decisions on pipe diameters have to be taken into account. Walski [W95] identifies the following situations:

(i) Master planning is the overall planning for constructions in a horizon of about 50 years. In this stage future municipal and industrial water use has a major influence on pipe sizes, and one major decision affecting the optimization of pipe diameters is the determination of the nominal head in pressure zones.

(ii) Preliminary engineering for transmission mains is important before the start of an installation project. It contains the reevaluation of demand, location and magnitudes of the pipes.

(iii) In subdivision layout the planning horizon only contains a few years resulting in clear estimates of demand. A main uncertainty is the pressure at the tie-in point of the subdivision, which can vary in future.

(iv) In rehabilitation of existing systems, the layout is known and the demand is quite clear, mostly driven by fire-flow requirements. The decision process of sizing new pipes should include consideration of alternatives like cleaning, lining or expanding pipes.

Due to different focus and quality of available input data, impact on engineering decisions as well as integration in the decision process will vary, but the underlying optimization problem remains unchanged.

2.1.2 Objectives and Parameters of Optimization

Sherali and Smith [S&S93] presented a lot of research on a holistic approach for optimal design and expansion plans. The least cost problem investigated in this chapter focuses on a single stage component of one submodel of this research, which provides optimal construction and replacement decisions for a prescribed reliability criterion. This component determines the least cost pipe sizing for a fixed network topology to meet given hydraulic pressure and flow requirements.

In the following, this model is referred to as the standard optimization model (NOP), and this chapter refers to its description and relaxations given by Sherali, Subramanian and Loganathan [S&a01].

Walski [W&a03] summarizes the main factors concerning optimization models for water distribution design from the engineering point of view as follows:

(i) Hydraulic feasibility

(ii) Satisfaction of demands

(iii) Meeting of pressure constraints

(iv) Budgetary constraints

(v) Reasonable levels of redundancy and reliability

(vi) Tradeoffs between different objectives (e.g. fire protection, water quality)

(vii) Uncertain parameters regarding future development

In the standard model the first three factors are described by constraints, and budgetary constraints are indirectly part of the objective function minimizing the cost.

Consideration of reasonable levels of redundancy and reliability is not part of the standard optimization model. Research on the design for meeting these requirements includes the implementation of parallel pipes or loops already in the layout of the network.

The last two factors are simplified in the model. In the United States fire standards are usually expressed in terms of minimum pressure that must be maintained in the system. Especially in residential areas, the capacity requirements of fire flows usually dominate those of customer demands. In case of long residence times, water quality problems can occur. Tradeoffs between fire protection and water quality as well as future developments are considered when estimating the demand. In the standard model, the demand is assumed to be a fixed input parameter. So the tasks of estimating are transferred to research on identification of the input data for the model.

2.1.3 Limits of Water Distribution Models

> "Prediction is very difficult,
> especially about the future."
> — Nils Bohr

When using the results of optimization as basis for design decisions, the following aspects illustrating the limits of the model should be born in mind:

One problem concerning **cost data** is the accuracy of the estimated costs. As supposed in [W&a03], they are often only accurate to $\pm 20\%$. As the relative costs for different diameters for pipe segments are consistent, this effect is not too significant when selecting diameter sizes.

Of more importance is the determination of the actual costs of the installed pipes. These costs depend not only on the diameter but on excavation and paving. Effects of different paving costs can be much greater than effects of diameters.

Sherali, Subramanian and Loganathan introduce a new real network design test problem based on the town of Blacksburg, Virginia, using unit pipe costs that "represent real-life values and include installation costs as well". No information about the amount of these costs or about the process of determination is available.

As discussed by Mays [M99], an engineering economic analysis as part of this process considers estimated life spans of pipes of different diameters, as e.g. pipes of cast iron with a diameter of 4 inch have an average life span of 50 years, whereas pipes with a diameter greater than 12 inch have an average life span of 100 years. Including estimates for maintenance costs introduces one further not negligible uncertainty in the discount factors used.

As already mentioned, no considerations for **reliability and redundancy** are included in the standard optimization models. The ability to respond to pipe breaks is of practical importance, but as stated in [W&a03], the main difficulty is the absence of a universally accepted quantifiable definition for redundancy and reliability.

This was already pointed out in [A&S77], considering reliability by forcing the network to have a fully looped configuration. As this was said to be not satisfactory, Alperovits and Shamir indicated that "a more intrinsic definition is needed, one which depends on a performance criterion for specified emergency situations." This "lack of reliability measure" is still part of Goulter's criticism on optimization models and is discussed in [G92] in detail. According to Geem [G00], networks designed on a purely cost-effective basis tend to be driven to a branched layout. This was already stated by Alperovits and Shamir [A&S77]:

> "Experience has shown that when a network is designed for a single loading, unless a minimum diameter is specified for all pipes, the optimal network will have a branching configuration: i.e., all loops will be opened in the process of the solution by deleting certain pipes."

Therefore minimum diameters constraints have often been used to ensure the retainment of loops. Obviously, the use of minimum diameters does not necessarily allow repairing of all segments.

In the context of risk analysis Mays [M99] provides a definition: *"Reliability* is defined as the complement of risk: i.e., the probability of nonfailure." Further research could focus on this aspect, that even in case of taking one pipe out of

Figure 2.1: Construction of an arch to prepare the tunnel to be lined with concrete[1]

Figure 2.2: This finished section of the tunnel, which is 24 feet in diameter, is under Shaft $13B$, near Central Reservoir[1]

[1]Courtesy of New York City Department of Environmental Protection

service for repairing or due to a break, the demand can still be met at least at a high percentage level. In the optimization problem introduced in the next section, this would increase the dimension of the decision vector as well as the number of equality constraints, since one could use new flow and head variables for every scenario.

Demand projections always are a great source of uncertainty. Here two aspects have to be considered: On the one hand changes in industry or real estate development can have great influence on demand and on the other hand location of piping capacity influences the real estate development.

According to [W95] sizing for pipes with a diameter larger than 16 inch is generally controlled by municipal and industrial water consumption and not by fire flows. When designing subdivisions most pipes will be of minimum diameter (6 inch in the United States) and the demand is driven by the fire-flow requirements. Pipes built in one of the test examples, the New York City Network, are shown in Figure 2.1 and 2.2.

In most models, **pumps and reservoirs** are not included. According to Walski [W&a95], optimization models should be able to handle more complicated problems to get closer to reality. As shown in [S&S93], enhancements of the model still include the originally formulated network optimization model as one component.

The current relevance of the problem is underlined by Loganathan, Sherali, Park and Subramanian [L&a02], where a subdivision of the water distribution system of Blacksburg, Virginia is calculated with realistic cost data. A second example is given by Kordab [K02], who designed the water distribution system of Escheburg, a small municipality located at the south east of Hamburg, with the EPANET software of the United States Environmental Protection Agency.

Finally, it has to be repeated that NOP and the expansions presented in this chapter are a model, i.e. an approximation of reality. The physical aspects of water distribution network design are quite well understood and the model is improved in Chapter 3 with the possibility to calculate the pressure head with the Darcy-Weisbach formula. One major challenge for further research includes economic aspects, especially the process of determining the cost functions.

Improvements of the first algorithms have "been obtained with sophisticated and elegant algorithms, which hold more attraction for theoreticians than for practitioners" [G92]. Even if Goulter may be right in saying this, the improvements in the algorithms – getting global optima in acceptable time – turns them into a factor of growing significance in all stages of the design process. The remainder of this study focuses upon the reliability of these calculations.

2.2 Network Optimization Problem (NOP)

"The least-cost pipe design problem is a hard nonconvex optimization problem having a number of local optima, ..." [S&a98]

2.2.1 Formulation of the Optimization Problem

The network optimization problem (NOP) first formulated by Schaake and Lai [S&L69] and later expanded is known in the formulation of Sherali, Subramanian and Loganathan ([S&a01] and [S01]) as

$$
\begin{cases}
\sum_{(i,j)\in P} \sum_{k=1}^{K} c_k\, x_{ijk} + \sum_{i\in S} c_{si}\, H_{si} \to \min & \\
 & \\
\Phi_{ij}(q_{ij},\, x_{ij}) = (H_i + E_i) - (H_j + E_j) & \forall (i,j) \in A \\
\sum_{k=1}^{K} x_{ijk} = L_{ij} & \forall (i,j) \in P \\
\sum_{j:(i,j)\in A} q_{ij} - \sum_{j:(j,i)\in A} q_{ji} = b_i & \forall i \in D \\
\sum_{j:(i,j)\in A} q_{ij} - \sum_{j:(j,i)\in A} q_{ji} \le b_i & \forall i \in S \\
q_{\min_{ij}} \le q_{ij} \le q_{\max_{ij}} & \forall (i,j) \in A \\
H_i + E_i \le F_i + H_{si} & \forall i \in S \\
H_{iL} \le H_i + E_i \le H_{iU} & \forall i \in D \\
H_{si} \ge 0 & \forall i \in S \\
x_{ijk} \ge 0 & \forall (i,j) \in P\,, \quad k = 1, \dots, K \\
x_{ijk} = \widetilde{x_{ijk}} & \forall (i,j) \in A \setminus P\,, \quad k = 1, \dots, K\,.
\end{cases}
$$

According to standard terminology, the interpretation of the variables are summarized in Table 2.1. The network layout is determined by a simple directed graph, represented by the set of nodes N, the set of arcs A and the length L_{ij} of the corresponding arc (i,j). An illustration for graphs of standard test problems are provided in the context of a detailed description of the corresponding networks in Figures 5.2, 5.3 and 5.4.

The objective is to determine the diameters of minimum cost for all pipes so that given water pressure requirements are satisfied in all nodes. The decision variables

N	Set of nodes of the network, $N = S \cup D$
S	Set of source nodes
D	Set of demand nodes
A	Set of directed arcs of the network
P	Set of arcs $P \subset A$ to be newly designed
d_k	Standard available diameters for segments of pipes, $k = 1, \dots, K$
c_k	Present value of construction and maintenance costs per unit length for a pipe of diameter d_k, including lining and cleaning
$\boldsymbol{x_{ijk}}$	**Length of segment of diameter $\boldsymbol{k}$ in link $\boldsymbol{(i,j) \in A}$**
x_{ij}	Vector with components x_{ijk}, $k = 1, \dots, K$
$\widetilde{x_{ijk}}$	Fixed value for length $x_{ijk} \in A \backslash P$
L_{ij}	Pipe length of a connection between nodes i and j
$\boldsymbol{q_{ij}}$	**Water flow rate in the pipe of link $\boldsymbol{(i,j)}$**
$q_{\min_{ij}}, q_{\max_{ij}}$	Analytically determined minimum and maximum bounds for q_{ij}
b_i	Net water supply ($b_i > 0$) or demand ($b_i \leq 0$) rate of node $i \in N$
E_i	Ground elevation at node $i \in N$
$\boldsymbol{H_i}$	**Established head above $\boldsymbol{E_i}$**
H_{iL}, H_{iU}	Bounds for the established head $H_i + E_i$ at demand nodes
F_i	Fixed maximum available energy head at source node S_i
c_{si}	Present value of costs per unit energy head for raising head above E_i
$\boldsymbol{H_{si}}$	**Additional head to be developed at each source node $\boldsymbol{i \in S}$**
Φ	Pressure head-loss in a given pipe segment due to friction
C_{HW}	Hazen-Williams coefficient based on roughness and diameter

Table 2.1: Meaning of variables of NOP, decision variables are marked bold.

include pipe segment x_{ijk} which represents the segment of the pipe (i, j) to be build of diameter k. A comprehensive summary of the model formulation is presented by Sherali, Totlani and Loganathan [S&a98] and is cited here to introduce the optimization problem:

> "Consider a distribution network comprised of a set of reservoirs or supply nodes and a set of consumption or demand nodes. Let these nodes be collectively identified by the index set $N = \{1, 2, \ldots, n\}$, where the set of source nodes is denoted by $S \subset N$ and the set of demand or transshipment nodes is denoted by $D \subset N$ such that $N = S \cup D$. Associate with each node a quantity b_i that represents the net water supply rate or demand rate corresponding to node i in the index set N. We will assume that $b_i > 0$ for $i \in S$ and $b_i \leq 0$ for $i \in D$. To ensure feasibility, we assume that the total supply rate is at least equal to the total demand rate.
>
> In standard terminology, sections of pipes are defined to be short (2–4m) lengths of pipe that are used to physically construct a pipeline. A segment is defined to be a length of pipe having constant properties of diameter, roughness, and annualized cost per unit length, perhaps composed of many sections. Links are defined as a collection of segments between two nodes, the lengths of which add up to the required length of the pipeline between the nodes.
>
> For each pipe connecting node pairs i and j, where $i, j \in N$, $i < j$, we create a (notationally directed) arc $(i, j) \in A$. For each $(i, j) \in A$, let L_{ij} denote the pipe length corresponding to an existing connection in the network between the nodes i and j. We will assume that each such link is constructed from segments of lengths having standard available diameters, chosen from the set $\{d_k,\ k = 1, \ldots, K\}$. Also, let us denote by c_k the cost per unit length for a pipe of diameter d_k.
>
> Associated with each link connecting node pairs $(i, j) \in A$ is the decision variable q_{ij} that represents the flow rate. Note that this variable may take any real value, thus permitting flow in either direction. A positive flow value means that flow is along the specified conventional direction of the arc.
>
> Our next set of decision variables relates to the lengths of segments having different standard diameters chosen from the set of available diameters that comprise each link of the network. Let x_{ijk} denote the length of segment of diameter k in the link $(i, j) \in A$, and let $\mathbf{x}_{ij}$ be the vector having components x_{ijk}, $(k = 1, \ldots, K)$. Now let us consider the heads at the various nodes $i \in N$ in the network. For each node $i \in S$ let E_i denote the ground elevation of node $i \in N$ and let H_i denote the established pressure head above E_i. Additionally, for the source nodes $i \in S$, let F_i denote the fixed maximum available head, and suppose

that there is an opportunity to further raise this head by an amount H_{si} at an annualized cost $c_{si} > 0$ per unit head. Correspondingly, for each demand node $i \in D$, suppose that there is the requirement that at a flow equilibrium; the established head $(H_i + E_i)$ at this node lies in the interval $[H_{iL}, H_{iU}]$, where $H_{iL} \leq H_{iU}$."

For the calculations in this thesis, the interpretation of all costs as annualized costs is changed to regard the present value of the overall costs, to coincide with Linaweaver and Clark, who investigated the correlation between capital costs and diameter for pipelines and tunnels. Their results have been used by Schaake and Lai [S&L69] when introducing the New York test example.

The constraints include the link length, the conservation of flow at all nodes and the conservation of energy. With the latter it is implicitly guaranteed that the hydraulic head loss over each loop in the pipe system is zero. Three formulae are commonly used to calculate the pressure head-loss due to friction in a pipe segment: Hazen-Williams, Chezy-Manning and Darcy-Weisbach. The Hazen-Williams formula is widely used in North America, Chezy-Manning is more commonly applied for open channel flow. For analyzing the water distribution system of Escheburg, Kordab used the Darcy-Weisbach formula and points out that it "is theoretically the most correct among the three and it is widely used in Europe" [K02].

According to mathematical publications of the standard problem (see e.g. [S&L69], [A&S77] or [S&a01]), the Hazen-Williams formula is used in the following. An expansion to the Darcy-Weisbach formula is presented in Chapter 3. Assuming smooth-flow conditions, the empirical Hazen-Williams equation is described by

$$\Phi(q,\, C_{HW},\, d,\, x) = c_{hl}\, \mathrm{sgn}\,(q) \left|\frac{q}{C_{HW}}\right|^{c_d} d^{-4.87}\, x\,. \tag{2.2.1}$$

For $k = 1, \ldots, K$ let α_k be defined by $\alpha_k := c_{hl}\, C_{HW}^{-c_d}\, d_k^{-4.87}$. To conform with calculations in literature, the units specified by Sherali, Subramanian and Loganathan [S&a01] are used throughout this thesis, i.e. instead of converting all input data to SI units, the original ones are kept. Hence, with cubic-meters per hour for the flow, meters for the pipe length and centimeters for the pipe diameter as standard units, it is

$$\begin{aligned}
\Phi_{ij}(q_{ij},\, x_{ij}) &= \sum_{k=1}^{K} \Phi(q_{ij},\, C_{HW},\, d_k,\, x_{ijk}) \\
&= c_{hl}\, \mathrm{sgn}\,(q_{ij}) \left|\frac{q_{ij}}{C_{HW}}\right|^{c_d} \sum_{k=1}^{K} d_k^{-4.87}\, x_{ijk} \\
&= \mathrm{sgn}\,(q_{ij})\, |q_{ij}|^{c_d} \sum_{k=1}^{K} \alpha_k\, x_{ijk}\,,
\end{aligned}$$

where the discharge coefficient $c_d = 1.852$ and the head-loss coefficient $c_{hl} = 15,200$ are used.

For the given New York test network, the flow is assumed to be in cubic-feet per second, the pipe length in feet and the pipe diameter in inches. So as in [S01], the adjusted parameter $c_{hl} = 851,500$ is obtained due to the corresponding units and for comparability reasons $c_d = 1.85$ is used.

C_{HW} is assumed to be constant for all pipes of the network due to the facts given by the test networks. Furthermore, if not otherwise specified, for all these networks no possibility to develop additional head at the source nodes is considered, i.e. $H_{si} = 0$ is assumed.

2.2.2 Standard Test Networks

In literature, three standard test cases have become widely accepted, one simple two-loop example as well as the networks of New York City and Hanoi. For all these problems Sherali, Subramanian and Loganathan [S&a01] present "proven global optimal solutions within a tolerance of 10^{-4} and / or within 1$ of optimality".

The **two-loop** test network was first introduced by Alperovits and Shamir [A&S77] in 1977 as example for their linear programming gradient method. This pipeline network operates under gravity for one loading. No possibility to develop additional head at the source node is regarded. The network topology consists of two loops.

Already in 1969, the **New York City** water supply system was introduced by Schaake and Lai [S&L69]. Probably because of its illustrative nature and its simplicity, it has become a benchmark for optimization approaches.

According to the NYC 2003 Drinking Water Supply and Quality Report [B&W03], this system supplies water for a daily consumption of approximately 1.2 billion gallons, i.e. about 4.5 billion liters. It consists of a system of wells in Queens, which provides less than 1 % of the total supply, and of three unfiltered surface water sources: The Croton provides about 10 %, the Catskill and the Delaware water collection systems approximately 90 %.

The optimization problem is an expansion network. The New York City Tunnel No. 1 connects the Hillview Reservoir to Brooklyn by way of Manhattan. City Tunnel No. 2 extends between Hillview Reservoir and Richmond by way of Queens. They were put into service 1917 and 1937, respectively. The construction of City Tunnel No. 3 began in 1970 to enhance and improve the water delivery system, work is currently proceeding. Ultimately, it will allow inspection and repair of City Tunnels No. 1 and 2.

The third example is the network of **Hanoi.** This planned water distribution trunk network in Hanoi, Vietnam, was introduced by Fujiwara and Khang [F&K90] in 1990 to illustrate a two-phase decomposition method described in their paper. It is a single source network with thirty-two nodes and thirty-four arcs.

The dimensions of all test networks are summarized in Table 2.2. The minimum costs known decreased from the first published solution by about 16% for the two-loop, about 51% for the New York and about 4% for the Hanoi test network. The network layouts are illustrated in Section 5.2 together with a detailed description of the input data used in WaTerInt.

2.2.3 Cost Function

For the costs of the New York test example, Schaake and Lai [S&L69] originally used results of a study of Linaweaver and Clark [LC64] of the factors that influence the cost of transmission facilities. According to Linaweaver and Clark, one basic unit to present data are dollars per mile for the investment cost of a given size facility. For these, they present data for 55 pipelines and 21 tunnels. As a result of a regression analysis for the variable diameter d in inches and the capital cost K they obtain for pipelines

$$K = 1,890\ d^{1.29}\ \frac{\text{US \$}}{\text{mile}} = 0.3580\ d^{1.29}\ \frac{\text{US \$}}{\text{ft}} \tag{2.2.2}$$

with a correlation coefficient of 0.98 , and for tunnels

$$K = 5,800\ d^{1.24}\ \frac{\text{US \$}}{\text{mile}} = 1.0985\ d^{1.24}\ \frac{\text{US \$}}{\text{ft}} \tag{2.2.3}$$

with a correlation coefficient of 0.77 .

The costs of Equation 2.2.3 are used for the New York networks, as part of Table A.1. Thus, as already stated, for the calculations part of this thesis overall costs are assumed, whereas Sherali, Totlani and Loganathan [S&a98] define their model as based on annualized costs.

Consequently, the costs of the optimization problem are on one hand the present value of construction and maintenance costs c_k per unit length for a pipe of diameter

	No. of nodes	No. of arcs	No. of available diameters	No. of nonlinear equalities	No. of linear equalities	No. of decision variables
Two-loop	7	8	14	8	14	127
New York (4")	26	33	43	33	58	1,478
New York (12")	26	33	15	33	58	554
Hanoi	32	34	6	34	65	270

Table 2.2: Dimension of standard test networks and number of linear and nonlinear equality constraints.

d_k, $k \in \{1, \ldots, K\}$ including lining and cleaning, and on the other hand the present value of the costs per unit energy c_{s1} for raising head above E_1 for the life span of the pipes. Simplified, it is assumed that this life span is the same as the planning horizon for the distribution network.

For all standard networks, the first are provided in Table A.1 and summarized in Figure 2.3. For all these networks only one source node exists, the head at the source node is fixed and the energy costs are set to zero. In reality, the energy costs are of great significance for water distribution design, and have therefore been included in the calculations in WaTerInt.

To add energy costs for the whole network, it suffices to consider the cost per meter head at the source node, as proposed by the formulation of NOP. Because of energy conservation, the head at the source node multiplied with the supply is obviously the same as the sum of head loss in all pipes multiplied with corresponding flow and the head at demand nodes multiplied with demand:

2.2.1 Lemma. For the single-loading nonlinear optimization problem (NOP) it holds

$$(H_1 + E_1)\, b_1 = \sum \Phi_{ij} q_{ij} - \sum_{k=2}^{n} (H_i + E_i)\, b_i \,.$$

Proof. Direct insertion results in

$$\begin{aligned}
\sum_{i=1}^{n} (H_i + E_i)\, b_i &= \sum_{i=1}^{n} (H_i + E_i) \left(\sum_{j:\,(i,j)\in A} q_{ij} - \sum_{j:\,(j,i)\in A} q_{ji} \right) \\
&= \sum_{i=1}^{n} \sum_{j:\,(i,j)\in A} (H_i + E_i)\, q_{ij} - \sum_{j=1}^{n} \sum_{i:\,(i,j)\in A} (H_j + E_j)\, q_{ij} \\
&= \sum_{(i,j)\in A} \left((H_i + E_i) - (H_j + E_j)\right) q_{ij} \quad = \quad \sum_{(i,j)\in A} \Phi_{ij}\, q_{ij} \,.
\end{aligned}$$

□

Figure 2.4 contains an estimation of realistic energy costs c_{s1} for the test networks. These depend on the assumed time period y for calculation in years, the discount factors used and the direct energy cost per kWh, c_{kWh} in US \$ per kWh, i.e.

$$c_{s1} = \frac{1 - q^y}{1 - q} \; \frac{24 \cdot 365 \cdot 9.81}{3,600} \; \frac{1}{\eta_P\, \eta_M} \; c_{\mathrm{kWh}} \, \frac{\mathrm{kWh}}{\mathrm{m}\, \frac{\mathrm{m}^3}{\mathrm{h}}} \,, \tag{2.2.4}$$

where $q = \frac{1}{1+\frac{p}{100}}$, with an underlying interest rate p, and with the assumption $1\,\mathrm{m}^3 = 1,000\,\mathrm{kg}$ for water. The efficiency rate of the pumps is described by η_P and the efficiency rate of the corresponding motor by η_M. Details can be found in Bollrich [B00]. According to engineering practice and to [D95], for the rest of this thesis values of $\eta_P = 0.7$ and $\eta_M = 0.6$ are assumed.

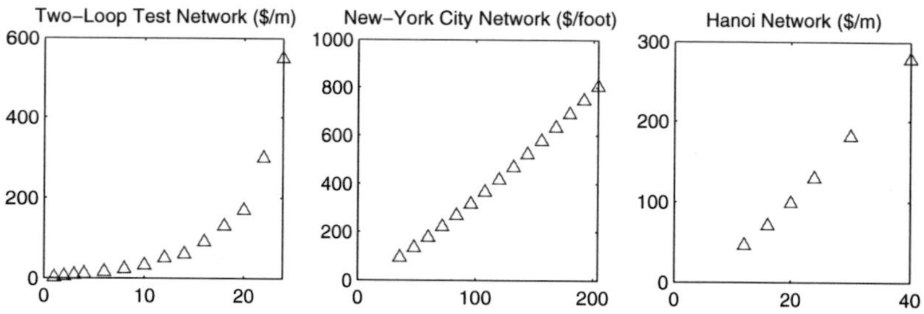

Figure 2.3: Costs per unit length subject to pipe diameters used in the test examples

For the range specified in Figure 2.4, this results in total energy cost of about 24 to 104 TUS$ per meter head for the two-loop network, 1,337 to 5,844 TUS$ per foot for New-York and about 425 to 1,859 TUS$ per meter for Hanoi.

First, the computational results discussed in Chapter 5.2 contain calculations with fixed head at the source node, for comparison with published results. Then, optimal solution with varying initial head are part of Tables D.50 to D.51 and with energy costs of Tables D.52 to D.53 and summarized in Section 5.3.2.

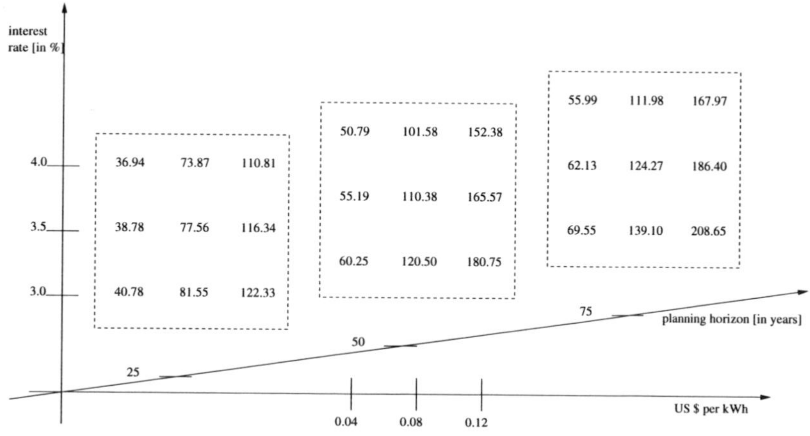

Figure 2.4: Energy costs in US$ per meter head subject to interest rate, cost in US$ per kWh and planning horizon of the network, determined with Equation 2.2.4.

2.2.4 Literature Survey of Optimization Techniques

"Very often, the so-called optimal design is only one of the local optima, ..." [S&L69]

Extensive research on the optimization problem for water distribution systems design has been done and is still going on as shown by recently published papers such as [W&S02], [L&a02], or [A&P05]. This research includes not only projects all over the world but also a wide range of optimization techniques.

A series of sessions held at a conference 1985 in Buffalo, New York, to bring together researches and practicing engineers was called "The Battle of the Network Models". For these sessions each participating group solved a water design problem for the same hypothetical water system. It took into account a lot more components than NOP, for instance a wide range of loadings, pump sizing, storage tank sizing and location. The results are summarized in [W&a95].

In [S&L69] Schaake and Lai present a **joint dynamic and linear programming** model for water distribution systems. The linear program is deduced from the nonlinear problem by transforming the nonlinearity to the objective function and using a linear approximation. The dynamic program regards capacity expansion. This model has been developed in a program of applied research in engineering systems analysis for the primary water distribution system of the City of New York.

Schaake and Lai considered only branched networks. As already shown, loops play an important role in the design when regarding reliability of the system. So the practical use for engineers was limited. Further on in their model the pipe diameters are continuous variables. Schaake and Lai already recommend for further investigations to use a discrete set of commercially available diameters. In the following this discrete problem is considered.

One approach for looped systems was proposed by Alperovits and Shamir [A&S77]. They developed a **linear programming gradient** method, where for a fixed flow the rest of the decision variables are determined with a linear optimization program and afterwards the gradient of the total costs with respect to changes in the flow distribution is determined. Using this gradient to improve the flow in subsequent iterations, a local optimum can be approached.

Quindry et al. [Q&a79] corrected this gradient in 1979, and this algorithm was expanded by improvements published e.g. by Fujiwara, Jenchaimahakoon and Edirisinghe [F&a87] or Kessler and Shamir [K&S89].

These early methods were not able to guarantee to find a global optimum as stated in [F&a87]: "The process is thus iterative and converges to a local optimal solution. It should be noted, however, that this local optimal solution may also be a global optimal solution."

Fujiwara and Khang ([F&K90] and [F&K91]) provide an extension of the linear programming gradient method and additionally allow, in a two-phase decomposition, to move from one local optimal solution to a different and better one.

Other approaches are developed e.g. by Rowell and Barnes [R&B82], who suggest a two level system including reliability criteria. Morgan and Goulter [M&G85] developed an iterative heuristic based on linear programming and the Hardy-Cross method, which is used to determine the flow for fixed diameters [C36].

Heuristic algorithms have been developed since 1970 and combine rules with randomness based on natural phenomena. With high probability these searching algorithms arrive at a point near the global optimum. Examples are simulated annealing, tabu search and evolutionary algorithms. Simulated annealing, based on the metallic annealing processes, was used to solve the water network design optimization problem as part of the model suggested by Loganathan, Greene and Ahn [L&a95] and of Cunha and Sousa 2001 [C&S01].

Genetic algorithms are a class of search algorithms based on artificial evolution. The first application of genetic algorithms to water distribution pipe network optimization has been developed 1992 by Murphy and Simpson according to [D&a96], where Dandy, Simpson and Murphy present an improved genetic algorithm for pipe network optimization with computational results for the New York test problem. The development of the computer model GANET by Savic and Walters involves the application of genetic algorithms [S&W95]. The New York City Water tunnel problem has been chosen by Wu and Simpson to investigate the performance of the self-adaptive boundary genetic algorithm search strategy published in [W&S02]. Recently, Awad and von Poser [A&P05] describe a genetic algorithm formulation and solve the two-loop and New York City problem as well.

Another approach is presented by Geem ([G00], [G&a00] and [G&a01]), where a meta-heuristic method called Harmony Search is developed. The algorithm is conceptualized from a musical improvisation process involving searching for a better harmony.

The genetic algorithms and harmony search consider only one diameter for each link. Thus, as summarized by Alperovits and Shamir [A&S77], the solution is not necessarily optimal for NOP: "In engineering practice it has been the custom to select a single diameter for the entire length of each link. If this is done, the design will not be optimal."

The first global optimization approaches for this problem are **branch and bound** methods. They were developed by Eiger, Shamir and Ben-Tal [E&a94] and Sherali, Subramanian and Loganathan ([S&S97], [S&a98], [S01] and [S&a01]) and are capable of providing solutions within a proven tolerance of a global optimum. To construct lower bounds needed for the branch and bound algorithm, Subramanian [S01] relaxed the nonlinear constraints in a transformed space via polyhedral outer approximations. This transformation and the respective polyhedral approximations form the basis of the verification procedure presented in Chapter 4.

2.3 Algorithm of Sherali, Subramanian and Loganathan

2.3.1 Transformation and Relaxations

Sherali, Subramanian and Loganathan [S&a01] successively branch the hyperrectangle spanned by the flow bounds. For each subproblem they determine a linear relaxation. Fixing in NOP the flow part of the solution of this lower bounding problem, they obtain a linear upper bounding problem.

According to [S&a01] and [S01], the problem is transformed to use monotonicity and a convex-concave nature of the nonlinear constraints for relaxations.

This transformation introduces the definition of

$$v(q) := \operatorname{sgn}(q)\,|q|^{c_d} \tag{2.3.1}$$

as well as $v_{\min} := v(q_{\min})$ and $v_{\max} := v(q_{\max})$.

Because of its monotone nature $v(q)$ can be written as

$$v(q) = \lambda v_{\min} + (1-\lambda)v_{\max} \quad \text{for } 0 \le \lambda \le 1\,,$$

which implies for a value of $c_d = 1.852$

$$q(\lambda) = \operatorname{sgn}\,[\lambda\, v_{\min} + (1-\lambda)v_{\max}]\ |\lambda\, v_{\min} + (1-\lambda)v_{\max}|^{\frac{1}{1.852}}$$

and its derivative

$$q'(\lambda) = \frac{v_{\min} - v_{\max}}{1.852}\,|\lambda\, v_{\min} + (1-\lambda)v_{\max}|^{\frac{-0.852}{1.852}}$$

for $\lambda \in [0\,,\ 1]$, $\lambda\, v_{\min} + (1-\lambda)v_{\max} \neq 0$.

By introducing the new decision variables λ_{ij} , x^1_{ijk} and x^2_{ijk} so that $0 \le \lambda_{ij} \le 1$,

$$x^1_{ijk} := \lambda_{ij}\, x_{ijk}\,, \quad x^2_{ijk} := (1-\lambda_{ij})\, x_{ijk} \quad \text{and} \quad x_{ijk} = x^1_{ijk} + x^2_{ijk} \tag{2.3.2}$$

for all $(i,j) \in A$, Sherali, Subramanian and Loganathan proved that the nonlinear equations of NOP can be written equivalently for each arc $(i,j) \in A$ as

$$\begin{cases} \sum_{k=1}^{K} v_{\min_{ij}} \alpha_k\, x^1_{ijk} + \sum_{k=1}^{K} v_{\max_{ij}} \alpha_k\, x^2_{ijk} = (H_i + E_i) - (H_j + E_j) \\ q_{ij} = \operatorname{sgn}\left[\lambda_{ij} v_{\min_{ij}} + (1-\lambda_{ij}) v_{\max_{ij}}\right] \left|\lambda_{ij} v_{\min_{ij}} + (1-\lambda_{ij}) v_{\max_{ij}}\right|^{\frac{1}{c_d}} \\ q_{\min_{ij}} \le q_{ij} \le q_{\max_{ij}} \\ 0 \le \lambda_{ij} \le 1 \\ v_{\min_{ij}} := v(q_{\min_{ij}})\,, \quad v_{\max_{ij}} := v(q_{\max_{ij}})\,. \end{cases}$$

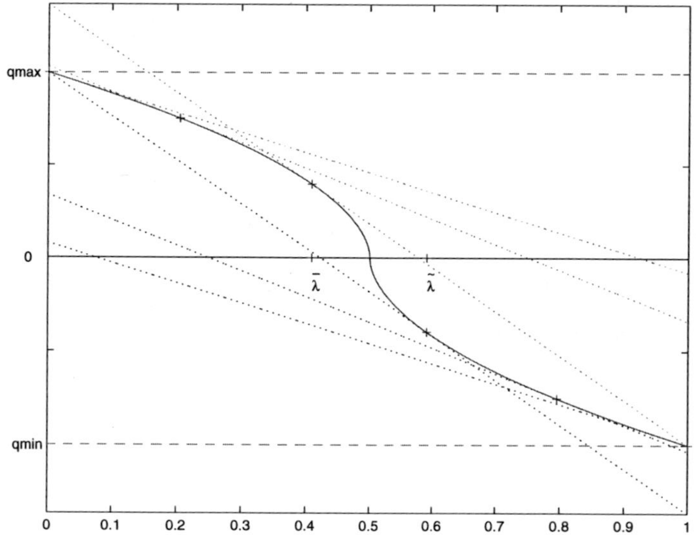

Figure 2.5: Illustration of the relaxations in the concave-convex case.

This yields an equivalent formulation of NOP. For deriving a relaxation, the bilinearities in Equation 2.3.2 are omitted, but the obvious aggregate relationship

$$\sum_{j=1}^{K} x^1_{ijk} = \lambda_{ij} L_{ij} \tag{2.3.3}$$

is added.

So with the definition of

$$\alpha_k = c_{hl}\, C_{HW}^{-c_d}\, d_k^{-4.87}$$

for all $k \in \{1, \dots, K\}$ and the decision variables x^1, x^2, λ, q, H_S and H the following **lower bounding** problem (LNOP) is obtained:

$$\begin{cases}
\displaystyle\sum_{(i,j)\in P} \sum_{k=1}^{K} c_k\,(x^1_{ijk} + x^2_{ijk}) + \sum_{i\in S} c_{si} H_{si} & \to \min \\[2em]
\displaystyle\sum_{k=1}^{K} v_{\min_{ij}} \alpha_k\, x^1_{ijk} + \sum_{k=1}^{K} v_{\max_{ij}} \alpha_k\, x^2_{ijk} & \\
\qquad\qquad = (H_i + E_i) - (H_j + E_j) & \forall (i,j) \in A \\
\displaystyle\sum_{k=1}^{K} (x^1_{ijk} + x^2_{ijk}) = L_{ij} & \forall (i,j) \in P \\
\displaystyle\sum_{k=1}^{K} x^1_{ijk} = \lambda_{ij} L_{ij} & \forall (i,j) \in P \\
x^1_{ijk} = \lambda_{ij} \widetilde{x_{ijk}} & \forall (i,j) \in A \setminus P,\ k = i, \dots, K \\
x^2_{ijk} = (1 - \lambda_{ij}) \widetilde{x_{ijk}} & \forall (i,j) \in A \setminus P,\ k = i, \dots, K \\
\displaystyle\sum_{j:(i,j)\in A} q_{ij} - \sum_{j:(j,i)\in A} q_{ji} = b_i & \forall i \in D \\
\displaystyle\sum_{j:(i,j)\in A} q_{ij} - \sum_{j:(j,i)\in A} q_{ji} \le b_i & \forall i \in S \\
\text{[polyhedral relaxation]} & \\
q_{\min_{ij}} \le q_{ij} \le q_{\max_{ij}} & \forall (i,j) \in A \\
H_i + E_i \le F_i + H_{si} & \forall i \in S \\
H_{iL} \le H_i + E_i \le H_{iU} & \forall i \in D \\
x^1_{ijk} \ge 0,\ x^2_{ijk} \ge 0 & \forall (i,j) \in P, \quad k = 1, \dots, K \\
0 \le \lambda_{ij} \le 1 & \forall (i,j) \in A \\
H_{si} \ge 0 & \forall i \in S\,.
\end{cases}$$

The **polyhedral relaxations** are developed in [S01]. They approximate the non-linear equalities

$$q_{ij} = \operatorname{sgn}\left[\lambda_{ij} v_{\min_{ij}} + (1-\lambda_{ij}) v_{\max_{ij}}\right] \left|\lambda_{ij} v_{\min_{ij}} + (1-\lambda_{ij}) v_{\max_{ij}}\right|^{\frac{1}{c_d}}$$

for all $(i\,,\,j) \in A$.

They are based on the idea to determine the points $\overline{\lambda}_{ij}$ and $\widetilde{\lambda}_{ij}$ with

$$q_{\min_{ij}} - q_{ij}(\overline{\lambda}_{ij}) - (1-\overline{\lambda}_{ij})\, q'_{ij}(\overline{\lambda}_{ij}) = 0 \quad \text{for} \quad \overline{\lambda}_{ij} \in (0\,,\,1)\,,$$

and

$$q_{\max_{ij}} - q_{ij}(\widetilde{\lambda}_{ij}) + \widetilde{\lambda}_{ij}\, q'_{ij}(\widetilde{\lambda}_{ij}) = 0 \quad \text{for} \quad \widetilde{\lambda}_{ij} \in (0\,,\,1)\,,$$

where the tangents are upper respectively lower bounds for the flow.

Then for all $(i, j) \in A$ the following relaxations are chosen according to case differentiation:

(i) $q_{\min_{ij}} < 0$ and $q_{\max_{ij}} > 0$ i.e. $q(\lambda_{ij})$ is concave-convex:

(a) $\overline{\lambda}_{ij}$ and $\widetilde{\lambda}_{ij}$ exist:

$$q_{ij} \leq q_{ij}(\hat{\lambda}) + (\lambda_{ij} - \hat{\lambda})\, q'_{ij}(\hat{\lambda}) \quad \text{for} \quad \hat{\lambda} = 0\,,\, \frac{\overline{\lambda}_{ij}}{2}\,,\, \overline{\lambda}_{ij}$$

and

$$q_{ij} \geq q_{ij}(\hat{\lambda}) + (\lambda_{ij} - \hat{\lambda})\, q'_{ij}(\hat{\lambda}) \quad \text{for} \quad \hat{\lambda} = \widetilde{\lambda}_{ij}\,,\, \frac{1+\widetilde{\lambda}_{ij}}{2}\,,\, 1\,,$$

for illustration see Figure 2.5.

(b) $\widetilde{\lambda}_{ij}$ does not exist:

$$q_{ij} \leq q_{ij}(\hat{\lambda}) + (\lambda_{ij} - \hat{\lambda})\, q'_{ij}(\hat{\lambda}) \quad \text{for} \quad \hat{\lambda} = 0\,,\, \frac{\overline{\lambda}_{ij}}{4}\,,\, \frac{\overline{\lambda}_{ij}}{2}\,,\, \frac{3\,\overline{\lambda}_{ij}}{4}\,,\, \overline{\lambda}_{ij}$$

and

$$q_{ij} \geq \lambda_{ij}\, q_{\min_{ij}} + (1-\lambda_{ij})\, q_{\max_{ij}}$$

(c) $\overline{\lambda}_{ij}$ does not exist:

$$q_{ij} \leq \lambda_{ij}\, q_{\min_{ij}} + (1-\lambda_{ij})\, q_{\max_{ij}}$$

and

$$q_{ij} \geq q_{ij}(\hat{\lambda}) + (\lambda_{ij} - \hat{\lambda})\, q'_{ij}(\hat{\lambda})$$

$$\text{for} \quad \hat{\lambda} = \widetilde{\lambda}_{ij}\,,\, \frac{1+\widetilde{\lambda}_{ij}}{4}\,,\, \frac{1+\widetilde{\lambda}_{ij}}{2}\,,\, \frac{3(1+\widetilde{\lambda}_{ij})}{4}\,,\, 1\,,$$

(ii) $q_{\min_{ij}} \geq 0$ i.e. $q(\lambda_{ij})$ is concave:

$$q_{ij} \geq \lambda_{ij}\, q_{\min_{ij}} + (1-\lambda_{ij})\, q_{\max_{ij}}$$

and

$$q_{ij} \leq q_{ij}(\hat{\lambda}) + (\lambda_{ij} - \hat{\lambda})\, q'_{ij}(\hat{\lambda}) \quad \text{for} \quad \hat{\lambda} = 0\,,\ 0.25\,,\ 0.5\,,\ 0.8\,,\ 0.9\,.$$

(iii) $q_{\max_{ij}} \leq 0$ i.e. $q(\lambda_{ij})$ is convex:

$$q_{ij} \leq \lambda_{ij}\, q_{\min_{ij}} + (1-\lambda_{ij})\, q_{\max_{ij}}$$

and

$$q_{ij} \geq q_{ij}(\hat{\lambda}) + (\lambda_{ij} - \hat{\lambda})\, q'_{ij}(\hat{\lambda}) \quad \text{for} \quad \hat{\lambda} = 0.1\,,\ 0.2\,,\ 0.5\,,\ 0.75\,,\ 1\,.$$

In the first case in (b) and (c) in [S&a01] only four equations are used. They are expanded to six for easier handling during the branch and bound algorithm.

During the determination of the relaxations, the derivative q' is not evaluated at a point where $\lambda\, v_{\min} + (1-\lambda) v_{\max} = 0$, therefore no exception concerning non-differentiability has to be considered when implementing q'.

Some of the computational results published in [S&a01] for the test networks are obtained with a different lower bounding scheme according to the reformulation-linearization technique (RLT) developed by Sherali and Tuncbilek. For these enhanced and stronger lower bounds additional constraints are generated, for detailed description see [S01]. Since the main purpose of WaTerInt is to investigate additional time for verification, it does not contain these alternative lower bounds.

2.3.2 Maximal Spanning Tree Based Branching

The branch and bound algorithm is based on branching the hyperrectangle spanned by the flow-bounds. Thus the number of branching variables is determined by the number of arcs in the network.

Some of the results for the test networks presented by Sherali, Subramanian and Loganathan [S&a01] are obtained using a maximal spanning tree based approach for reducing the candidate set of branching variables (MSTR).

Using MSTR for the smaller two-loop network, the computation time published by Subramanian [S01] decreases about a factor between 4 and 10 depending on use of RLT and number of relaxations. For the larger Hanoi test network no results are published by Subramanian without this additional procedure.

Loganathan, Sherali, Park and Subramanian [L&a02] emphasize that the computational experience "clearly indicated that the maximal spanning tree reduction procedure (MSTR) is an indispensable strategy. In fact, when we suppressed the procedure MSTR for the two larger New York and Blacksburg test problems, we were unable to obtain good quality feasible solutions within the time limit of 10 CPU hours."

For being able to get results in acceptable time this procedure is part of WaTerInt. In the following the definitions of graph theory according to Nemhauser and Wolsey [NW88] are used for description of MSTR.

At the beginning of the branch and bound procedure, a maximal spanning tree for the distribution network is constructed with Kruskal's algorithm using arc weights

$$q_{\max_{ij}} - q_{\min_{ij}} \text{ for all arcs } (i,j) \in A$$

and regarding the graph as undirected.

In Kruskal's algorithms an edge is colored blue to indicate acceptance, and red to indicate rejection. During the algorithm, the edges are colored while maintaining color invariance, which means that there exists a maximal spanning tree containing all of the blue and none of the red edges. This algorithm is summarized in Table 2.3.

The maximal spanning tree is not necessarily unique, but Kruskal's algorithm guarantees to obtain one of the maximal spanning trees. Let T denote the set of arcs which are elements of this tree. The remaining arcs $A \setminus T$ are called non-tree arcs.

Sherali, Subramanian and Loganathan propose that "the supply nodes are connected to a dummy sink via slack arcs having a large weight for this purpose, in order to balance supply and demand, and thereby obtain an equality flow conservation system" [S&a01]. As in the implementation of WaTerInt only one source node is regarded, no additional slack arcs have to be concerned.

The constructed spanning tree "yields a valid basis for the underlying network flow problem, and given the flow on the independent non-basic arcs, the corresponding flows on the dependent basic arcs are uniquely determined" [S&a01]. This is explained in the derivation of the network simplex [NW88] as the flow equations are a non-singular linear system of equations, and is shown directly in the following lemma explaining the update of the flow bounds.

2.3.1 Lemma. Let (N, A) be a simple directed graph, and let $T \subset A$ be a spanning tree of the corresponding undirected graph, i.e. a tree that connects all of the vertices N. Let the source node $s \in N$ be assigned as root of this tree. On all edges (k,l) element of $A \setminus T$ let the flow be determined as q_{kl}.

Additionally, let the flow equations

$$\sum_{j:(i,j)\in A} q_{ij} - \sum_{j:(j,i)\in A} q_{ji} = b_i \quad \text{for all } i \in N \setminus \{s\}$$

hold true.

Then the flow is determined for all $(i,j) \in A$.

Determination of Maximum Spanning Tree

Initially each vertex is regarded as a separate blue tree.
 	
for all edges E element of A in decreasing order by costs
    if both ends of E are element of the same blue tree, color E red,
    else color E blue
    fi
rof

Table 2.3: Kruskal's Algorithm

Proof. With complete induction it is shown that q_{ij} is determined for all $(i,j) \in T$. The flow equation implies

$$\sum_{j:(i,j)\in A\setminus T} q_{ij} + \sum_{j:(i,j)\in T} q_{ij} - \sum_{j:(j,i)\in A\setminus T} q_{ji} - \sum_{j:(j,i)\in T} q_{ji} = b_i \quad \text{for all } i \in N \setminus \{s\}\,.$$

According to the definition of a leaf $l \in T$ there exists only one edge $(u,v) \in T$ with $u = l$ or $v = l$. W.l.o.g. let this edge be (k,l). This implies the determination of q_{kl} with

$$q_{kl} = \sum_{j:(l,j)\in A\setminus T} q_{lj} - \sum_{j:(j,l)\in A\setminus T} q_{jl} - b_l\,.$$

Assuming as induction hypotheses that the flow is determined for all children of $l \in T$, it has to be shown that the flow on edge (k,l) or respectively (l,k) is determined for the parent $k \in T$. W.l.o.g. (k,l) is regarded, then all $c \in T$ with $c \neq k$ and $(l,c) \in T$ or $(c,l) \in T$ are children of l. So again the flow equation implies the determination of q_{kl} with

$$q_{kl} = \sum_{j:(l,j)\in A\setminus T} q_{lj} + \sum_{c:(l,c)\in T} q_{lc} - \sum_{j:(j,l)\in A\setminus T} q_{jl} - \sum_{\substack{c:(c,l)\in T\\ c\neq k}} q_{cl} - b_l\,,$$

which also holds true for the source node $k = s$ and therefore proofs the assumption. □

So only the non-tree arcs have to be bisected, which extremely reduces the number of branching variables.

Additionally, the flow bounds are adjusted during the branch and bound procedure before evaluating the lower bounding problem. For all tree-arcs (i,j) ordered according to a post-order tree traversal, as direct consequence of the flow equations the following update is used:

$$
\begin{aligned}
q_{\max_{ij}} &= \min\left\{ q_{\max_{ij}} \quad , \quad -b_j - \sum_{\substack{k:(k,j)\in A \\ k\neq i}} q_{\min_{kj}} + \sum_{k:(j,k)\in A} q_{\max_{jk}} \ , \right. \\
&\qquad \left. b_i + \sum_{k:(k,i)\in A} q_{\max_{ki}} - \sum_{\substack{k:(i,k)\in A \\ k\neq j}} q_{\min_{ik}} \right\} \\
q_{\min_{ij}} &= \max\left\{ q_{\min_{ij}} \quad , \quad -b_j - \sum_{\substack{k:(k,j) \\ k\neq i}} q_{\max_{kj}} + \sum_{k:(j,k)\in A} q_{\min_{jk}} \ , \right. \\
&\qquad \left. b_i + \sum_{k:(k,i)\in A} q_{\min_{ki}} - \sum_{\substack{k:(i,k)\in A \\ k\neq j}} q_{\max_{ik}} \right\} .
\end{aligned}
$$

This procedure extremely reduces the branching variables, as according to Subramanian the structure of most water distribution networks is almost “tree-like”, i.e. they have a value of ρ close to 1 , where ρ is defined as the arc-to-node ratio of a network, i.e. $\rho = \frac{|A|}{|N|}$ (see [S01]).

2.3.3 Branch and Bound Algorithm

Sherali, Subramanian and Loganathan [S&a01] use a standard branch and bound algorithm, summarized in Table 2.4. Again, a comprehensive introduction is provided by Sherali, Totlani and Loganathan [S&a98]:

> “Each branch-and bound node principally differs in the specification of the hyperrectangle Ω . The hyperrectangle associated with note t of the branch-and-bound tree at the main iteration or stage S of the procedure is denoted by
>
> $$\Omega^{S,t} = \{q \ : \ q_{\min}^{S,t} \leq q \leq q_{\max}^{S,t}\} .$$
>
> In our implementation of the branch-and-bound procedure, we successively partition the hyperrectangle defined by the initial bounds $\Omega^{1,1} \equiv \Omega$ on the flow variables into smaller and smaller hyperrectangles. At any stage S of the branch-and-bound algorithm, we have a set of active or nonfathomed nodes denoted as T_S . We select an active nodes t^* in T_S that has the least lower bound, breaking ties arbitrarily, and we partition the hyperrectangle associated with this node according to a suitable branching variable selection strategy. [...] The selection of a branching

variable according to these strategies ensures convergence of the overall procedure to a global optimum for NOP (Ω). The solution process continues by solving the bounding problems for the resulting two node subproblems, and then fathoming nodes as permissible on the basis of this analysis. Whenever the set of active nodes is empty, the algorithm terminates."

For each subproblem, for the optimal solution of the corresponding lower bounding problem the distance to feasibility Δ_A is defined by

$$\Delta_A := \max_{(i,j)\in A} \left| (H_i + E_i) - (H_j + E_j) - \operatorname{sgn}(q_{ij}) \, |q_{ij}|^{c_d} \sum_{k=1}^{K} \alpha_{ijk} x_{ijk} \right| .$$

So an optimal solution of NOP is obtained if $\Delta_A = 0$, and analogous to [S&a01] a vector with $\Delta_A < 10^{-6}$ (`DELTA_ACC`) is accepted to be admissible in WaTerInt.

According to the structure of the optimization problem the upper bounds can be obtained by solving a linear problem. Therefore the flow obtained with the lower bounding problem is regarded as fixed and the original problem NOP is solved.

This results in the following linear **upper bounding problem** for the decision variables x_{ijk} and H_i:

$$\begin{cases}
\sum_{(i,j)\in P} \sum_{k=1}^{K} c_k \, x_{ijk} + \sum_{i\in S} c_{si} \, H_{si} \to \min & \\[2em]
v(q_{ij}) \sum_{k=1}^{K} \alpha_k \, x_{ijk} = (H_i + E_i) - (H_j + E_j) & \forall (i,j) \in A \\
\sum_{k=1}^{K} x_{ijk} = L_{ij} & \forall (i,j) \in P \\
H_i + E_i \le F_i + H_{si} & \forall i \in S \\
H_{iL} \le H_i + E_i \le H_{iU} & \forall i \in D \\
H_{si} \ge 0 & \forall i \in S \\
x_{ijk} \ge 0 & \forall (i,j) \in P, \quad k = 1, \dots, K \\
x_{ijk} = \widetilde{x_{ijk}} & \forall (i,j) \in A \setminus P, \; k = 1, \dots, K .
\end{cases}$$

```
Branch & Bound Loop

for 1 to max-iterations
    for both subproblems
        Computation of flow bounds according to MSTR
        Determination of adjusted bounds: lb, ub
        Computation of relaxations: Aeq, A, b
        Solution of lower bounding problem: sub_lb
        if (admissible region of subproblem is empty) prune subproblem

        Determination of branching index and Δ_A
        if (Δ_A < DELTA_ACC) update incumbent solution, prune subproblem

        Determination of upper bounding problem for fixed flow
        Solution of upper bounding problem: fval_ub
        if (fval_ub < global_ub) update incumbent solution
    rof

    for all subproblems
        if (sub_lb ≥ (1-accuracy)*global_ub) prune subproblem
    rof

    if (list of subproblems is empty) stop: optimal solution found

    Selection of node to be branched
    Branching of subproblem: two new subproblems
    Removal of branched subproblem
rof
```

Table 2.4: Branch and Bound Algorithm

For the subproblem with the least lower bound, the index for bisecting the flow interval is chosen according to one of the following **branching variable selection strategies** recommended in [S&a01]. For description and comparison of further strategies see [S01].

Let $\widehat{q}$ denote the optimal flow determined with the lower bounding problem at iteration step $S+1$ of the branch and bound procedure.

If an improvement in the global lower bound (GLB) is obtained, i.e.

$$0.9 \cdot \mathrm{GLB}_{S+1} > \mathrm{GLB}_S$$

then the branching variable (r,s) is selected as

$$(r,s) \in \operatorname{argmax}\left\{\min\left(q_{\max_{ij}} - \widehat{q_{ij}}\ ,\ \widehat{q_{ij}} - q_{\min_{ij}}\right) : \quad (i,j) \in A \setminus T\right\}\ ,$$

and the corresponding bisected intervals are $[q_{\min_{rs}}\ ,\ \widehat{q_{rs}}]$ and $[\widehat{q_{rs}}\ ,\ q_{\max_{rs}}]$,

else

$$(r,s) \in \operatorname{argmax}\left\{q_{\max_{ij}} - q_{\min_{ij}} : \quad (i,j) \in A \setminus T\right\}$$

is used and the subintervals are $[q_{\min}, 0]$ and $[0, q_{\max}]$ if $q_{\min} \le 0 \le q_{\max}$, otherwise the interval is cut into halves.

A second branching strategy is based on the maximal distance to feasibility. As long as the length of the non-tree arc flow bounds are greater than a fixed value, e.g. 100 in WaTerInt, the branching variable is chosen as

$$(r,s) \in \operatorname{argmax}\left\{q_{\max_{ij}} - q_{\min_{ij}} : \quad (i,j) \in A \setminus T\right\}\ ,$$

afterwards the branching variable (r,s) is chosen as argument of $\Delta_{A\setminus T}$. The subintervals in both cases are again $[q_{\min}, 0]$ and $[0, q_{\max}]$ if $q_{\min} \le 0 \le q_{\max}$ and otherwise halves of the interval to be bisected.

The results for both branching strategies, the first identified by the parameter 1 and the second by 2, are compared in Tables D.37 and D.38. For the New York networks, within the maximum number of iterations no solution could be obtained with the second strategy, therefore the first one is preferred and used as default.

2.4 Head Constraint Propagation

The constraint propagation technique introduced by Sherali, Subramanian and Loganathan [S&a01] identifies parts of the flow bounds that cannot contain feasible points due to the Kirchhoff's equations. In addition, the head bounds can be used to reduce the hyperrectangle spanned by the flow bounds.

In essence, this relies on the fact that the function for determining the head loss in a pipe for given flow $\Phi(q)$ is strictly monotone and invertible, and that this function is monotone with respect to d, when regarded for fixed flow as function in the diameter, i.e. $\Phi(d)$.

Thus, in **a first step** for the incumbent subproblem, the new flow bounds obtained after the flow constraint propagation are used to improve the head bounds. Bounds for the admissible head loss in the pipe (i, j) can be determined for newly to be designed arcs by

$$[\widetilde{\phi}_{\min_{ij}}, \widetilde{\phi}_{\max_{ij}}] = \begin{cases} [\Phi_K(q_{\min_{ij}}, L_{ij}), \ \Phi_1(q_{\max_{ij}}, L_{ij})] & \text{for } q_{\min_{ij}} > 0 \\ [\Phi_1(q_{\min_{ij}}, L_{ij}), \ \Phi_K(q_{\max_{ij}}, L_{ij})] & \text{for } q_{\max_{ij}} < 0 \\ [\Phi_1(q_{\min_{ij}}, L_{ij}), \ \Phi_1(q_{\max_{ij}}, L_{ij})] & \text{else,} \end{cases}$$

and for fixed arcs by

$$[\widetilde{\phi}_{\min_{ij}}, \widetilde{\phi}_{\max_{ij}}] = \sum_{k=1}^{K} \left[\Phi_k(q_{\min_{ij}}, \widetilde{x_{ijk}}), \ \Phi_k(q_{\max_{ij}}, \widetilde{x_{ijk}})\right] ,$$

where according to the Hazen-Williams formula

$$\Phi_k(q, x) = \operatorname{sgn}(q) \, |q|^{c_d} \, \alpha_k \, x .$$

Obviously

$$[\phi_{\min_{ij}}, \phi_{\max_{ij}}] = [\widetilde{\phi}_{\min_{ij}}, \widetilde{\phi}_{\max_{ij}}] \cap ([H_{iL}, H_{iU}] - [H_{jL}, H_{jU}]) ,$$

and thus for every arc (i, j) the new head bounds at node i and j are

$$[H^a_{iL}, H^a_{iU}] = [H_{iL}, H_{iU}] \cap \Big([H_{jL}, H_{jU}] + [\phi_{\min_{ij}}, \phi_{\max_{ij}}]\Big)$$

and

$$[H^a_{jL}, H^a_{jU}] = [H_{jL}, H_{jU}] \cap \Big([H_{iL}, H_{iU}] - [\phi_{\min_{ij}}, \phi_{\max_{ij}}]\Big) .$$

In WaTerInt first all non-tree arcs, and then the tree arcs in the post-tree order determined for the MSTR are passed to adjust the head bounds.

Then, in **a second step,** these head bounds imply a minimum and maximum flow for every arc. Again, first the maximum and minimum head loss for every arc is

$$[\phi_{\min_{ij}}, \phi_{\max_{ij}}] = [H_{iL}, H_{iU}] - [H_{jL}, H_{jU}] .$$

Furthermore, it is

$$[\widetilde{q}_{\min_{ij}}, \widetilde{q}_{\max_{ij}}] = \begin{cases} [\Phi_1^{-1}(\phi_{\min_{ij}}, L_{ij}) \,,\, \Phi_K^{-1}(\phi_{\max_{ij}}, L_{ij})] & \text{for } \phi_{\min_{ij}} > 0 \\ [\Phi_K^{-1}(\phi_{\min_{ij}}, L_{ij}) \,,\, \Phi_1^{-1}(\phi_{\max_{ij}}, L_{ij})] & \text{for } \phi_{\max_{ij}} < 0 \\ [\Phi_K^{-1}(\phi_{\min_{ij}}, L_{ij}) \,,\, \Phi_K^{-1}(\phi_{\max_{ij}}, L_{ij})] & \text{else,} \end{cases}$$

where

$$\Phi_k^{-1}(\phi \,,\, L_{ij}) = \operatorname{sgn}(\phi) \left(\frac{|\phi|}{\alpha_k \, L_{ij}} \right)^{\frac{1}{c_d}}$$

and for fixed arcs

$$[\widetilde{q}_{\min_{ij}}, \widetilde{q}_{\max_{ij}}] = [\operatorname{sgn}(\phi_{\min_{ij}}) \left(\frac{|\phi_{\min_{ij}}|}{\sum\limits_{k=1}^{K} \alpha_k \, \widetilde{x_{ijk}}} \right)^{\frac{1}{c_d}} , \operatorname{sgn}(\phi_{\max_{ij}}) \left(\frac{|\phi_{\max_{ij}}|}{\sum\limits_{k=1}^{K} \alpha_k \, \widetilde{x_{ijk}}} \right)^{\frac{1}{c_d}}] \,.$$

This implies for the adjusted flow bounds

$$[q^a_{\min_{ij}}, q^a_{\max_{ij}}] \;=\; [q_{\min_{ij}}, q_{\max_{ij}}] \,\cap\, [\widetilde{q}_{\min_{ij}}, \widetilde{q}_{\max_{ij}}] \,.$$

This constraint propagation is summarized in Table 2.6. The first step is part of the WaTerInt procedure `adjust_head` and the second of `adjust_flow`. Obviously, this reduction is essential for fixed arcs, as the known diameter allows to use tight bounds for the head-loss. For the expansion test networks the computational time is reduced for most networks by about one third to one half, as summarized in Table 2.5.

Test Network	CPU Time (s)	# Sub-Problems	Lower Bound (US$)	Upper Bound (US$)
Two-loop Expansion	1	79	65,013.2711	65,013.2992
	1	9	65,013.2990	65,013.3393
New York (12")	541	14,523	38,049,833.1417	38,049,864.9609
	420	10,929	38,049,839.4065	38,049,864.8659
New York (4")	2,078	34,671	37,896,865.3229	37,896,903.2120
	1,173	19,387	37,896,865.4104	37,896,903.2702

Table 2.5: Comparison of computational times for the expansion test networks as part of Table D.38. For every test example the first row contains the results without and the second with head constraint propagation, i.e. for the first branching strategy 1 and 5 for the second.

Head Constraint Propagation

Adjust head bounds:
for all pipes $(i,j) \in A$:
Determination of $[\phi_{ij_{\min}}, \phi_{ij_{\max}}]$ according to flow bounds
Intersection of $[\phi_{ij_{\min}}, \phi_{ij_{\max}}]$ with difference obtained from head bounds
rof
for all non-tree arcs and then all post-ordered tree arcs (i,j)
$H_{iL} = \max\left(H_{iL}\ ,\ H_{jL} + \phi_{ij_{\min}}\right)$
$H_{jL} = \max\left(H_{jL}\ ,\ H_{iL} - \phi_{ij_{\max}}\right)$
$H_{iU} = \min\left(H_{iU}\ ,\ H_{jU} + \phi_{ij_{\max}}\right)$
$H_{jU} = \min\left(H_{jU}\ ,\ H_{iU} - \phi_{ij_{\min}}\right)$
rof
if ($i \in N$ exists with $H_{iL} > H_{iU}$) prune subproblem
 	
Adjust flow bounds:
for all pipes $(i,j) \in A$:
Determination $[\phi_{ij_{\min}}, \phi_{ij_{\max}}]$ as difference from head bounds
Determination of $\widetilde{q}_{\min_{ij}}, \widetilde{q}_{\max_{ij}}$ for all pipes according to $[\phi_{ij_{\min}}, \phi_{ij_{\max}}]$
Intersection of actual flow bounds with $\widetilde{q}_{\min_{ij}}, \widetilde{q}_{\max_{ij}}$
rof
if (at least one pipe $(i,j) \in A$ exists with $q_{\min_{ij}} > q_{\max_{ij}}$) prune subproblem

Table 2.6: Overview of head constraint propagation.

This procedure can be selected by the branching strategies 3 and 4 in WaTerInt, where 3 refers to the branching strategy 1 as described in Section 2.3.3 together with this head constraint propagation and 4 to the second one.

Due to small rounding errors in floating point calculation when applying this procedure, the two-loop expansion network is identified to be infeasible, see Table D.2. Thus, the case where lower and upper flow bounds for a fixed arc are already the same has to be treated separately. Therefore the branching strategies 5 and 6 are added for floating point calculation. For these, pipes (i,j) where $q_{\min_{ij}} = q_{\max_{ij}}$ are omitted in the adjustment of flow bounds. Again, strategy 5 corresponds to 1 and 6 to 2.

2.5 Properties of the Standard Problem (NOP)

"LP ist the most important development not only for OR but also for mathematics." – M. Grötschel[2]

2.5.1 Adjacency Property

The cost functions used in the standard test examples are plotted in Figure 2.3. As already explained in Section 2.1.3, originally Schaake and Lai used continuous variables to represent pipe diameters.

Alperovits and Shamir used discrete variables. They already stated in 1977 that it "can be shown that at the optimum, each link will contain at most two segments, their diameters being adjacent on the candidate list for that link" and took advantage of this adjacency property to reduce the set of variables in each iteration step.

Fujiwara and Khang again used continuous ones in 1990. But for converting their solution to a solution of the discrete problem, they used this property. Fujiwara and Dey [F&D87] present conditions for the pipe diameters of an optimal solution of NOP, which is based on head-loss calculations due to Hazen-Williams.

This property can also be shown for the Darcy-Weisbach optimization problem, which will be introduced in Chapter 3. It is directly used for the verification of calculations in Chapter 4. The proof of Theorem 2.5.7 is analogous to the one provided by Fujiwara and Dey, except for extracting all parts related to the choice of the Hazen-Willams equation.

The proof is based on the fact that for a fixed flow and fixed head-loss in one pipe, the length of the segments to be chosen of one diameter can be obtained with a linear optimization problem. Certain convexity arguments and linear programming theory then guarantee that any optimal solution of this problem fulfills the adjacency property, the adjacent indices are denoted with k^* and k^*+1. Therefore the following definitions are introduced.

2.5.1 Definition. For all $k \in \{1, \dots K\}$ let $\hat{J}_k$ denote the *unit head-loss due to friction* in a pipe of diameter d_k, i.e. the overall pressure head-loss in a pipe due to friction can be determined as

$$\Phi_{ij}(q_{ij}, x_{ij}) = \sum_{k=1}^{K} \hat{J}_k \, x_{ijk} \, . \tag{2.5.1}$$

[2] H. -J. Lüthi. *Die Jahrestagung 2004 der GOR und NGB.* OR News, 22, (11) 2004.

2.5.2 Definition and Lemma. Let $K \in I\!N$ with $K \geq 3$ and for $k \in \{1, \ldots, K\}$ let x_k, $y_k \in I\!R$.

(i) Suppose $x_1 < x_2 < \ldots < x_K$, then the set of points $\{(x_k, y_k)\}_{k=1}^{K}$ is said to form *a discrete, strictly convex function,* iff

$$\frac{y_{k+1} - y_k}{x_{k+1} - x_k} < \frac{y_{k+2} - y_{k+1}}{x_{k+2} - x_{k+1}} \tag{2.5.2}$$

for all $k \in \{1, \ldots, K-2\}$.

(ii) Let $x_1 > x_2 > \ldots > x_K$, then a simple reordering implies the set of points $\{(x_k, y_k)\}_{k=1}^{K}$ to form *a discrete, strictly convex function,* iff

$$\frac{y_{k+1} - y_k}{x_{k+1} - x_k} > \frac{y_{k+2} - y_{k+1}}{x_{k+2} - x_{k+1}} \tag{2.5.3}$$

for all $k \in \{1, \ldots, K-2\}$.

2.5.3 Lemma. Let $K \in I\!N$ with $K \geq 3$ and for $k \in \{1, \ldots, K\}$ let x_k, $y_k \in I\!R$. Suppose $x_1 > x_2 > \ldots > x_K$ and the set of points $\{(x_k, y_k)\}_{k=1}^{K}$ to form a discrete, strictly convex function.

Then for arbitrary but fixed $k^* \in \{1, \ldots, K\}$ it holds

$$y_k \; > \; y_{k^*} + (x_k - x_{k^*}) \, \frac{y_{k^*} - y_{k^*+1}}{x_{k^*} - x_{k^*+1}} \, .$$

for all $k \in \{1, \ldots, K\}$, with $k \notin \{k^*, \; k^*+1\}$.

Proof: Since the set of points $\{(x_k, y_k)\}_{k=1}^{n}$ forms a strictly convex function, the expression $\dfrac{y_{k+1} - y_k}{x_{k+1} - x_k}$ is strictly monotone decreasing in k.

Therefore for $k \leq k^* - 1$ it is

$$\frac{y_{k^*+1} - y_{k^*}}{x_{k^*+1} - x_{k^*}} < \frac{y_{k^*} - y_k}{x_{k^*} - x_k}$$

and for $k \geq k^* + 2$

$$\frac{y_{k^*+1} - y_{k^*}}{x_{k^*+1} - x_{k^*}} > \frac{y_k - y_{k^*}}{x_k - x_{k^*}} \, ,$$

as illustrated in Figure 2.6. Therefore the assertion follows for all $k \in \{1, \ldots, K\}$, with $k \notin \{k^*, \; k^*+1\}$. □

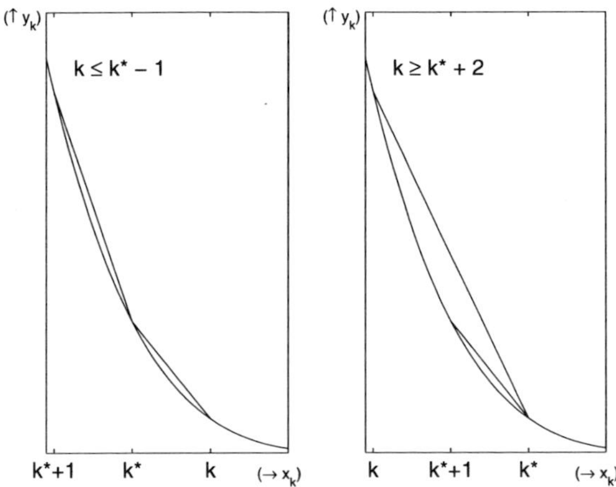

Figure 2.6: Illustration of slopes used in the proof of Lemma 2.5.3.

2.5.4 Lemma. Let $K \in \mathbb{N}$ with $K \geq 3$. For $k \in \{1, \ldots, K\}$ let x_k, $y_k \in \mathbb{R}$ with $y_1 < y_2 < \ldots < y_K$ and suppose the function $f : [x_1, x_K] \to \mathbb{R}$ is strictly monotone and convex.

Let either

(i) $x_1 < x_2 < \ldots < x_K$ and the function f being strictly monotone decreasing

(ii) or $x_1 > x_2 > \ldots > x_K$ and f being strictly monotone increasing.

If the set of points $\{(x_k, y_k)\}_{k=1}^{n}$ forms a discrete, strictly convex function then the set of points $\{(f(x_k), y_k)\}_{k=1}^{n}$ also forms a strictly convex function.

Proof: Assume that $x_1 < x_2 < \ldots < x_K$ and that f is strictly monotone decreasing. As the set of points $\{(x_k, y_k)\}_{k=1}^{K}$ forms a discrete, strictly convex function, for all $k = 1, \ldots, K-2$ it is

$$\frac{y_{k+2} - y_{k+1}}{x_{k+2} - x_{k+1}} > \frac{y_{k+1} - y_k}{x_{k+1} - x_k} .$$

The function $f : D \to \mathbb{R}$ being strictly monotone decreasing and convex, implies for the slopes that

$$\frac{f(x_{k+1}) - f(x_k)}{x_{k+1} - x_k} < \frac{f(x_{k+2}) - f(x_{k+1})}{x_{k+2} - x_{k+1}} < 0$$

for $k = 1, \ldots, K-2$.

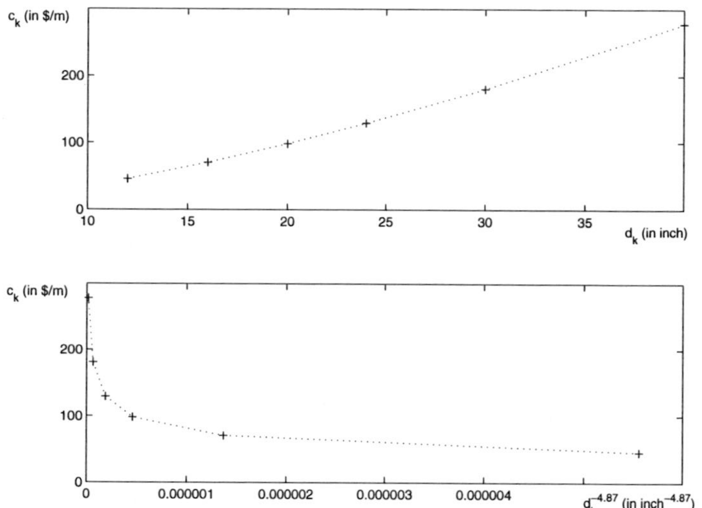

Figure 2.7: Costs per unit length c_k subject to pipe diameters d_k , and subject to a power of diameters $d_k^{-4.87}$ for the Hanoi network $(k = 1, \dots, 6)$.

So, for all $k = 1, \dots, K-2$ direct insertion results in

$$\frac{y_{k+2} - y_{k+1}}{f(x_{k+2}) - f(x_{k+1})} = \frac{y_{k+2} - y_{k+1}}{x_{k+2} - x_{k+1}} \frac{1}{\frac{f(x_{k+2}) - f(x_{k+1})}{x_{k+2} - x_{k+1}}} < \frac{y_{k+1} - y_k}{f(x_{k+1}) - f(x_k)} .$$

As $f(x_1) > f(x_2) > \ldots > f(x_K)$ this proves the first part of the lemma, and an analogous argument proves the second part. □

2.5.5 Remarks. The available diameters and corresponding unit costs of the standard test examples are part of Table A.1, the values of d_k and $d_k^{-4.87}$ for the Hanoi network are illustrated in Figure 2.7. Verified calculations for the following properties can be provided with the toolbox INTLAB.

(i) For the New York and the Hanoi network the strict convexity of $\{(d_k, c_k)\}_{k=1}^{K}$ can easily be shown by verifying the monotonicity of the corresponding slopes.

(ii) The points $\{(d_k, c_k)\}_{k=1}^{14}$ of the two-loop network do not form a strictly convex function. But the pipe costs are a strictly convex function of a power of pipe diameters, more precisely the set of $\{(d_k^{-4.87}, c_k)\}_{k=1}^{14}$ forms a strictly convex function.

2.5.6 Lemma. Let $K \in \mathbb{N}$, $K \geq 3$ and for $k \in \{1, \ldots, K\}$ let c_k, d_k, $\hat{J}_k \in \mathbb{R}$. Suppose $c_1 < c_2 < \ldots < c_K$ and $\hat{J}_1 > \hat{J}_2 > \ldots > \hat{J}_k$.

Assume the following properties are fulfilled:

(i) There exists one and only one $k^* \in \{1, \ldots, K\}$ with

$$\hat{J}_{k^*} \hat{L} \geq \hat{\Phi} \geq \hat{J}_{k^*+1} \hat{L}.$$

(ii) For all $k \in \{1, \ldots, K\}$, with $k \notin \{k^*, k^*+1\}$

$$c_k > c_{k^*} + (\hat{J}_k - \hat{J}_{k^*}) \frac{c_{k^*} - c_{k^*+1}}{\hat{J}_{k^*} - \hat{J}_{k^*+1}}.$$

Then linear optimization problem

$$\begin{cases} \sum_{k=1}^{K} c_k \hat{x}_k \to \min \\ A \hat{x} = \begin{pmatrix} \hat{L} \\ \hat{\Phi} \end{pmatrix} \\ \hat{x} \geq 0, \end{cases} \tag{2.5.4}$$

where $A \in \mathbb{R}^{2 \times K}$,

$$A = \begin{pmatrix} 1 & 1 & \ldots & 1 \\ \hat{J}_1 & \hat{J}_2 & \ldots & \hat{J}_K \end{pmatrix}$$

fulfills the property that at any optimal solution only one or two adjacent x_k will be non-zero.

Proof. As $\hat{J}_1 > \hat{J}_2 > \ldots > \hat{J}_K$, the matrix A is non-singular. Since any two columns of A are linearly independent, choose $B^* = \{k^*, k^*+1\}$ as basic index set. Then the corresponding matrices are

$$A_{B^*} = \begin{pmatrix} 1 & 1 \\ \hat{J}_{k^*} & \hat{J}_{k^*+1} \end{pmatrix} \quad \text{and} \quad A_{B^*}^{-1} = \frac{1}{\hat{J}_{k^*+1} - \hat{J}_{k^*}} \begin{pmatrix} \hat{J}_{k^*+1} & -1 \\ -\hat{J}_{k^*} & 1 \end{pmatrix}.$$

Therefore it is

$$\hat{x}_{B^*} = \begin{pmatrix} \hat{x}_{k^*} \\ \hat{x}_{k^*+1} \end{pmatrix} = A_{B^*}^{-1} \begin{pmatrix} \hat{L} \\ \hat{\Phi} \end{pmatrix} = \frac{1}{\hat{J}_{k^*} - \hat{J}_{k^*+1}} \begin{pmatrix} -\hat{J}_{k^*+1} \hat{L} + \hat{\Phi} \\ \hat{J}_{k^*} \hat{L} - \hat{\Phi} \end{pmatrix}.$$

The inequalities of assumption (i) imply

$$\hat{\Phi} - \hat{J}_{k^*+1}\hat{L} \geq 0 \qquad \text{and} \qquad \hat{J}_{k^*}\hat{L} - \hat{\Phi} \geq 0\,.$$

Since $\hat{J}_{k^*} - \hat{J}_{k^*+1} > 0$ these inequalities are equivalent to $\hat{x}_{B^*} \geq 0$. Therefore this basic solution is an admissible one.

So the feasible area is non-empty. As additionally the objective function is bounded below, there exists an optimal solution.

The components k of the reduced costs for the basic solution $\hat{x}_{B^*}$ are for $1 \leq k \leq K$ and $k \notin \{k^*,\ k^*+1\}$

$$\begin{aligned} c_k - t_k &= c_k - \frac{1}{\hat{J}_{k^*+1} - \hat{J}_{k^*}} \left(\begin{pmatrix} \hat{J}_{k^*+1} & -1 \\ -\hat{J}_{k^*} & 1 \end{pmatrix} \begin{pmatrix} 1 \\ \hat{J}_k \end{pmatrix} \right)^{\top} \begin{pmatrix} c_{k^*} \\ c_{k^*+1} \end{pmatrix} \\ &= c_k - \left(c_{k^*} + \frac{\hat{J}_k - \hat{J}_{k^*}}{\hat{J}_{k^*} - \hat{J}_{k^*+1}} (c_{k^*} - c_{k^*+1}) \right)\,. \end{aligned}$$

Using assumption (ii) for the reduced costs for all non-basic indices k one obtains

$$c_k - t_k > 0\,.$$

Therefore the vector $\hat{x}$ with $x_{k^*},\ x_{k^*+1} \geq 0$ and else $x_k = 0$, is optimal. As the reduced costs are positive for all $k \notin \{k^*,\ k^*+1\}$ this vector is unique. □

2.5.7 Theorem. Let $K \in \mathbb{N}$, $K \geq 3$ and for $k \in \{1,\ldots,K\}$ let c_k, d_k, $J_k \in \mathbb{R}$. Let $d_1 < d_2 < \ldots < d_K$ and $c_1 < c_2 < \ldots < c_K$ and suppose $|\hat{J}_1| > |\hat{J}_2| > \ldots > |\hat{J}_k|$.

Then at any optimal solution of NOP each link will consist of either one pipe segment or two pipe segments of adjacent diameters if the set of data points $\{(|\hat{J}_k|, c_k)\}_{k=1}^{K}$ forms a discrete, strictly convex function.

Proof (see [F&D87]): Let the optimal solution vector be $\left((\hat{x}_{ijk}), (\hat{q}_{ij}), (\hat{H}_l)\right)^{\top}$ for $(i,j) \in A$, $k = 1,\ldots,K$ and $l \in N$. Suppose one pipe (i,j) is regarded and assume w.l.o.g. for the optimal flow $\hat{q}_{ij} \geq 0$, otherwise the orientation of the original problem for this pipe could be changed. If $\hat{q}_{ij} = 0$ obviously $\hat{x}_{ij1} = L_{ij}$ and $\hat{x}_{ijk} = 0$ for $k \in \{2,\ldots,K\}$ is optimal and therefore the adjacency property is fulfilled.

Hence, $\hat{q}_{ij} > 0$ can be assumed. According to the definition of $\hat{J}_k$ as unit head-loss due to friction this implies $\hat{J}_k > 0$ for all $k \in \{1,\ldots,K\}$.

For better readability the indices i and j for this pipe are omitted in the following considerations, and the head-loss in this pipe is defined as $\hat{\Phi}$, i.e.

$$\hat{q} := \hat{q}_{ij}\,, \qquad \hat{\Phi} := (\hat{H}_i - E_i) - (\hat{H}_j - E_j)\,, \qquad \hat{L} := L_{ij}$$

and

$$\hat{x} = (\hat{x}_1\,,\, \hat{x}_2, \ldots, \hat{x}_K) := (\hat{x}_{ij1}\,,\, \hat{x}_{ij2}, \ldots, \hat{x}_{ijK})\,.$$

The optimal length $\hat{x}$ for the pipe (i,j) is a solution of the linear optimization problem 2.5.4 part of Lemma 2.5.6.

As the optimal solution is feasible and as $\hat{L} \geq 0$ it follows from Equation 2.5.1 that

$$\hat{J}_1\,\hat{L} \;\geq\; \hat{\Phi} = \sum_{i=1}^{K} \hat{J}_i\,\hat{x}_i \;\geq\; \hat{J}_K\,\hat{L}\,,$$

and there exists one unique $k^* \in \{1, \ldots, K\}$ with

$$\hat{J}_{k^*}\,\hat{L} \;\geq\; \hat{\Phi} \;\geq\; \hat{J}_{k^*+1}\,\hat{L}\,.$$

Since the set of points $\{(\hat{J}_k, c_k)\}_{k=1}^{n}$ forms a strictly convex function, Lemma 2.5.3 guarantees for all $k \in \{1, \ldots, K\}$, with $k \notin \{k^*,\, k^*+1\}$

$$c_k \;>\; c_{k^*} + (\hat{J}_k - \hat{J}_{k^*})\,\frac{c_{k^*} - c_{k^*+1}}{\hat{J}_{k^*} - \hat{J}_{k^*+1}}\,.$$

Hence, Lemma 2.5.6 implies the adjacency property. □

2.5.8 Corollary. For the standard test networks, NOP with head-loss calculations based on Hazen-Williams equation fulfills the adjacency property.

Proof: It suffices to show that the assumptions of Theorem 2.5.7 are fulfilled:

(i) Obviously for all standard test networks the assumptions $c_1 < c_2 < \ldots < c_K$ and $d_1 < d_2 < \ldots < d_K$ are fulfilled.

(ii) According to the Hazen-Williams equation one obtains

$$\hat{J}_k = c_{hl}\,\mathrm{sgn}\,(q)\left|\frac{q}{C_{HW}}\right|^{c_d} d_k^{-4.87}\,,$$

which implies for fixed $q \in I\!R \setminus \{0\}$ that $|\hat{J}_1| > |\hat{J}_2| > \ldots > |\hat{J}_k|$.

(iii) According to Remark 2.5.5 (ii) the points $\{(d_k^{-4.87}, c_k)\}_{k=1}^{K}$ form a strictly convex function for all standard test networks. As $c_{hl} > 0$ for arbitrary but fixed $q \in I\!R \setminus \{0\}$ the points $\{(|\hat{J}_k|, c_k)\}_{k=1}^{K}$ form a strictly convex function as well. □

2.5.9 Remarks.

(i) For the verification in Chapter 4, the fact that all test networks fulfill the adjacency property will be used.

(ii) Fujiwara and Dey [F&D87] additionally state that at "any optimal solution each link will consist of either one pipe segment or two pipe segments of adjacent diameters if and only if the plot of data points $\{(J_j, c_j)\}_{j=1}^{n}$ is a discrete, strictly convex function."

Nevertheless if the adjacency property is fulfilled for the linear program defined in Lemma 2.5.6, the set of points $\{(\hat{J}_j, c_j)\}_{k=1}^{K}$ does not necessarily form a discrete, strictly convex function.

This can be shown by the example of choosing $A = \begin{pmatrix} 1 & 1 & 1 & 1 \\ 4 & 3 & 2 & 1 \end{pmatrix}$ and $c = (1,3,4,10)^\top$. For a right hand side $(\hat{\Phi}, \hat{L})^\top = (20,40)^\top$ the unique optimal solution is $(0,0,20,0)^\top$, i.e. just one x_k is nonzero and for $(\hat{\Phi}, \hat{L})^\top = (30,40)^\top$ the optimal solution is $(0,0,10,20)^\top$, where just two adjacent x_k are nonzero.

From Lemma 2.5.4 and the example of Remark 2.5.5 (ii) it is obvious that if the set of points $\{(d_k, c_k)\}_{k=1}^{n}$ forms a discrete, strictly convex function then the set of points $\{(d_k^{-4.87}, c_k)\}_{k=1}^{n}$ also forms a strictly convex function but not vice versa.

In summary, the strict convexity of $\{(d_k, c_k)\}_{k=1}^{n}$ is a sufficient but not necessary condition for $\{(\hat{J}_k, c_k)\}_{k=1}^{n}$ forming a strictly convex function. And $|\hat{J}_k|$ being strictly monotone decreasing in k together with the strict convexity of $\{(|\hat{J}_k|, c_k)\}_{k=1}^{n}$ is a sufficient but not necessary condition for the adjacency property.

2.5.2 Elimination of Inequalities

In this section the flow conservation constraints

$$\begin{cases} \sum\limits_{j:(i,j)\in A} q_{ij} - \sum\limits_{j:(j,i)\in A} q_{ji} \leq b_i & \forall i \in S \\ \sum\limits_{j:(i,j)\in A} q_{ij} - \sum\limits_{j:(j,i)\in A} q_{ji} = b_i & \forall i \in D \end{cases}$$

are investigated in detail. In all standard test networks just one source node exists, therefore this special case of NOP is considered and the inequalities can be omitted:

2.5.10 Lemma. Let the water distribution network contain only one source node with net water supply b_1 and $n-1$ nodes with demand rate b_i for $i=2,\dots,n$. Suppose $\sum_{i=2}^{n} b_i = -b_1$ and suppose that for all $(i,j) \in A$ a flow is denoted by $q_{ij} \in I\!R$ which fulfills the equalities of the flow conservation constraints, i.e.

$$\sum_{j:(i,j)\in A} q_{ij} - \sum_{j:(j,i)\in A} q_{ji} = b_i \quad \text{for } i=2,\dots,n\,.$$

Then it is implied that

$$\sum_{j:(1,j)\in A} q_{1j} - \sum_{j:(j,1)\in A} q_{j1} = b_1\,.$$

Proof. With simple insertion one obtains

$$\begin{aligned} b_1 = -\sum_{i=2}^{n} b_i &= -\sum_{i=2}^{n}\left(\sum_{j:(i,j)\in A} q_{ij} - \sum_{j:(j,i)\in A} q_{ji}\right) \\ &= \sum_{j:(1,j)\in A} q_{1j} - \sum_{j:(j,1)\in A} q_{j1} - \left(\sum_{i=1}^{n}\sum_{j:(i,j)\in A} q_{ij} - \sum_{i=1}^{n}\sum_{j:(j,i)\in A} q_{ji}\right). \end{aligned}$$

Because of a reordering of the sum, the last term is zero, therefore the assertion follows. □

2.5.3 Equivalent Formulation of NOP

An equivalent formulation of NOP is developed in this section, to explain – at least heuristically – the fact that some components of the optimal head will attain the simple bounds, as this turns out to be one of the main difficulties when verifying feasibility of the upper bound in the branch and bound algorithms of Sherali, Subramanian and Loganathan. Further on this formulation could explain the assumption that the optimal solution tends to be driven to a branched one.

First a linear optimization problem similar to the one regarded in the proof of Theorem 2.5.7 is investigated:

2.5.11 Lemma. Suppose $\alpha_1 > \alpha_2 > \dots > \alpha_K > 0$ and $0 < c_1 < c_2 < \dots < c_K$, and that the set of points $\{(\alpha_k, c_k)\}_{k=1}^{K}$ forms a discrete, strictly convex function.

Let the linear problem (LP_y) be of the form

$$\begin{cases} \sum_{k=1}^{K} c_k \, x_k \to \min \\ \sum_{k=1}^{K} \alpha_k \, x_k = y \\ \sum_{k=1}^{K} x_k = 1 \\ x_k \geq 0 \quad \text{for } 1 \leq k \leq K \,. \end{cases} \qquad (LP_y)$$

Then for $y \in [\alpha_K \,,\, \alpha_1]$ there exists one and only one optimal solution vector with optimal value p_y^* of (LP_y).

Proof. Setting $\hat{\Phi} = y$, $\hat{L} = 1$ and $\hat{J}_k = \alpha_k$ for all $k \in \{1, \ldots, K\}$, this is a direct consequence from the linear problem defined in Lemma 2.5.6 having a unique solution. □

2.5.12 Lemma.

(i) With the notation of Lemma 2.5.11 let $f : [\alpha_1 \,,\, \alpha_K] \to [c_1 \,,\, c_K]$ be defined by $f(y) = p_y^*$.

Then f is well-defined.

(ii) The function f defined in (i) is piecewise linear, strictly monotone decreasing, continuous and convex.

Proof. The first part follows directly from Lemma 2.5.11, as $c_1 \leq p_y^* \leq c_K$. From the proof of Theorem 2.5.7 follows that the optimal solution of (LP_y) fulfills the adjacency property. There exists $k^* \in \{1, \ldots, K\}$ so that $\alpha_{k^*+1} \leq y \leq \alpha_{k^*}$. Then the optimal value p_y^* is expressed by

$$p_y^* = \frac{c_{k^*+1}\,\alpha_{k^*} - c_{k^*}\,\alpha_{k^*+1}}{\alpha_{k^*} - \alpha_{k^*+1}} + \frac{c_{k^*} - c_{k^*+1}}{\alpha_{k^*} - \alpha_{k^*+1}}\, y \,,$$

which implies f to be piecewise linear.

Since $\frac{c_k - c_{k+1}}{\alpha_k - \alpha_{k+1}} < 0$, the function f is strictly monotone decreasing.

Furthermore $f(\alpha_{k^*}) = c_{k^*}$ and $f(\alpha_{k^*+1}) = c_{k^*+1}$, so f is continuous.

As the set of the points $\{(\alpha_k, c_k)\}_{k=1}^{K}$ forms a discrete, strictly convex function, the slopes of f are strictly monotone increasing and therefore f is convex. □

2.5.13 Lemma. Suppose the linear program (LP_y) satisfies the assumptions of Lemma 2.5.11. Let $p^* \in [c_1\,,\,c_K]$.

Then there exists one and only one y so that p^* is an optimal solution p_y^* of (LP_y).

Proof. Regarding the function f defined in Lemma 2.5.12, it is

$$\min_{y \in [\alpha_1\,,\,\alpha_K]} f(y) = c_1 \qquad \text{and} \qquad \max_{y \in [\alpha_1\,,\,\alpha_K]} f(y) = c_K\,.$$

Since f is continuous and strictly monotone decreasing the theorem for monotone functions implies f to be bijective and guarantees the existence of

$$f^{-1} : [c_1\,,\,c_K] \to [\alpha_1\,,\,\alpha_K]\,,$$

which proves the lemma. □

For an equivalent formulation of NOP new decision variables y_{ij} are introduced for all $(i,j) \in P$ with $\alpha_k = c_{hl}\, C_{HW}^{-c_d}\, d_k^{-4.87}$ for $k \in \{1, \dots, K\}$ by

$$y_{ij} := \sum_{k=1}^{K} \alpha_k \, \frac{x_{ijk}}{L_{ij}}\,.$$

Analogously for all fixed arcs $(i,j) \in A \setminus P$ let $\widetilde{y_{ij}}$ be defined by

$$\widetilde{y_{ij}} := \sum_{k=1}^{K} \alpha_k \, \frac{\widetilde{x_{ijk}}}{L_{ij}}\,.$$

Let the vector $y = (y_{ij})_{(i,j) \in P}$. Regarding just one source node, NOP can be formulated equivalently with the function f of Lemma 2.5.12 and the reduced decision vector $(y, q, H)^\top$ as

$$\begin{cases} \displaystyle\sum_{(i,j) \in P} L_{ij}\, f\,(y_{ij}) \to \min & \\[2ex] \displaystyle y_{ij}\, \mathrm{sgn}\,(q_{ij})\, |q_{ij}|^{c_d} = \frac{(H_i + E_i) - (H_j + E_j)}{L_{ij}} & \forall (i,j) \in A \\ \displaystyle\sum_{j:(i,j) \in A} q_{ij} - \sum_{j:(j,i) \in A} q_{ji} = b_i & \forall i \in D \\ y_{ij} = \widetilde{y_{ij}} & \forall (i,j) \in A \setminus P \\ \alpha_K \le y_{ij} \le \alpha_1 & \forall (i,j) \in P \\ q_{\min_{ij}} \le q_{ij} \le q_{\max_{ij}} & \forall (i,j) \in A \\ H_{iL} \le H_i + E_i \le H_{iU} & \forall i \in D\,. \end{cases}$$

2.5.14 Remarks.

(i) Even if this formulation does not explicitly contain the set of indices for pipe segments of diameter k, it does not reduce the dimension of the problem. Implicitly they are part of the piecewise linear function f, which probably has to be split by using the original vector (x_{ijk}) for $(i,j) \in P$ and $k \in [1,\dots,K]$, for calculating a solution.

(ii) For an equivalent formulation the flow can be assumed to be non-zero for all pipes. This is no restriction to the engineering problem:

To ensure looped networks the assumption of a minimum diameter is already introduced to NOP. From the engineering point of view it would not make sense to build a pipe through which no or nearly no water will flow. Therefore the nonlinear constraints can be written for all $(i,j) \in A$ as

$$y_{ij} = \frac{(H_i + E_i) - (H_j + E_j)}{L_{ij}\ \mathrm{sgn}\,(q_{ij})\ |q_{ij}|^{c_d}} \tag{2.5.5}$$

with the additional constraint $q_{ij} \neq 0$.

(iii) During the verification of feasibility one has to determine an interval containing a zero of the nonlinear or linear constraints, which has to be within simple bounds. With the Brouwer fixed point theorem, existence of a zero is guaranteed when mapping an interval into itself. In the procedures provided in INTLAB (cf. Section 1.1.4) it is required that this interval is mapped into its interior. Then the zero will be part of the interior as well, and no inclusion can be determined for a solution being part of the boundary.

Suppose $(\hat{y},\ \hat{q},\ \hat{H})$ is an optimal solution of this formulation of NOP. Assume a sink is part of this solution, as for example node 17 in the New York City network. Then considering the nonlinear constraint 2.5.5 one can expect the head-loss in the pipes connecting node 17 to be maximal, which implies the head at this node to take the value of the lower bound H_{17L}.

These heuristic considerations may explain one of the main difficulties in the verification of feasibility of an optimal solution vector for NOP.

(iv) Further on, considering this formulation of NOP, one could imagine that an optimal solution will tend to be a branched one because of the structure of the objective function.

Chapter 3

NOP with Darcy-Weisbach Friction Loss Formula

"Thus, because of its general accuracy and complete range of application, the Darcy-Weisbach equation should be considered the standard [...]"[1]

In Europe, a theoretically more exact calculation of head-loss due to friction is used, based on Darcy-Weisbach instead of the Hazen-Williams formula. The hydraulic basics for the Darcy-Weisbach formula include factors concerning the roughness of the pipes. The Colebrook-White equation for determining the friction factors is introduced and hydraulic coefficients to be used for the standard test networks are discussed.

The differences between these formulae are analyzed and a corresponding optimization model to NOP is formulated. The adjacency property is proved for this optimization problem and the branch and bound algorithm introduced in Chapter 2 is adjusted for solving this new optimization problem, while investigating the increased nonlinearity in detail.

To combine the advantages of Darcy-Weisbach with the more simple structure of Hazen-Williams, an adjustment to the latter is proposed to obtain a closer approximation to Darcy-Weisbach. Again, details for the standard networks are provided.

3.1 Basic Hydraulic Principles

For description of variables and constants used in the context of head-loss according to Darcy-Weisbach see Table 3.1. In this section, pipes are assumed to be circular.

[1]Glenn O. Brown. *The History of the Darcy-Weisbach Equation for Pipe Flow Resistance.* Proceedings of the 150th Anniversary Conference of ASCE, Washington, D.C., A. Fredrich, and J. Rogers eds., American Society of Civil Engineers, Reston, VA., Environmental and Water Resources History, pp. 34 - 43, 2002.
`http://biosystems.okstate.edu/darcy/DarcyWeisbach/HistoryoftheDarcyWeisbachEq.pdf`

c_{DW}	Constant for unit conversion
c_{CW}	Constant for Colebrook-White equation
q	Pipeline flow rate $(\frac{L^3}{T})$
f	Darcy-Weisbach friction factor
d	Pipeline diameter (L)
l	Pipe length (L)
A	Cross-sectional area of pipeline (L^2)
g	Gravitational acceleration constant $(\frac{L}{T^2})$ $g = 9.81\frac{m}{s^2}$ respectively $g = 32.2\frac{ft}{s^2}$
ϵ	Equivalent sand roughness (L)
$\mathcal{R}$	Reynolds number
V	Average fluid velocity $(\frac{L}{T})$
ν	Kinematic viscosity $(\frac{L^2}{T})$

Table 3.1: Variables and constants of the Darcy-Weisbach formula with units, where for example L stands for meter or inch and T for seconds.

Then for full flowing circular pipes, the average fluid velocity can be computed with

$$V = \frac{q}{A} = \frac{4\,q}{\pi\,d^2}\,.$$

As noted in Chapter 2, the nonlinear optimization problem (NOP) can be expanded to the calculation of the head-loss in a pipe with the use of Darcy-Weisbach formula

$$\Phi(q,\ f,\ d,\ l) = \operatorname{sgn}(V)\; c_{DW}\, f\,\frac{l\,V^2}{d\,2\,g} = \operatorname{sgn}(q)\; c_{DW}\,\frac{8\,f\,l\,q^2}{g\,d^5\,\pi^2}\,, \tag{3.1.1}$$

which is predominant for calculations in Europe.

The constant c_{DW} is added due to unit conversion. Regarding the units as proposed by Sherali, Subramanian and Loganathan [S&a01], i.e. the flow in cubic-meters per hour, the pipe length in meter and the pipe diameter in centimeters, the constant has a value of $c_{DW} = \frac{10^6}{1296}\,\frac{h^2\,cm^5}{s^2\,m^5}$, and for cubic-feet per second, feet and inch it is $c_{DW} = 248,832\;\frac{inch^5}{ft^5}$.

For turbulent flow, the Darcy-Weisbach friction factor[2] f can be determined according to Prandtl-Kármán for hydraulically smooth pipes with

$$\frac{1}{\sqrt{f}} = 2\,\log\frac{\mathcal{R}\,\sqrt{f}}{2.51}$$

and for flow in rough pipes as

$$\frac{1}{\sqrt{f}} = -2\,\log\frac{3.71\,d}{\epsilon}\,.$$

According to Damrath and Cord-Landwehr [DC98], the usual diameters and velocities in water supply systems imply flows predominantly in the transition region. In the Colebrook-White equation it is implicitly described for turbulent flow as

$$\frac{1}{\sqrt{f}} = -2\,\log\left(\frac{2.51}{\mathcal{R}\,\sqrt{f}} + \frac{\epsilon}{3.71\,d}\right)\,. \tag{3.1.2}$$

According to Colebrook, who presented this theoretical formula for flow in transition zones [C38] "this transition-curve merges asymptotically into the smooth- and rough-law curves." Thus, they can be used in general.

WaTerInt is based on this version of the formula, which is predominant in German literature and corresponds to [DC98], [B00] and [M&S02], whereas in [S&J76], [M99] and [K00] a value of 3.7 instead of 3.71 is used. Walski et al. [W&a03] introduce further constants for the Colebrook-White formula slightly differing from these values, which are based on the natural logarithm.

[2]The corresponding German notation for the friction factor f is λ, which is called "Rohrreibungsbeiwert" [B00], "Reibungszahl" [DC98] or "Widerstandsbeiwert" ([B00], [DC98]).

The value of 3.7 was used in the original publication of this formula by Colebrook [C38]. But he derived it while combining experimental results from Nikuradse with an extension of the Prandtl-von-Kármán laws for smooth and for rough pipes. Following his derivation of the factor, it is the rounded value of the product of $0.113 \cdot 33$.

In 1944, Moody [M44] used the value 3.7 for the development of his diagrams, which he derives directly from the results from von Kármán as result of the expression $\frac{10^{0.87}}{2}$. Rounding this value up to three digits implies 3.71 , the factor used for the following calculations.

Further on, in this section the temperature of the water in the pipes is first of all assumed to have a constant value of $10°$ C. Then according to [W&a03] the Reynolds number can be determined with

$$\mathcal{R} = \frac{|V|\, d}{\nu} = \frac{4\, |q|}{\pi\, d\, \nu},$$

where ν represents the kinematic viscosity.

In WaTerInt, rounded values according to [W&a03] are used, i.e. for water of $10°$ C a value of $\nu = 1.306 \cdot 10^{-6} \frac{m^2}{s}$ and correspondingly $\nu = 1.407 \cdot 10^{-5} \frac{ft^2}{s}$. For the New York network, additionally water of $4°$ C with a viscosity value of $\nu = 1.687 \cdot 10^{-5} \frac{ft^2}{s}$ and of $20°$ C with $\nu = 1.088 \cdot 10^{-5} \frac{ft^2}{s}$ is to be considered. This corresponds to values of $\nu = 1.586 \cdot 10^{-6} \frac{m^2}{s}$ for water of $4°$ C and to $\nu = 1.003 \cdot 10^{-6} \frac{m^2}{s}$ for $20°$ C.

The Colebrook-White Equation 3.1.2 can be written as

$$\frac{1}{\sqrt{f}} = -2 \log \left(\frac{c_{CW}\, d}{|q|\, \sqrt{f}} + \frac{\epsilon}{3.71\, d} \right) .$$

The constant c_{CW} can easily be determined with the kinematic viscosity: Regarding the same units as in the Darcy-Weisbach equation and the viscosity in $\frac{m^2}{s}$ or respectively $\frac{ft^2}{s}$ values of

$$c_{CW} = 22.59\, \pi\, \nu\, \frac{m\, s}{cm\, h} \quad \text{and} \quad c_{CW} = \frac{2.51}{48}\, \pi\, \nu\, \frac{ft}{inch}$$

are used.

According to Walski et al. [W&a03], laminar flow is indicated by a Reynolds number less than $2,000$, transitional flow by a range from $2,000$ to $4,000$ and a Reynolds number greater than $4,000$ describes turbulent flow. Damrath and Cord-Landwehr [DC98] explicity note the transition point to have a Reynolds-number of 2,320. For laminar flow the friction factor has to be correctly determined according to Hagen-Poiseuille with

$$f_{\text{lam}} = \frac{64}{\mathcal{R}},$$

which implies for the head-loss $\Phi(q, d, l) = \frac{128\, \nu}{\pi\, g\, d^4}\, q\, l$.

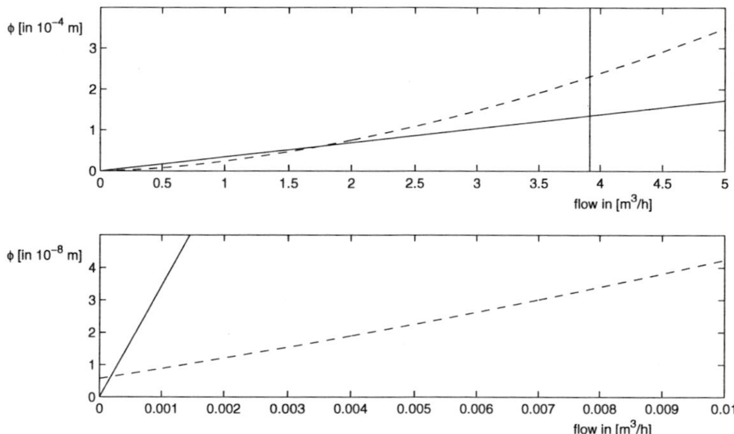

Figure 3.1: Laminar flow is defined by $\mathcal{R} \leq 2.320$, the corresponding flow is marked by the vertical line. For different scale, the head-loss correctly calculated with Hagen-Poiseuille is drawn solid and the one used in DNOP dashed. Basis are the constants used for the two-loop network and an 18"-pipe of length $1,000$ m.

Regarding the constants of the two-loop network and an 18"-pipe, $\mathcal{R} = 2,320$ corresponds to a flow of approximately $3.92\frac{m^3}{h}$ and for a 1"-pipe of $0.22\frac{m^3}{h}$.

Using the Hagen-Poiseuille for laminar flow and the Colebrook-White equation, also known as Prandtl-Colebrook law, for turbulent flow, Burgschweiger, Gnädig and Steinbach [B&a04] describe that this "model is highly accurate but has jump discontinuities at the transitions between laminar and turbulent flow." Thus, they developed an explicit smooth approximation to these formulae for the friction factor f , which is part of an optimization module for minimum cost operative planning in operation at Berliner Wasserbetriebe (BWB).

The differences between Hagen-Poiseuille and Colebrook-White are illustrated for the constants of the two-loop network and an 18" pipe of length $1,000$ m in Figure 3.1. This discontinuity mentioned by Burgschweiger, Gnädig and Steinbach is obvious, but a head-loss lower than 10^{-4} is small compared to a maximum head-loss of approximately 30 m in the whole network when regarding the optimal solution obtained with Hazen-Williams for the two-loop network. The differences therefore are negligible from the engineering point of view.

Additionally, water flow in distribution pipe systems can usually be expected to be turbulent. Thus, in DNOP, at slow flow speeds, where laminar flow can occur, the Colebrook-White equation is still used.

The network optimization design problem is based on the assumption that pipes of the same diameter have the same cost independently of the part of the network they are used in. Therefore the value of the equivalent sand roughness[3] ϵ can as well be assumed to be fixed for all pipes in the network. The model is simplified to not contain increasing roughness over time, which is normally observed for aging pipes.

When interpreting the solutions obtained, it should be noted that over time the equivalent sand roughness generally increases. Karney [K02] emphasizes that this is the only part of the Darcy-Weisbach formula that remains empirical:

> "Although it is true that the functional relationship of the Darcy-Weisbach formula reflects logical associations implied by the dimensions of the various terms, determination of the equivalent uniform sand-grain size is essentially experimental."

According to the materials used for the standard networks, the following values are chosen as default in WaTerInt:

According to the diameters used for the two-loop test example ranging from 1" to 24", PVC pipes are assumed. According to Bollrich [B00] an equivalent sand roughness of $\epsilon = 0.003$ mm should be used. Walski et al. [W&a03] suggest a value of 0.0015 mm, however they also suggest a Hazen-Williams coefficient of $140 - 153$, which is larger than the value of $C_{HW} = 130$ as used in the two-loop example. Mutschmann and Stimmelmayr [M&S02] propose a value of 0.01. Thus, in WaTerInt a value of $0.003\,\mathrm{mm}$ is used.

As shown in Figure 2.2, the pipes for the NYC water supply system are lined with concrete. For this material Bollrich suggests an equivalent sand roughness ϵ of $6\,\mathrm{mm} = 0.02\,\mathrm{ft}$, the value used in WaTerInt.

According to [B03], in Hanoi pipes of diameter up to 75 mm are of PVC or galvanized iron, from 75 to 250 mm of asphalted cast iron or steel and greater than 250 mm of ductile cast iron. As the pipes for the Hanoi test example are designed with a minimum diameter of 12 inch, the equivalent sand roughness for cast iron should be regarded. For this, Bollrich [B00] gives a value of $0.25\,\mathrm{mm}$, which is also proposed by Mutschmann and Stimmelmayr [M&S02].

[3] The German notation for ϵ is k, called "k-Wert" ([B00], [DC98], [M&S02]), "absolute hydraulische Rauheit" [B00], "absolute Rauheit" [DC98] or "Rauhigkeitswert" [M&S02].

3.2 Comparison of Hazen-Williams and Darcy-Weisbach Formula

"Thus one must be cautious when using the Hazen-Williams equation." [L98]

The Hazen-Williams and the Darcy-Weisbach formula are introduced in water distribution network design for the determination of the head-loss due to friction in a pipe. Therefore the main question in a comparison of these formulae focuses on differences in the head-loss obtained, as summarized by Karney: [K02]

> "Different equations should still produce similar head-discharge behavior. That is, the physical relation between head-loss and flow for a physical segment of pipe should be predicted well by any practical loss relation."

When the head-loss for a given flow is compared between the two formulae, a good match only exists for a small parameter range of flow velocity, diameter and equivalent sand roughness. For the default values for the equivalent sand roughness, the deviations in head-loss calculations according to Darcy-Weisbach or Hazen-Williams subject to the flow are significant. Figure 3.2 contains an overview of the absolute differences subject to the flow, more detailed plots containing boundary lines for velocities of $1\,\frac{\mathrm{m}}{\mathrm{s}}$, $2.5\,\frac{\mathrm{m}}{\mathrm{s}}$ and $5\,\frac{\mathrm{m}}{\mathrm{s}}$ are part of the Appendix, Figures E.1 to Figure E.9. Obviously the difference attains its maximum for small diameter and large flow.

This coincides with Liou [L98], who demonstrate on the basis of historic experimental data that C_{HW} "is a strong function of Reynolds number and pipe size and that the Hazen-Williams equation has narrow applicable ranges for Reynolds numbers and pipe sizes" and that the "level of error when the Hazen-Williams equations is used outside its data ranges is significant."

Both head-loss equations can be solved for the flow q. Then for values of

$$\Phi_{DW} > \frac{8\, c_{DW}\, c_{CW}^2\, l}{\pi^2\, g\, d^3 \left(1 - \frac{\epsilon}{3.71\, d}\right)^2},$$

i.e. for the values of ϵ used in WaTerInt, approximately $\Phi_{DW} > 4 \cdot 10^{-5}$ meter for the two-loop network, $\Phi_{DW} > 8 \cdot 10^{-10}$ foot for the New York network and $\Phi_{DW} > 2 \cdot 10^{-8}$ meter for Hanoi, the relation of head-loss due to Hazen-Williams to head-loss due to Darcy-Weisbach is independent of the flow and can be obtained as

$$\Phi_{HW} = c_{hl}\, d^{-4.87}\, l \left(\frac{-\pi}{c_{HW}} \sqrt{\frac{\Phi_{DW}\, g\, d^5}{2\, c_{DW}\, l}}\ \log\left(\frac{c_{CW}}{\pi} \sqrt{\frac{8\, c_{DW}\, l}{\Phi_{DW}\, g\, d^3}} + \frac{\epsilon}{3.71\, d} \right) \right)^{c_d}.$$

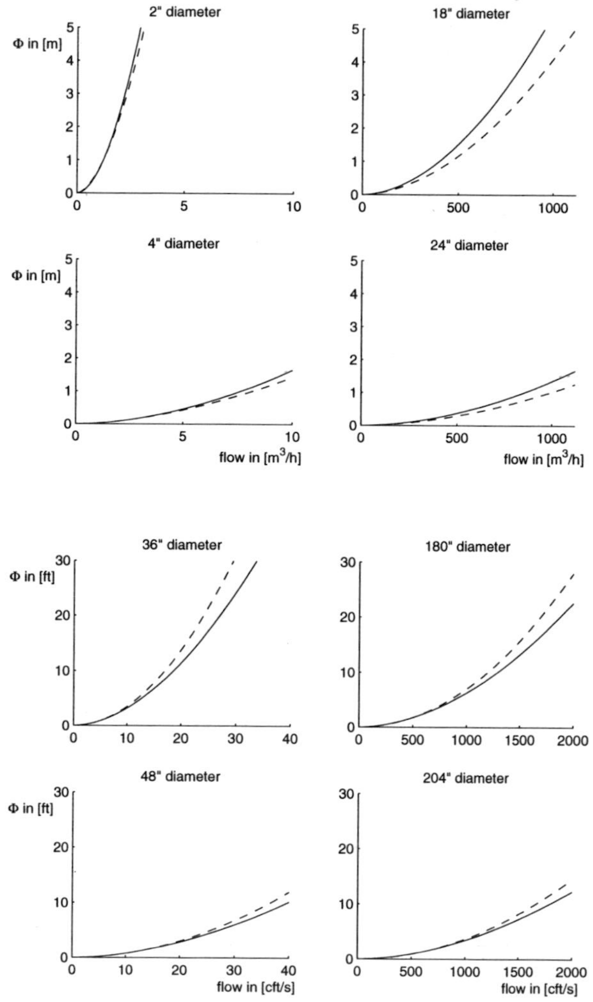

Figure 3.2: Comparison of head-loss calculated with the Darcy-Weisbach and Hazen-Williams formulae. The first four figures are based on the default values of the two-loop network and a length of 1,000 m, and the second four on the default values for the New York networks and a length of 10,000 ft. The head-loss according to Hazen-Williams is plotted solid and according to Darcy-Weisbach dashed.

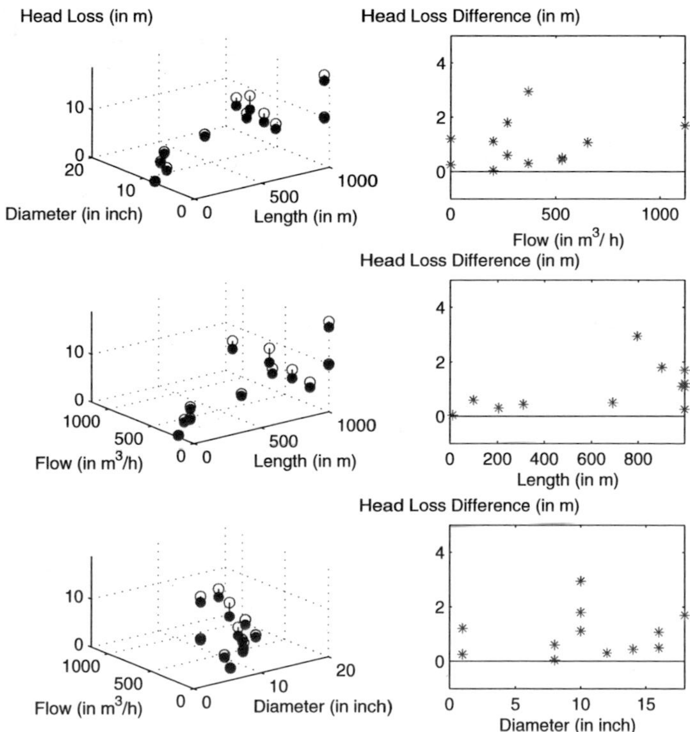

Figure 3.3: Comparison of head-loss calculations with the Darcy-Weisbach and Hazen-Williams formulae, head-loss difference is head-loss according to Hazen-Williams minus head-loss according to Darcy-Weisbach. Basis is the optimal solution for the two-loop test network containing optimal flow, diameter and length, see Table D.8. For all pipe segments the head-loss according to Hazen-Williams is plotted with a circle and according to Darcy-Weisbach with a filled one.

The relative differences of head-loss, i.e. $\Delta\Phi = (\Phi_{HW} - \Phi_{DW})/\Phi_{DW}$, subject to the head-loss according to Darcy-Weisbach is illustrated in Figures E.10 to E.12.

The solution of the two-loop network, as part of Table D.1, and the solution for the New York network published by Sherali, Subramanian and Loganathan is investigated in more detail. The differences in the head-loss for all pipe segments larger than one meter, or for the New York network larger than one foot, are used as basis for the comparison of head-loss calculation. Comparisons of these solutions are summarized in Figure 3.3 and 3.4 respectively.

In all test networks the Reynolds numbers of the optimal solutions, as presented in Tables D.8, D.10, D.11 and D.12, indicate turbulent flow in all pipes. For the segments of the two-loop network the values range from $7.39 \cdot 10^3$ to $6.63 \cdot 10^5$, for the New York networks from $1.45 \cdot 10^5$ to $6.06 \cdot 10^6$ and for the Hanoi network from $4.83 \cdot 10^4$ to $5.31 \cdot 10^6$.

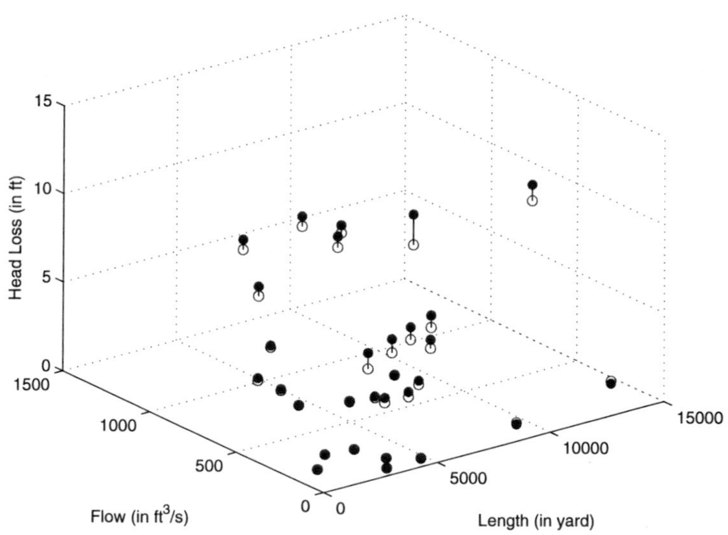

Figure 3.4: Comparison of head-loss calculations with the Darcy-Weisbach and Hazen-Williams formulae. Basis is the optimal solution for the New York (4") network published in [S&a01]. For all pipe segments the head-loss according to Hazen-Williams is plotted with a circle and according to Darcy-Weisbach with a filled one.

This corresponds to modeling theory, as according to [W&a03] "the flow of water through municipal water systems is almost always turbulent, except in the periphery where water demand is low and intermittent, and may result in laminar and stagnant flow conditions."

Regarding the head-loss as a function of the pipeline flow rate, the difference between the formulae is obviously based on the exponents 1.852 , or 1.85 respectively, for the Hazen-Willams equation and 2 for Darcy-Weisbach.

This is underlined by Williams and Hazen [WH05], who describe how the exponents were obtained:

> "The exponents in the formula used were selected as representing as nearly as possible average conditions, as deduced from the best available records of experiments upon the flow of water in such pipes and channels as most frequently occur in water-works practice."

The friction factor used in the Darcy-Weisbach equation depends on the flow. As will be shown in Section 3.4.1, it can be expressed by a function depending on the flow and the diameter, which for all diameters is strictly monotone decreasing with respect to the flow. So obviously this equation provides more detailed calculations, as it includes an observation of Hazen and Williams that the exponent may vary for different diameters.

3.3 Darcy-Weisbach Optimization Problem

The Darcy-Weisbach network optimization problem (DNOP) is obtained from NOP by substituting the Hazen-Williams formula with the Darcy-Weisbach formula and adding the Colebrook-White equation.

The simplifications resulting from the properties of the optimization problem discussed in Chapter 2 are included. So just one source node is considered and, according to the test examples to be calculated, first no additional head is considered for the source node, i.e. it is assumed that $\sum_{i \in S} c_{si} H_{si} = 0$ and $F_i = 0$ for $i \in S$. Thus, according to Lemma 2.5.10 the inequality for the source node is eliminated.

Assume the available diameters are ordered, i.e. $d_1 < d_2 < \ldots < d_K$. For guaranteeing a solution for the friction factor of the Colebrook-White equation $\epsilon < 3.71\, d_1$ is assumed, which is obviously fulfilled for all reasonable test networks.

The interpretation of the variables corresponds to the one used in NOP, as summarized in Table 2.1. The network layout is again determined by N, A and the length L_{ij} representing a simple directed graph.

Let $\mathcal{K} = \{1, \ldots, K\}$ and let for $k \in \mathcal{K}$ the factor β_k be defined as $\beta_k := \dfrac{8\, c_{DW}}{g\, \pi^2\, d_k^5}$.

Then instead of NOP, the following problem (DNOP) is obtained, where the decision vector is again $(x,\, q,\, H)^\top$:

$$
\begin{cases}
\displaystyle\sum_{(i,j)\in P} \sum_{k=1}^{K} c_k\, x_{ijk} + c_{s1}\, H_{s1} \to \min & \\[2ex]
\displaystyle\frac{1}{\sqrt{f_{ijk}}} + 2\log\left(\frac{c_{CW}\, d_k}{|q_{ij}|\,\sqrt{f_{ijk}}} + \frac{\epsilon}{3.71\, d_k}\right) = 0 & \forall (i,j) \in A,\ \forall\, k \in \mathcal{K} \\[2ex]
\displaystyle\operatorname{sgn}(q_{ij})\, q_{ij}^2 \sum_{i=1}^{K} f_{ijk}\, \beta_k\, x_{ijk} = (H_i + E_i) - (H_j + E_j) & \forall (i,j) \in A \\[2ex]
\displaystyle\sum_{k=1}^{K} x_{ijk} = L_{ij} & \forall (i,j) \in P \\[2ex]
\displaystyle\sum_{j:(i,j)\in A} q_{ij} - \sum_{j:(j,i)\in A} q_{ji} = b_i & \forall\, i \in D \\
q_{\min_{ij}} \le q_{ij} \le q_{\max_{ij}} & \forall (i,j) \in A \\
H_{iL} \le H_i + E_i \le H_{iU} & \forall\, i \in N \\
H_{s1} \ge 0 & \\
x_{ijk} \ge 0 & \forall (i,j) \in P,\ \forall\, k \in \mathcal{K} \\
x_{ijk} = \widetilde{x_{ijk}} & \forall (i,j) \in A \setminus P,\ \forall\, k \in \mathcal{K}\,.
\end{cases}
$$

3.4 Properties of the Darcy-Weisbach Problem

> "... the transition curves are somewhat complex and are not, therefore, easy to use, ..."
>
> C. F. Colebrook, [C38]

3.4.1 Nonlinearity of the Colebrook-White Equation

For solving the increased nonlinearity of the optimization problem, first of all the friction factors f_{ijk} are eliminated. Therefore, a number of auxiliary functions is introduced.

As already explained, the equivalent sand roughness ϵ is assumed to have a fixed value for all pipes in the network. Thus for $k \in \mathcal{K} = \{1, \dots, K\}$ and fixed diameter d_k the friction factor f_{ijk} depends only on the flow, and the function ψ_k can be defined.

3.4.1 Definition. Let the auxiliary function $\psi_k : \mathbb{R}_+ \setminus \{0\} \times \mathbb{R} \setminus \{0\} \to \mathbb{R}$ be defined by

$$\psi_k(f, q) = \frac{1}{\sqrt{f}} + 2 \log \left(\frac{c_{CW}\, d_k}{|q|\, \sqrt{f}} + \frac{\epsilon}{3.71\, d_k} \right) .$$

Then for fixed flow $\bar{q}$, the friction factor f can be determined by finding a root of the auxiliary function ψ_k, i.e. a point f with $\psi_k(f, \bar{q}) = 0$. An example of $\psi_k(f, \bar{q})$ for the first pipe of the two-loop network is plotted in Figure 3.5.

3.4.2 Lemma.

(i) For arbitrary but fixed $\bar{q} \in \mathbb{R} \setminus \{0\}$ the function $\psi_k(f, \bar{q})$ is strictly monotone deceasing in f and a unique f_{k_0} exists with $\psi_k(f_{k_0}, \bar{q}) = 0$.

(ii) For arbitrary but fixed $\bar{f} \in \mathbb{R}_+ \setminus \{0\}$ the function $\psi_k(\bar{f}, q)$ is strictly monotone deceasing in q for $q \geq 0$, and increasing for $q \leq 0$.

Proof. As $c_{CW} > 0$, $d_k > 0$ and $\epsilon > 0$, the function ψ_k is strictly monotone decreasing in f. Regarding

$$\lim_{f \to 0} \psi_k(f, q) = \infty \quad \text{and} \quad \lim_{f \to \infty} \psi_k(f, q) = 2 \log \left(\frac{\epsilon}{3.71\, d_k} \right) ,$$

it follows existence and uniqueness of $f_{k_0} \in \mathbb{R}_+ \setminus \{0\}$ with $\psi_k(f_{k_0}, q) = 0$, iff $\epsilon < 3.71\, d_k$. The last is fulfilled according to the definition of DNOP.

Obviously, the second part is again a direct consequence of the monotonicity of the logarithm. □

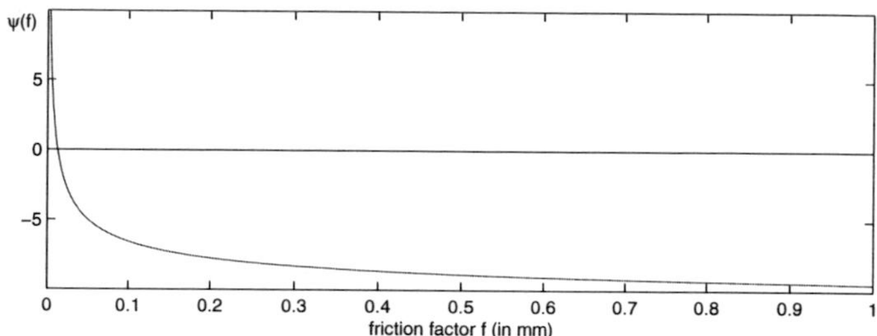

Figure 3.5: Auxiliary function $\psi(f) = \psi_k(f, \bar{q})$ for determining the Darcy-Weisbach friction factor using a flow of $\bar{q} = 1120\frac{m^3}{h}$, an equivalent sand roughness of 0.003 mm and a pipe diameter of 18 inch.

Therefore the following function is well-defined:

3.4.3 Definition. Let for $k \in \mathcal{K}$ the function $u_k : I\!R_+ \setminus \{0\} \to I\!R$ be defined by

$$u_k(q) = f \quad \text{iff} \quad \psi_k(f, q) = 0 \ .$$

3.4.4 Lemma. The function u_k has the following properties:

(i) u_k is strictly monotone decreasing.

(ii) The range of u_k is the open interval $(\frac{1}{4}\left[\log(\frac{\epsilon}{3.71\,d_k})\right]^{-2}, \infty)$ and the function u_k is invertible with $u^{-1} : (\frac{1}{4}\left[\log(\frac{\epsilon}{3.71\,d_k})\right]^{-2}, \infty) \to I\!R_+ \setminus \{0\}$ and

$$u_k^{-1}(f) = \frac{c_{CW}\, d_k}{\sqrt{f}\left(10^{-\frac{1}{2\sqrt{f}}} - \frac{\epsilon}{3.71\,d_k}\right)} \ .$$

(iii) The function u_k^{-1} is strictly monotone decreasing, and both u_k^{-1} and the function u_k are strictly convex.

Proof.

(i) Assume there exists $q_1,\, q_2 \in I\!R_+ \setminus \{0\}$, $q_1 < q_2$ and $u_k(q_1) = f_1$, $u_k(q_2) = f_2$ with $f_1 \le f_2$. Then the function $\psi_k(f, q)$, being strictly monotone decreasing in q for fixed f, and strictly monotone decreasing in f for fixed q according to Lemma 3.4.2, implies

$$0 = \psi_k(f_1, q_1) > \psi_k(f_1, q_2) \ge \psi_k(f_2, q_2) = 0 \,,$$

a contradiction. Therefore $f_1 > f_2$ must hold true.

(ii) The strict monotonicity of u_k guarantees the existence of u_k^{-1}, and that u_k^{-1} is strictly monotone decreasing. The equation $\psi_k(f,q)=0$ can be solved for q to obtain the term for the inverse function $q=u_k^{-1}(f)$.

Regarding the pole of the expression, it is

$$\lim_{f \searrow \frac{1}{4}\left[\log(\frac{\epsilon}{3.71\,d_k})\right]^{-2}} u^{-1}(f) = \infty \quad \text{and} \quad \lim_{f \nearrow \infty} u^{-1}(f) = 0\,,$$

which implies the lemma.

(iii) The monotonicity of u_k^{-1} is a direct consequence of (i), i.e. u_k being strictly monotone decreasing.

The second derivative of $u_k^{-1}(f)$ is

$$u_k^{-1\prime\prime}(f) = T_{u1}(f) + T_{u2}(f) + T_{u3}(f) - T_{u4}(f)\,,$$

with

$$T_{u1}(f) = \frac{3\,c_{CW}\,d_k}{4\,f^{\frac{5}{2}}\left(10^{-\frac{1}{2\sqrt{f}}} - \frac{\epsilon}{3.71\,d_k}\right)}\,, \qquad T_{u2}(f) = \frac{5\,c_{CW}\,d_k\left(10^{-\frac{1}{2\sqrt{f}}}\right)\ln(10)}{8\,f^3\left(10^{-\frac{1}{2\sqrt{f}}} - \frac{\epsilon}{3.71\,d_k}\right)^2}\,,$$

$$T_{u3}(f) = \frac{c_{CW}\,d_k\left(10^{-\frac{1}{2\sqrt{f}}}\right)^2[\ln(10)]^2}{8\,f^{\frac{7}{2}}\left(10^{-\frac{1}{2\sqrt{f}}} - \frac{\epsilon}{3.71\,d_k}\right)^3}\,, \quad T_{u4}(f) = \frac{c_{CW}\,d_k\left(10^{-\frac{1}{2\sqrt{f}}}\right)[\ln(10)]^2}{16\,f^{\frac{7}{2}}\left(10^{-\frac{1}{2\sqrt{f}}} - \frac{\epsilon}{3.71\,d_k}\right)^2}\,.$$

One obtains $T_{u2}>0$ and, as $f > \frac{1}{4}\left(\log(\frac{\epsilon}{3.71\,d_k})\right)^{-2}$, it follows

$$10^{-\frac{1}{2\sqrt{f}}} - \frac{\epsilon}{3.71\,d_k} > 0\,,$$

which implies $T_{u1}>0$. For the last two terms it is

$$T_{u3} - T_{u4} = \frac{c_{CW}\,d_k\,10^{-\frac{1}{2\sqrt{f}}}\,[\ln(10)]^2}{16\,f^{\frac{7}{2}}\left(10^{-\frac{1}{2\sqrt{f}}} - \frac{\epsilon}{3.71\,d_k}\right)^2}\left(\frac{2\cdot 10^{-\frac{1}{2\sqrt{f}}}}{10^{-\frac{1}{2\sqrt{f}}} - \frac{\epsilon}{3.71\,d_k}} - 1\right) > 0\,,$$

as

$$\frac{2\cdot 10^{-\frac{1}{2\sqrt{f}}}}{10^{-\frac{1}{2\sqrt{f}}} - \frac{\epsilon}{3.71\,d_k}} - 1 = \frac{10^{-\frac{1}{2\sqrt{f}}} + \frac{\epsilon}{3.71\,d_k}}{10^{-\frac{1}{2\sqrt{f}}} - \frac{\epsilon}{3.71\,d_k}} > 0\,.$$

So, $u_k^{-1\prime\prime}(f)>0$ for all $f \in (\frac{1}{4}\left[\log(\frac{\epsilon}{3.71\,d_k})\right]^{-2}, \infty)$, which implies the strict convexity of u_k^{-1}.

Using the derivation rule for the inverse function, for $q \in \mathbb{R}_+ \setminus \{0\}$ one obtains

$$u_k''(q) = -\frac{1}{\left[u_k^{-1'}(u_k(q))\right]^3} \, u_k^{-1''}(u_k(q)) \ ,$$

which proves the strict convexity of u_k. □

An illustration of the monotonicity of u_k is provided in Figure 3.6, where the zeros of ψ_k are illustrated for different flows q.

3.4.5 Remarks.

(i) In the context of the adjacency property for DNOP, monotonicity with respect to d_k has to be investigated. Therefore the function u^{-1} can be defined as a function in d and f, i.e. $u^{-1} : [d_1, d_K] \times (\frac{1}{4}\left(\log(\frac{\epsilon}{3.71\,d})\right)^{-2}, \infty) \to \mathbb{R}_+ \setminus \{0\}$ with

$$u^{-1}(d, f) = \frac{c_{CW}\, d}{\sqrt{f}\left(10^{-\frac{1}{2\sqrt{f}}} - \frac{\epsilon}{3.71\,d}\right)} \ ,$$

as illustrated in Figure 3.7.

(ii) When assuming the stronger condition $\epsilon < 1.885\,d_1$ instead of $\epsilon < 3.71\,d_1$ for DNOP, for fixed values of f with $f > (\frac{1}{4}\left(\log(\frac{2\epsilon}{3.71\,d_k})\right)^{-2}$ the function $u^{-1}(d, f)$ is strictly monotone increasing in d. This lower bound of f corresponds to a value for the flow of $q = -\frac{7.42\, c_{CW}\, d^2}{\epsilon} \log(\frac{2\epsilon}{3.71\,d})$.

For the two-loop network data and a diameter of 1 inch, this results in a flow of approximately $5\,\frac{m^3}{h}$, and for a diameter of 24 inch in a flow of $3,898\,\frac{m^3}{h}$. So within the simple bounds of DNOP no monotonicity of u^{-1} in d can be assumed.

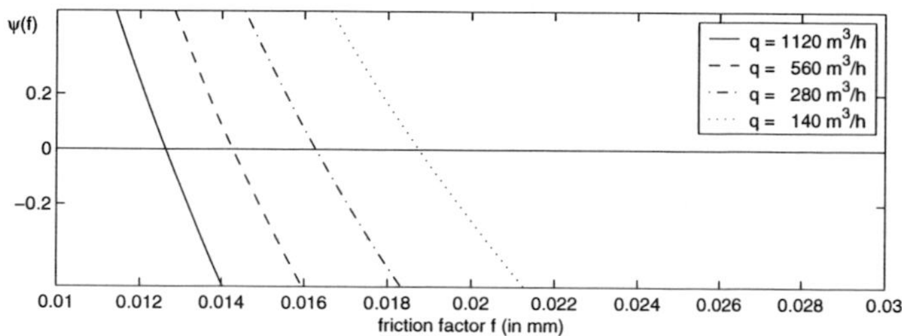

Figure 3.6: Auxiliary function $\psi(f) = \psi(f, q)$ for determining the Darcy-Weisbach friction factor for different flows q, using an equivalent sand roughness of 0.003 mm and a pipe diameter of 18 inch.

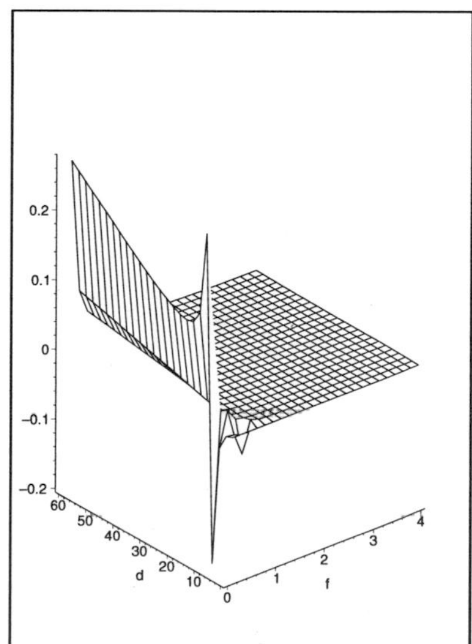

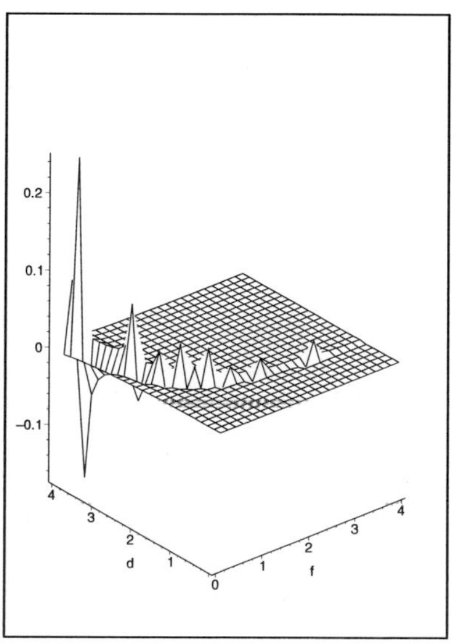

Figure 3.7: Continuation of the auxiliary function u^{-1} regarded as function of two parameters, i.e. depending on d and f. The plot is generated by Maple and based on the constants of the two-loop network except that for better illustration an equivalent sand roughness of 19 mm is used, the diameter is measured in centimeters. The domain of the lower bound of the diameter is set to 0 and different ranges for d are plotted.

To obtain relaxations for the nonlinearity of DNOP similar to those used for NOP, the term of the head-loss formula which corresponds to the expression q^{c_d} has to be investigated.

Thus, according to the transformation of NOP introduced by Sherali, Subramanian and Loganathan, auxiliary functions w_k are introduced in the following definition.

3.4.6 Definition.

(i) Let the function $w_k : \mathbb{R} \to \mathbb{R}$ be defined by

$$w_k(q) := \begin{cases} \operatorname{sgn}(q)\, q^2\, u_k(|q|) & \text{for } q \neq 0 \\ 0 & \text{else.} \end{cases}$$

(ii) Let $w_{k_{\min}}$ and $w_{k_{\max}}$ be defined as

$$w_{k_{\min}} := w_k(q_{\min}) \quad \text{and} \quad w_{k_{\max}} := w_k(q_{\max})\,.$$

(iii) Let the set $D_{w_k^{-1}}$ denote the union of the following open intervals and $\{0\}$, i.e.

$$D_{w_k^{-1}} := (-\infty\,,\, -\left(\frac{c_{CW}\, d_k}{1 - \frac{\epsilon}{3.71\, d_k}}\right)^2)\, \cup\, \{0\}\, \cup\, (\left(\frac{c_{CW}\, d_k}{1 - \frac{\epsilon}{3.71\, d_k}}\right)^2\,,\, \infty)\,.$$

Heuristically, the function w_k is similar to the function $v(q) = \operatorname{sgn}(q)\, |q|^{c_d}$ defined for the relaxations of Sherali, Subramanian and Loganathan in Equation 2.3.1 on page 49, as it describes the part of the head-loss that depends on the flow.

Essentially, these functions incorporate the main difference of the optimization problems: Whereas the head-loss in one pipe (i,j) according to Hazen-Williams,

$$\Phi_{ij}(q_{ij}, x_{ij}) = v(q_{ij}) \sum_{k=1}^{K} \alpha_k\, x_{ijk}\,, \tag{3.4.1}$$

can be separated in the product of two functions, one depending only on the flow q and one only on the diameters used, this is not possible for the head-loss according to Darcy-Weisbach,

$$\Phi_{ij}(q_{ij}, x_{ij}) = \sum_{k=1}^{K} w_k(q_{ij})\, \beta_k\, x_{ijk}\,. \tag{3.4.2}$$

For investigating DNOP, first the properties of the functions w_k are summarized in the following lemma and the family of curves is plotted in Figure 3.8.

3.4.7 Lemma.

(i) The function w_k of Definition 3.4.6 is strictly monotone increasing in q.

(ii) The function w_k is invertible and w_k^{-1} is strictly monotone increasing with $w_k^{-1} : D_{w_k^{-1}} \to I\!R$,

$$w_k^{-1}(\phi) = \begin{cases} -2\,\mathrm{sgn}\,(\phi)\,\sqrt{|\phi|}\,\log\left(\frac{c_{CW}\,d_k}{\sqrt{|\phi|}} + \frac{\epsilon}{3.71\,d_k}\right) & \text{for } \phi \neq 0 \\ 0 & \text{else.} \end{cases}$$

(iii) The function w_k^{-1} is strictly convex for $\phi < 0$ and strictly concave for $\phi > 0$.

(iv) The function w_k is strictly concave for $q < 0$ and strictly convex for $q > 0$.

Proof.

(i) The derivative of w_k is for all $q \in I\!R \setminus \{0\}$

$$w_k'(q) = 2\,|q|\,u_k(|q|) + q^2\,u_k'(|q|) = 2\,|q|\,u_k(|q|) + q^2 \frac{1}{u_k^{-1\prime}(u_k(|q|))}\,.$$

With the derivative of u_k^{-1},

$$u_k^{-1\prime}(f) = -\,\frac{u_k^{-1}(f)}{2\,f}\left(1 + \frac{\ln(10)}{2\,c_{CW}\,d_k} 10^{-\frac{1}{2\sqrt{f}}}\,u_k^{-1}(f)\right)\,,$$

one obtains

$$w_k'(q) = \frac{q^2\,u_k(|q|)\,\ln(10)}{c_{CW}\,d_k\,10^{\frac{1}{2\sqrt{u_k(|q|)}}} + \frac{\ln(10)}{2}\,|q|} > 0\,.$$

As $w_k(q) < 0$ for $q < 0$ and $w_k(q) > 0$ for $q > 0$, the monotonicity follows for all $q \in I\!R$.

(ii) As the function $w_k : I\!R \to I\!R$ is strictly monotone increasing, the inverse function w_k^{-1} exists and is strictly monotone increasing. The function w_k is continuous except in $q = 0$. Therefore, the range of w_k can be obtained when regarding the limites

$$\lim_{q \searrow 0} w_k(q) = \lim_{f \to \infty} w_k\left(u_k^{-1}(f)\right) = \lim_{f \to \infty} \frac{(c_{CW}\,d_k)^2}{\left(10^{-\frac{1}{2\sqrt{f}}} - \frac{\epsilon}{3.71\,d_k}\right)^2} = \frac{(c_{CW}\,d_k)^2}{\left(1 - \frac{\epsilon}{3.71\,d_k}\right)^2}$$

and

$$\lim_{q \nearrow \infty} w_k(q) = \lim_{f \to \frac{1}{4}\left(\log\left(\frac{\epsilon}{3.71\,d_k}\right)\right)^{-2}} \frac{(c_{CW}\,d_k)^2}{\left(10^{-\frac{1}{2\sqrt{f}}} - \frac{\epsilon}{3.71\,d_k}\right)^2} = \infty\,.$$

As w_k is symmetric with respect to the origin it is

$$\lim_{q\nearrow 0} w_k(q) = -\frac{(c_{CW}\, d_k)^2}{\left(1-\frac{\epsilon}{3.71\, d_k}\right)^2}, \quad \text{and} \quad \lim_{q\searrow -\infty} w_k(q) = -\infty .$$

So the range of w_k , which is the range of the domain of w_k^{-1} , is the union of two open intervals and $\{0\}$, i.e.

$$(-\infty\, ,\, -\left(\frac{c_{CW}\, d_k}{1-\frac{\epsilon}{3.71\, d_k}}\right)^2) \cup \{0\} \cup (\left(\frac{c_{CW}\, d_k}{1-\frac{\epsilon}{3.71\, d_k}}\right)^2, \infty) .$$

According to Definition 3.4.3 it is $\psi_k(u_k(|q|)\, , |q|) = 0$, therefore direct insertion results in $w_k^{-1}(w_k(q)) = q$, which proves the assertion.

(iii) For $\phi \in D_{w_k^{-1}} \setminus \{0\}$ the second derivative $w^{-1''}$ of w^{-1} is

$$w^{-1''}(\phi) = \frac{\operatorname{sgn}(\phi)}{2\ln(10)}\left(T_{w1}(\phi) + T_{w2}(\phi)\right) ,$$

where

$$\begin{aligned} T_{w1}(\phi) &= \frac{\ln\left(\frac{c_{CW}\, d_k}{\sqrt{|\phi|}} + \frac{\epsilon}{3.71\, d_k}\right)}{|\phi|^{\frac{3}{2}}} , \\ T_{w2}(\phi) &= -\frac{c_{CW}\, d_k}{\phi^2\left(\frac{c_{CW}\, d_k}{\sqrt{|\phi|}} + \frac{\epsilon}{3.71\, d_k}\right)} + \frac{c_{CW}^2\, d_k^2}{|\phi|^{\frac{5}{2}}\left(\frac{c_{CW}\, d_k}{\sqrt{|\phi|}} + \frac{\epsilon}{3.71\, d_k}\right)^2} . \end{aligned}$$

For $\phi \in D_{w_k^{-1}} \setminus \{0\}$ it is $\frac{c_{CW}\, d_k}{\sqrt{|\phi|}} + \frac{\epsilon}{3.71\, d_k} < 1$, and therefore $T_{w1}(\phi) \le 0$. Furthermore it is

$$T_{w2}(\phi) = \frac{c_{CW}\, d_k}{\phi^2\left(\frac{c_{CW}\, d_k}{\sqrt{|\phi|}} + \frac{\epsilon}{3.71\, d_k}\right)}\left(-1 + \frac{c_{CW}\, d_k}{\sqrt{|\phi|}\left(\frac{c_{CW}\, d_k}{\sqrt{|\phi|}} + \frac{\epsilon}{3.71\, d_k}\right)}\right) \le 0 ,$$

if and only if

$$\frac{c_{CW}\, d_k}{\sqrt{|\phi|}\left(\frac{c_{CW}\, d_k}{\sqrt{|\phi|}} + \frac{\epsilon}{3.71\, d_k}\right)} \le 1 ,$$

which is equivalent to $\frac{\epsilon\sqrt{|\phi|}}{3.71\, d_k} \ge 0$. The last is obviously fulfilled, as all constants are greater than zero. So $w^{-1''}(\phi) > 0$ for $\phi < 0$ and $w^{-1''}(\phi) < 0$ for $\phi > 0$, which proves the assertion.

(iv) The derivation rule for the inverse function implies

$$w_k''(q) = -\frac{1}{[w_k^{-1'}(w_k(q))]^3} w_k^{-1''}(w_k(q)) \ .$$

As $w(q) > 0$ iff $q > 0$, the lemma follows from w_k and w_k^{-1} being monotone increasing and from (iii). □

3.4.8 Corollary. Regarding the Darcy-Weisbach formula for a vanishing flow, Lemma 3.4.7 implies for the head-loss

$$\lim_{q \searrow 0} \Phi(q) = \lim_{q \searrow 0} \frac{8\, c_{DW}\, l}{g\, d_k^5\, \pi^2}\, q^2\, u(q) = \frac{8\, c_{DW}\, c_{CW}{}^2\, l}{g\, d_k^3\, \pi^2 \left(1 - \frac{\epsilon}{3.71\, d_k}\right)^2} \ ,$$

and analogously

$$\lim_{q \nearrow 0} \Phi(q) = -\frac{8\, c_{DW}\, c_{CW}{}^2\, l}{g\, d_k^3\, \pi^2 \left(1 - \frac{\epsilon}{3.71\, d_k}\right)^2} \ .$$

Regarding the constants used for the three standard test examples and a length of $1,000$ m, the absolute value of this bound has a range from $4 \cdot 10^{-12}$ m for the New-York network and a diameter of 204 inch, and $3 \cdot 10^{-5}$ m for the two-loop network and a diameter of 1 inch. Obviously theses values are negligible and result from not using the original formula for laminar flow.

3.4.2 Equivalent Formulation of DNOP

With the auxiliary functions w_k of Definition 3.4.6, the nonlinearity of DNOP can be written for all $(i,j) \in A$ as

$$\sum_{i=1}^{K} \beta_k\, w_k(q_{ij})\, x_{ijk} = (H_i + E_i) - (H_j + E_j) \tag{3.4.3}$$

and therefore DNOP can equivalently be formulated as

$$
\begin{cases}
\sum\limits_{(i,j)\in P} \sum\limits_{k=1}^{K} c_k \, x_{ijk} + c_{s1} \, H_{s1} \to \min & \\
\sum\limits_{i=1}^{K} \beta_k \, w_k(q_{ij}) \, x_{ijk} = (H_i + E_i) - (H_j + E_j) & \forall (i,j) \in A \\
\sum\limits_{k=1}^{K} x_{ijk} = L_{ij} & \forall (i,j) \in P \\
\sum\limits_{j:(i,j)\in A} q_{ij} - \sum\limits_{j:(j,i)\in A} q_{ji} = b_i & \forall \, i \in D \\
q_{\min_{ij}} \leq q_{ij} \leq q_{\max_{ij}} & \forall (i,j) \in A \\
H_{iL} \leq H_i + E_i \leq H_{iU} & \forall \, i \in N \\
H_{s1} \geq 0 & \\
x_{ijk} \geq 0 & \forall (i,j) \in P \,, \, \forall \, k \in \mathcal{K} \\
x_{ijk} = \widetilde{x_{ijk}} & \forall (i,j) \in A \setminus P \,, \, \forall \, k \in \mathcal{K} \,.
\end{cases}
$$

3.4.3 Adjacency Property for DNOP

Analogous to the Theorem 2.5.7 of Fujiwara and Dey, first the adjacency property of DNOP is shown for the natural case that the set of points $\{(d_k, c_k)\}_{k=1}^{K}$ forms a strictly convex function. Lemma 3.4.10 as essential part of this proof is further needed for the head constraint propagation.

As this case does not include the theoretically based cost function of the two-loop network, an auxiliary function $\widetilde{J}$ is introduced to prove the more general case that it is sufficient for the adjacency property to hold true, if the set of points $\left\{(d_k^{-3}, c_k)\right\}_{k=1}^{K}$ forms a strictly convex function. Obviously, Theorem 3.4.11 is a special case and direct consequence of Theorem 3.4.15.

3.4.9 Definition. Define $\gamma = \frac{8 c_{DW}}{g \, \pi^2}$, i.e. $\beta_k = \gamma \, d_k^{-5}$.

(i) The unit head-loss according to Darcy-Weisbach is given by

$$\hat{J}_k(q) = \gamma \, d_k^{-5} \, w_k(q)$$

for $k \in \mathcal{K}$.

(ii) Regarding the continuation $w : I\!R \times I\!R_+ \to I\!R$ of w_k , the auxiliary function $J : I\!R \times I\!R_+ \to I\!R$ is defined by

$$J(q,d) := \gamma \, d^{-5} \, w(q,d) \,.$$

3.4.10 Lemma.

(i) For fixed $d \in [d_1, d_K]$ the auxiliary functions J of Definition 3.4.9 is strictly monotone increasing in q, strictly concave in q for $q < 0$ and strictly convex in q for $q > 0$.

(ii) For fixed $q < 0$ the function J is strictly monotone increasing in d and strictly concave in d.

For fixed $q > 0$ the function J is strictly monotone decreasing in d and strictly convex in d.

Proof.

(i) As $\gamma\, d_k^{-5} > 0$, Lemma 3.4.7 (i) implies the monotonicity and (iv) the concave-convex structure.

(ii) The monotonicity and convexity can be shown by direct determination of the corresponding derivatives of the function J, for details see Appendix F.1. □

First, the adjacency property is shown as direct consequence of the convexity of the continuation of the unit head-loss J with respect to d:

3.4.11 Theorem. Let $K \in \mathbb{N}$, $K \geq 3$ and for $k \in \{1, \dots, K\}$ let c_k, $d_k \in \mathbb{R}$ with $d_1 < d_2 < \dots < d_K$ and $c_1 < c_2 < \dots < c_K$. Assume the set of points $\{(d_k, c_k)\}_{k=1}^{K}$ forms a strictly convex function.

Then the optimal solution of DNOP with head-loss calculations based on Darcy-Weisbach equation fulfills the adjacency property.

Proof: Assume an arbitrary but fixed flow $q_{ij} \in \mathbb{R} \setminus \{0\}$ is part of an optimal solution of DNOP. Analogous to Theorem 2.5.7, $q_{ij} \neq 0$ can be assumed since otherwise choosing the minimum diameter for the whole pipe will be optimal. Then it suffices to show that the assumptions of Theorem 2.5.7 are fulfilled:

(i) According to the definition of the unit head-loss according to Darcy-Weisbach it is

$$\hat{J}_k(q_{ij}) = \gamma\, d_k^{-5} \operatorname{sgn}(q)\, q^2\, u_k(|q|)\ .$$

Since Lemma 3.4.10 (ii) implies for fixed $q < 0$ that the function $\hat{J}$ is strictly monotone increasing and for $q > 0$ strictly monotone decreasing, it follows for $q \neq 0$ that

$$|\hat{J}_1(q)| > |\hat{J}_2(q)| > \dots > |\hat{J}_K(q)|\ .$$

(ii) Additionally Lemma 3.4.10 (ii) implies for fixed $q < 0$ that the function J is strictly concave and for $q > 0$ strictly convex. Therefore according to Lemma 2.5.4 (i) for arbitrary but fixed $q \neq 0$ the set of points

$$\left\{(|\hat{J}_k(q)|\,,\ c_k)\right\}_{k=1}^{K} = \{(|J(q, d_k)|\,,\ c_k)\}_{k=1}^{K}$$

forms a strictly convex function. □

3.4.12 Corollary. For the networks of Hanoi and New-York the set of points $\{(d_k, c_k)\}_{k=1}^{K}$ forms a strictly convex function according to Remark 2.5.5. So these fulfill the adjacency property.

3.4.13 Definition. Let the auxiliary functions $\widetilde{J} : I\!R \times I\!R_+ \to I\!R$ be defined by

$$\widetilde{J}(q, \delta) := \gamma\, \delta^{\frac{5}{3}}\; w(q, \delta^{-\frac{1}{3}})\ .$$

3.4.14 Lemma. The function $\widetilde{J}$ of Definition 3.4.13 is strictly monotone increasing and strictly convex in δ for fixed $q > 0$, and strictly monotone decreasing and strictly concave in δ for fixed $q < 0$.

Proof. Direct determination of the partial derivatives implies the lemma, for details see Appendix F.2. □

3.4.15 Theorem. Let $K \in I\!N$, $K \geq 3$ and for $k \in \{1, \ldots, K\}$ let c_k, $d_k \in I\!R$ with $d_1 < d_2 < \ldots < d_K$ and $c_1 < c_2 < \ldots < c_K$. Assume the set of points $\left\{(d_k^{-3}, c_k)\right\}_{k=1}^{K}$ forms a strictly convex function.

Then the optimal solution of DNOP with head-loss calculations based on Darcy-Weisbach equation fulfills the adjacency property.

Proof: Analogous to Theorem 3.4.11, the assumptions of Theorem 2.5.7 are fulfilled for $q \neq 0$:

(i) According to the definition of the unit head-loss according to Darcy-Weisbach it is

$$\hat{J}_k(q_{ij}) = J(q_{ij}, d_k) = \widetilde{J}(q_{ij}, d_k^{-3})\ .$$

Obviously $d_1^{-3} > d_2^{-3} > \ldots > d_K^{-3}$, so Lemma 3.4.14 implies for fixed $q \neq 0$

$$|\hat{J}_1(q)| > |\hat{J}_2(q)| > \ldots > |\hat{J}_K(q)|\ .$$

(ii) Additionally, according to Lemma 3.4.14 the function $|\widetilde{J}|$ is strictly monotone increasing and strictly convex for fixed $q \neq 0$. Therefore Lemma 2.5.4 (ii) implies the set of points $\left\{\widetilde{J}(q_{ij}, d_k^{-3})\ ,\ c_k)\right\}_{k=1}^{K}$ to form a strictly convex function, which proves the theorem. □

3.4.16 Corollary. All standard test networks fulfill the adjacency property.

Proof. For the two-loop and the two-loop expansion network, it can be verified that the points $\left\{(d_k^{-3}, c_k)\right\}_{k=1}^{K}$ form a strictly convex function, therefore Theorem 3.4.15 proves the corollary. □

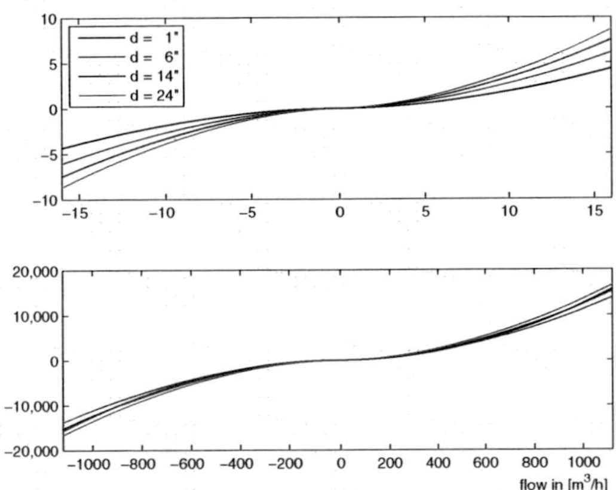

Figure 3.8: Family of functions w_k of Definition 3.4.6 (i) in small and large scale. Basis are the constants used for the two-loop network, and the second plot illustrates that they are not monotone in k for all $q \in I\!R$.

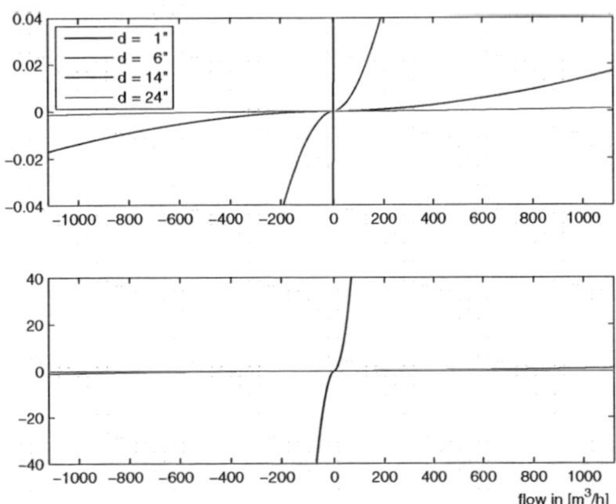

Figure 3.9: Family of functions $\hat{J}_k(q)$ of Definition 3.4.9 (i) in small and large scale, based on default values of the two-loop network. Lemma 3.4.10 (ii) shows monotonicity with respect to d_k, as well for $q < 0$ as for $q > 0$.

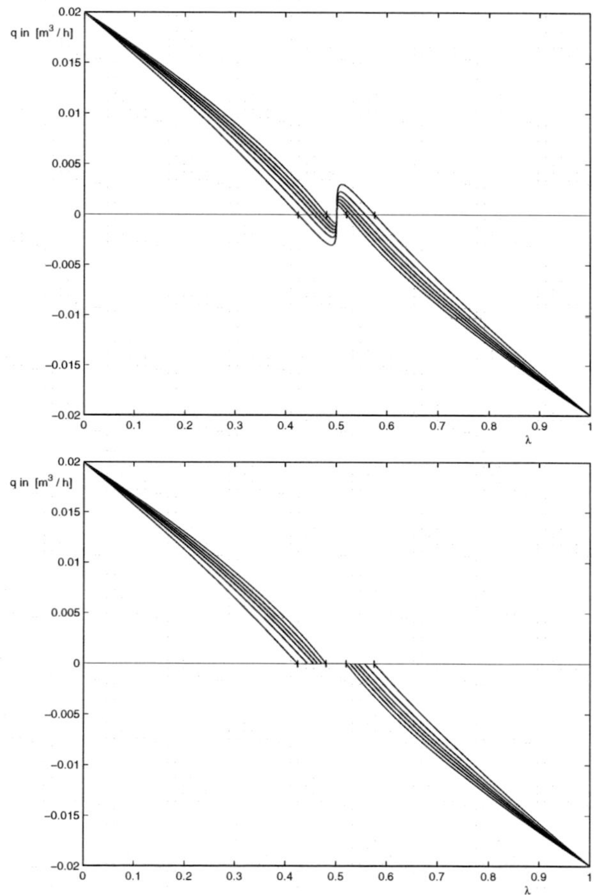

Figure 3.10: Auxiliary functions of DNOP, plotted with default data of the Hanoi network for all available diameters $k \in \mathcal{K}$. In the first plot, the continuations $w_k^{-1}(\lambda w_{k_{\min}} + (1-\lambda)\, w_{k_{\max}})$ are plotted, and the boundaries of the domains of w_1^{-1} and w_K^{-1} are marked at the λ-axis. The second figure contains $p_k(\lambda)$ for all $k \in \mathcal{K}$, which restricts w_k^{-1} to its domain, p_1 is plotted red and p_K green.

3.5 Determination of Global Minimum Costs

3.5.1 Branch and Bound Algorithm

For solving the global optimization problem, the branch and bound algorithm of Sherali, Subramanian and Loganathan [S&a01] is basically used and adjusted where necessary. For example, the distance to feasibility Δ_A is redefined by

$$\Delta_A := \max_{(i,j)\in A} \left| (H_i + E_i) - (H_j + E_j) - \sum_{k=1}^{K} w_k(q_{ij})\, \beta_{ijk}\, x_{ijk} \right| .$$

Analogous to Section 2.3.3, a vector with $\Delta_A < 10^{-6}$ (DELTA_ACC) is accepted to be admissible with respect to the nonlinear constraints in WaTerInt.

The branching strategies are analogous as for NOP, and in the next sections the main changes to subprocedures are summarized.

3.5.2 Constraint Propagation

The flow constraint propagation of Sherali, Subramanian and Loganathan [S&a01] can be directly applied to Kirchhoff's equations.

The head constraint propagation can be used analogously to Section 2.4, as the monotonicity arguments are provided in Lemma 3.4.10.

The calculation for fixed arcs is simplified to regard only one of the fixed diameters. Let $k_{l_{ij}}$ be the lowest diameter with segment length $x_{ijk} \neq 0$ for pipe $(i,j) \in A$ and correspondingly $k_{u_{ij}}$ the largest. Then in the first part it is

$$[\widetilde{\phi}_{\min_{ij}}, \widetilde{\phi}_{\max_{ij}}] = \begin{cases} [\Phi_{k_{u_{ij}}}(q_{\min_{ij}}, L_{ij})\,,\ \Phi_{k_{l_{ij}}}(q_{\max_{ij}}, L_{ij})] & \text{for } q_{\min_{ij}} > 0 \\ [\Phi_{k_{l_{ij}}}(q_{\min_{ij}}, L_{ij})\,,\ \Phi_{k_{u_{ij}}}(q_{\max_{ij}}, L_{ij})] & \text{for } q_{\max_{ij}} < 0 \\ [\Phi_{k_{l_{ij}}}(q_{\min_{ij}}, L_{ij})\,,\ \Phi_{k_{l_{ij}}}(q_{\max_{ij}}, L_{ij})] & \text{else} \end{cases}$$

and in the second

$$[\widetilde{q}_{\min_{ij}}, \widetilde{q}_{\max_{ij}}] = \begin{cases} [\Phi^{-1}_{k_{l_{ij}}}(\phi_{\min_{ij}}, L_{ij})\,,\ \Phi^{-1}_{k_{u_{ij}}}(\phi_{\max_{ij}}, L_{ij})] & \text{for } \phi_{\min_{ij}} > 0 \\ [\Phi^{-1}_{k_{u_{ij}}}(\phi_{\min_{ij}}, L_{ij})\,,\ \Phi^{-1}_{k_{l_{ij}}}(\phi_{\max_{ij}}, L_{ij})] & \text{for } \phi_{\max_{ij}} < 0 \\ [\Phi^{-1}_{k_{u_{ij}}}(\phi_{\min_{ij}}, L_{ij})\,,\ \Phi^{-1}_{k_{u_{ij}}}(\phi_{\max_{ij}}, L_{ij})] & \text{else,} \end{cases}$$

where

$$\Phi_k(q\,,\,L_{ij}) = \hat{J}_k(q)\,L_{ij} \qquad \text{and} \qquad \Phi_k^{-1}(\phi\,,\,L_{ij}) = w_k^{-1}\left(\frac{\phi}{\beta_k\,L_{ij}}\right) .$$

To avoid getting numerical errors when determining the friction factor f for flow with values close to zero, in the first part an obtained maximum flow out of the interval $(0, \texttt{HEPS})$ is set to HEPS and a minimum flow out of $(-\texttt{HEPS}, 0)$ to -HEPS. In WaTerInt this constant has a value of HEPS $= 0.01$.

Analogously, in the second part an obtained head loss with $\phi \leq \texttt{MEPS}$ is regarded as zero. To avoid getting tiny absolute values for the flow bounds implying numerical difficulties, an obtained minimum flow out of the interval $(0, \texttt{FEPS})$ or a maximum flow out of $(-\texttt{FEPS}, 0)$ is set to zero. In WaTerInt values of MEPS $= 10^{-6}$ and FEPS $= 0.01$ are used.

3.5.3 Lower Bounding Problem

For obtaining a lower bounding problem and for developing polyhedral relaxations the following function is needed.

3.5.1 Definition.
(i) Let $\lambda_k^* = \frac{w_{k_{\max}}}{w_{k_{\max}} - w_{k_{\min}}}$, and let $\lambda_k^{\triangleleft}$ and $\lambda_k^{\triangleright}$ be defined by

$$\lambda_k^{\triangleleft} = \frac{w_{k_{\max}} - \left(\frac{c_{CW}\, d_k}{1 - \frac{\epsilon}{3.71\, d_k}}\right)^2}{w_{k_{\max}} - w_{k_{\min}}} \quad \text{and} \quad \lambda_k^{\triangleright} = \frac{w_{k_{\max}} + \left(\frac{c_{CW}\, d_k}{1 - \frac{\epsilon}{3.71\, d_k}}\right)^2}{w_{k_{\max}} - w_{k_{\min}}} \,.$$

(ii) Let $p_k : [0,1] \to I\!R$ be defined by

$$p_k(\lambda) = \begin{cases} w_k^{-1}\left(\lambda\, w_{k_{\min}} + (1-\lambda)\, w_{k_{\max}}\right) & \text{for } 0 \leq \lambda < \lambda^{\triangleleft} \text{ or } \lambda^{\triangleright} < \lambda \leq 1 \\ 0 & \text{else.} \end{cases}$$

3.5.2 Lemma. For $0 \leq \lambda < \lambda^{\triangleleft}$ and for $\lambda^{\triangleright} < \lambda \leq 1$ the derivative can be determined with

$$p_k'(\lambda) = (w_{k_{\max}} - w_{k_{\min}}) \left(\frac{\log(T_p)}{\sqrt{|z|}} - \frac{c_{CW}\, d_k}{|z|\, T_p\, \ln 10} \right),$$

where $T_p = \frac{c_{CW}\, d_k}{\sqrt{|z|}} + \frac{\varepsilon}{3.71\, d_k}$ and $z = \lambda\, w_{k_{\min}} + (1-\lambda)\, w_{k_{\max}}$.

Proof. For arbitrary λ with either $0 \leq \lambda < \lambda^{\triangleleft}$ or $\lambda^{\triangleright} < \lambda \leq 1$ it follows

$$\lambda\, w_{k_{\min}} + (1-\lambda)\, w_{k_{\max}} \in D_{w_k^{-1}},$$

therefore p_k is well-defined, the derivative can be directly obtained. □

The set

$$[-\left(\frac{c_{CW}\, d_k}{1 - \frac{\epsilon}{3.71\, d_k}}\right)^2, \left(\frac{c_{CW}\, d_k}{1 - \frac{\epsilon}{3.71\, d_k}}\right)^2] \setminus \{0\}$$

is not part of the domain of the function w_k^{-1}.

For illustration purposes, Figure 3.10 contains, in the first plot, the continuations of the functions $w_k^{-1}(\lambda w_{k_{\min}} + (1-\lambda) w_{k_{\max}})$, where the symbolic expression of w_k^{-1} is evaluated for this interval as well. Obviously the functions p_k used for the lower bounding problem are more natural, where the value of w_k^{-1} is assigned to zero in this interval.

As explained in Section 3.4.1, the main difference of DNOP to NOP is the fact that the nonlinear part $w_k(q_{ij})$ depends on the diameter k as well. Therefore, the additionally introduced variables λ_{ij} are to be substituted by λ_{ijk}.

For simplicity, polyhedral relaxations are only developed for fixed arcs. Hence, for obtaining a lower bounding problem analogous to LNOP, for all $k \in \mathcal{K}$ new decision variables x^1_{ijk} and x^2_{ijk} are introduced for all $(i,j) \in A$, with

$$x_{ijk} = x^1_{ijk} + x^2_{ijk} \,.$$

Let $\mathcal{L} = \{(i,j,k) \mid (i,j) \in A \backslash P \text{ and } \widetilde{x_{ijk}} \neq 0\}$. Then further new decision variables are λ_{ijk} for all $(i,j,k) \in \mathcal{L}$ such that

$$x^1_{ijk} := \lambda_{ijk}\, x_{ijk}\,, \quad x^2_{ijk} := (1-\lambda_{ijk})\, x_{ijk} \tag{3.5.1}$$

with $0 \leq \lambda_{ijk} \leq 1$.

Let x^1 denote the vector $(x^1_{ijk})_{(i,j)\in A,\, k\in\mathcal{K}}$, analogously let $x^2 := (x^2_{ijk})_{(i,j)\in A,\, k\in\mathcal{K}}$, and $\lambda := (\lambda_{ijk})_{(i,j,k)\in\mathcal{L}}$.

For the fixed arcs of an expansion network the values $\widetilde{x_{ijk}}$ are known, and for these Equations 3.5.1 result in linear constraints.

If one would expand Equations 3.5.1 to newly to be designed arcs, the bilinearities could not be analogously relaxed with an aggregate relationship like Equation 2.3.3 on page 51. But, as the constraint propagation used in the branch and bound algorithm tightens the flow bounds quite well, one possibility is to simply relax the corresponding nonlinear constraints by omission. Obviously this increases the computational time needed and further research for other relaxations is recommended.

Lemma 3.4.7 (i) implies the function $w_k : I\!R \to I\!R$ to be monotone increasing, hence

$$w_k([q_{\min}, q_{\max}]) \subset [w_{k_{\min}}, w_{k_{\max}}]\,,$$

if $q_{\min} < 0$ and $q_{\max} > 0$, the inclusion is strict, as

$$[-\left(\frac{c_{CW}\, d_k}{1-\frac{\epsilon}{3.71\, d_k}}\right)^2, \left(\frac{c_{CW}\, d_k}{1-\frac{\epsilon}{3.71\, d_k}}\right)^2] \quad \not\subset \quad w_k([q_{\min}, q_{\max}])\,.$$

So, for all $(i,j) \in A$, the nonlinear equations of DNOP,

$$\begin{cases} \sum_{k=1}^{K} \beta_k \, w_k(q_{ij}) \, x_{ijk} = (H_i + E_i) - (H_j + E_j) \\ q_{\min_{ij}} \leq q_{ij} \leq q_{\max_{ij}} \,, \end{cases}$$

can be relaxed by

$$\begin{cases} \sum_{k=1}^{K} \beta_k \, w_{k_{\min_{ij}}} \, x^1_{ijk} + \sum_{k=1}^{K} \beta_k \, w_{k_{\max_{ij}}} \, x^2_{ijk} = (H_i + E_i) - (H_j + E_j) \\ q_{ij} = p_k\,(\lambda_{ijk}) & \forall \, (i,j,k) \in \mathcal{L} \\ q_{\min_{ij}} \leq q_{ij} \leq q_{\max_{ij}} & \forall (i,j) \in A \setminus P \\ 0 \leq \lambda_{ijk} \leq 1 & \forall \, (i,j,k) \in \mathcal{L} \,. \end{cases}$$

And in summary, for the decision variables x^1 , x^2 , λ , q and H the following **lower bounding** problem is obtained:

$$\begin{cases} \sum_{(i,j) \in P} \sum_{k=1}^{K} c_k \, (x^1_{ijk} + x^2_{ijk}) \; + \; c_{s1} \, H_1 & \to \min \\ & \\ \sum_{k=1}^{K} \beta_k \, w_{k_{\min_{ij}}} \, x^1_{ijk} + \sum_{k=1}^{K} \beta_k \, w_{k_{\max_{ij}}} \, x^2_{ijk} & \\ \qquad\qquad = (H_i + E_i) - (H_j + E_j) & \forall (i,j) \in A \\ \sum_{k=1}^{K} (x^1_{ijk} + x^2_{ijk}) = L_{ij} & \forall (i,j) \in P \\ x^1_{ijk} = \lambda_{ijk} \, \widetilde{x_{ijk}} & \forall (i,j) \in A \setminus P, \; k = 1, \ldots, K \\ x^2_{ijk} = (1 - \lambda_{ijk}) \, \widetilde{x_{ijk}} & \forall (i,j) \in A \setminus P, \; k = 1, \ldots, K \\ \sum_{j:(i,j) \in A} q_{ij} - \sum_{j:(j,i) \in A} q_{ji} = b_i & \forall i \in D \\ \text{[polyhedral relaxation for } q_{ij} = p_k(\lambda_{ijk})\text{]} & \forall (i,j) \in A \setminus P, \; k = 1, \ldots, K \\ q_{\min_{ij}} \leq q_{ij} \leq q_{\max_{ij}} & \forall (i,j) \in A \\ H_{iL} \leq H_i + E_i \leq H_{iU} & \forall i \in N \\ x^1_{ijk} \geq 0 \,, \; x^2_{ijk} \geq 0 & \forall (i,j) \in P \,, \quad k = 1, \ldots, K \\ 0 \leq \lambda_{ijk} \leq 1 & \forall (i,j) \in A \setminus P, \; k = 1, \ldots, K \,. \end{cases}$$

3.5.4 Linear Polyhedral Relaxations for Expansion Networks

In this section, relaxations are developed analogous to those of Sherali, Subramanian and Loganathan [S&a01]. Currently, only the nonlinearities of fixed arcs are relaxed by tangents, the others are simply relaxed by omission. Due to the constraint propagation techniques used, computational results can be obtained and are part of Appendix D.2.

As already explained, adding further relaxations may decrease the computational time, but instead, approximations to the Darcy-Weisbach formula may be used for water distribution design, as suggested by the adjusted Hazen-Williams formula at the end of this chapter.

For every pipe (i,j) and $k \in \mathcal{K}$ with $\widetilde{x_{ijk}} \neq 0$, the function p_k of Definition 3.5.1 (ii) has to be relaxed. An example is part of the second plot of Figure 3.10.

3.5.3 Lemma.

(i) The functions p_k of Definition 3.5.1 are strictly monotone decreasing in λ.

(ii) They are concave for $\lambda < \lambda^{\triangleleft}$ and convex for $\lambda > \lambda^{\triangleright}$.

Proof. According to Lemma 3.4.7 (ii) w_k^{-1} is strictly monotone increasing. Obviously $\lambda w_{k_{\min}} + (1-\lambda)\, w_{k_{\max}}$ is strictly monotone decreasing in λ for $w_{k_{\min}} < w_{k_{\max}}$. Therefore the composition of $p_k^{-1}(\lambda)$ is strictly monotone decreasing. An analogous argument, based on Lemma 3.4.7 (iii) proves the second part. □

Analogous to Sherali, Subramanian and Loganathan [S&a01], the linear relaxations yield an approximation for a concave-convex function. They are based on the idea to determine points $\overline{\lambda}_{ijk}$ and $\widetilde{\lambda}_{ijk}$ with

$$q_{\min_{ij}} - p_{k_{ij}}(\overline{\lambda}_{ijk}) - (1 - \overline{\lambda}_{ijk})\, p'_{k_{ij}}(\overline{\lambda}_{ijk}) = 0 \quad \text{for} \quad \overline{\lambda}_{ijk} \in (0\,,\ 1)\,,$$

and

$$q_{\max_{ij}} - p_{k_{ij}}(\widetilde{\lambda}_{ijk}) + \widetilde{\lambda}_{ijk}\, p'_{ijk}(\widetilde{\lambda}_{ijk}) = 0 \quad \text{for} \quad \widetilde{\lambda}_{ijk} \in (0\,,\ 1)\,,$$

where the tangents are upper respectively lower bounds for the flow.

Then for all $(i,j) \in A \setminus P$ and all $k \in \mathcal{K}$ with $\widetilde{x_{ijk}} \neq 0$ the same relaxations as presented on page 52 to 53 are chosen, except that q_{ij} is substituted by $p_{k_{ij}}$ and λ_{ij} by λ_{ijk}.

Analogous to Section 2.3.1, the derivative of p'_k and p_k are not evaluated at a point where $\lambda\, w_{\min} + (1 - \lambda) w_{\max} = 0$, therefore no exception concerning non-differentiability has to be considered when implementing p_k in WaTerInt.

In the first case, i.e. for $p_{k_{ij}}$ being concave-convex, it has to be ensured that no points are cut off due to the continuation of p_k in the interval $[\lambda^{\triangleleft}_{ijk}, \lambda^{\triangleright}_{ijk}]$ for $(i,j) \in N$ and $k \in \mathcal{K}$. Thus, the following inequalities have to hold true:

(i) $q_{\min_{ij}} < 0$ and $q_{\max_{ij}} > 0$ i.e. $q(\lambda_{ijk})$ is concave-convex:

(a) $\overline{\lambda}_{ijk}$ and $\widetilde{\lambda}_{ijk}$ exist:

$$p(\lambda^{\triangleright}_{ijk}) \leq p(\overline{\lambda}_{ijk}) + (\lambda^{\triangleright}_{ijk} - \overline{\lambda}_{ijk})\, p'(\overline{\lambda}_{ijk})$$

and

$$p(\lambda^{\triangleleft}_{ijk}) \geq p(\widetilde{\lambda}_{ijk}) + (\lambda^{\triangleleft}_{ijk} - \widetilde{\lambda}_{ijk})\, p'(\widetilde{\lambda}_{ijk})$$

(b) $\widetilde{\lambda}_{ijk}$ does not exist:

$$p(\lambda^{\triangleleft}_{ijk}) \geq \lambda^{\triangleleft}_{ijk}\, q_{\min} + (1 - \lambda^{\triangleleft}_{ijk})\, q_{\max}$$

and

$$p(\lambda^{\triangleright}_{ijk}) \leq q(\overline{\lambda}_{ijk}) + (\lambda^{\triangleright}_{ijk} - \overline{\lambda}_{ijk})\, p'(\overline{\lambda}_{ijk})$$

(c) $\overline{\lambda}_{ijk}$ does not exist:

$$p(\lambda^{\triangleright}_{ijk}) \leq \lambda^{\triangleright}_{ijk}\, q_{\min} + (1 - \lambda^{\triangleright}_{ijk})\, q_{\max}$$

and

$$p(\lambda^{\triangleleft}_{ijk}) \geq p(\widetilde{\lambda}_{ijk}) + (\lambda^{\triangleleft}_{ijk} - \widetilde{\lambda}_{ijk})\, p'(\widetilde{\lambda}_{ijk})\,.$$

In most cases they can be expected to be fulfilled. Therefore, in the current implementation of WaTerInt it is checked if these are fulfilled, and if not, again the corresponding pipe and diameter is just relaxed by omission.

3.5.5 Determination of Upper Bound

With the equivalent formulation of DNOP developed in Section 3.4.1, one can easily see that for a fixed flow vector q this optimization problem is linear for the remaining decision variables x_{ijk} and H_i for $(i,j) \in P$, $k \in \mathcal{K}$ and $i \in N$.

Therefore again, the flow obtained with the lower bounding problem is used for the incumbent upper bounding problem.

To calculate the values $u_k(q_{ij})$, the zero of the function ψ_k, i.e.

$$\psi_k(u_k(q_{ij}), q_{ij}) = 0\,,$$

is determined with the MATLAB function `fzero` in WaTerInt, which is based on bisection methods.

This is time consuming and further improvement may be possible, if already calculated values are tabulated or even the expected range is tabulated at the beginning and then appropriate interpolations are used during the branch and bound algorithm.

3.6 Adjusted Hazen-Williams Formula

> "... the values of the exponents vary with different surfaces, and also their values may not be exactly the same for large diameters and for small ones, ... " [WH05]

The computational time to solve DNOP exceeds the time needed for NOP significantly. As the Hazen-Williams formula is already a good approximation of Darcy-Weisbach, it is likely that a better fit can be obtained by improving the coefficients used in this formula. When comparing head-loss calculations due to both formulae, the main differences are obvious:

(i) It is possible to factorize the Hazen-Williams formula, so that one part depends only on the flow and the other on the diameter, whereas the implicitly defined term $u(q, d)$ in Darcy-Weisbach contains both, as shown by Equations 3.4.1 and 3.4.2.

(ii) For the Darcy-Weisbach formula, two parameters have to be specified, the equivalent sand roughness and viscosity, whereas Hazen-Williams contains only one hydraulic parameter C_{HW} .

According to the introductory citation of Williams and Hazen, further improvement may be obtained when considering different exponents for every diameter in the original formulation of their equation [WH05],

$$V = C_{HW} \left(\frac{d}{4}\right)^{0.63} \phi^{0.54} \, 0.001^{-0.04} ,$$

here within the notation analogous to Chapter 2. But for an exponent of ϕ depending on the diameter, the first advantage would not remain. Thus, first the results of NOP is improved just by using different Hazen-Williams coefficients for different diameters, i.e. changing C_{HW} to C_{HW_k} , $k \in \{1, \dots, K\}$.

Second, the more natural input parameters, equivalent sand roughness ϵ and viscosity ν , can be incorporated into Hazen-Williams. The Hazen-Williams coefficients C_{HW_k} and the discard coefficient c_d are to be adjusted, such that the head-loss obtained with both formulae is as close as possible in the flow and diameter range of the network calculated.

Therefore, all available diameters are used and the flow is discretized at `NMAX` points, i.e. in WaTerInt a value of `NMAX = 100` is used. This grid is chosen linearly between 1 m/h, or 1 ft/s for the New York networks, and the absolute maximum of the flow bounds, i.e. for the two-loop and two-loop expansion network 1,120 m/h, for Hanoi 19,940 m/h and for the New York networks 2,017.5 ft/s.

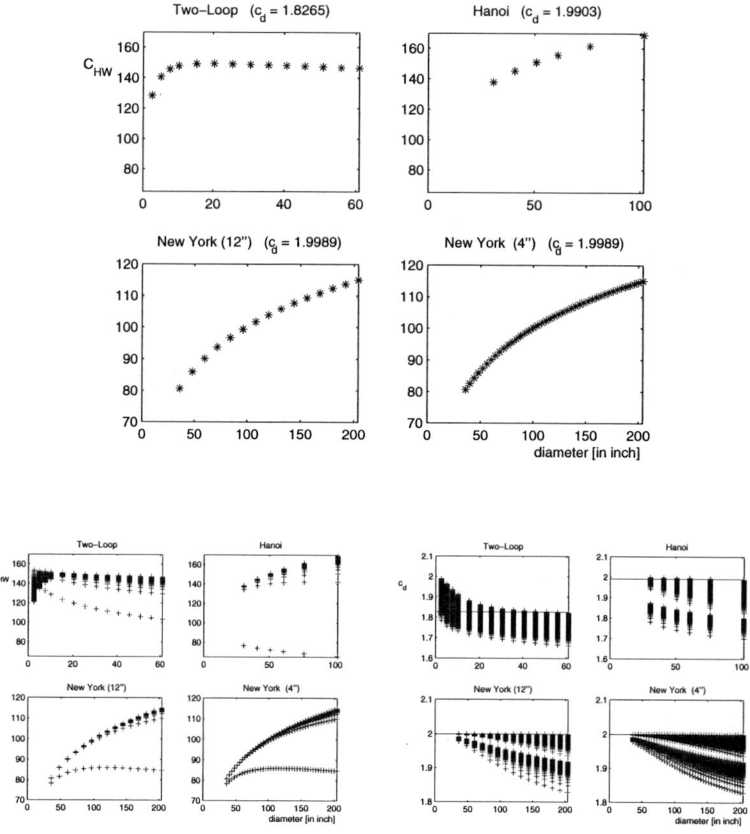

Figure 3.11: Determination of AHW-Coefficients with minimum flow of 1 m/h or for the New York networks of 1 ft/s, as implemented in WaTerInt. The first figure contains the coefficients that are selected for the default values which are the median of all data points shown in the second figure.
The third figure shows all values calculated for the discard coefficient c_d, where the median is indicated by the horizontal line.

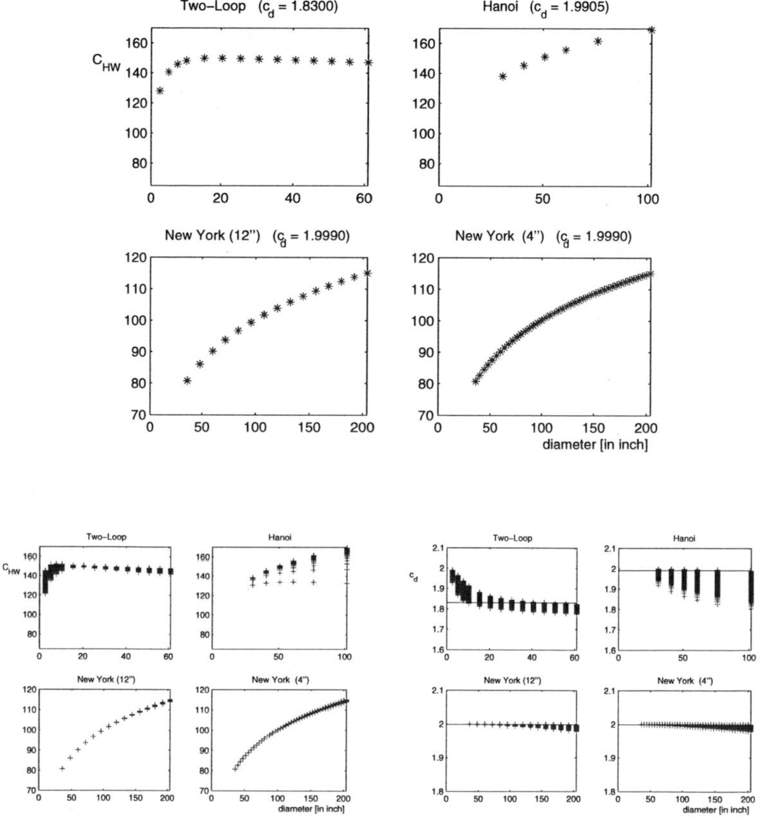

Figure 3.12: Determination of AHW-Coefficients with minimum flow of 100 m/h or 100 ft/s, respectively. The first figure contains the coefficients that would be selected as median when using this minimum flow, the second contains all data points considered.
Again, the third figure shows all values calculated for the discard coefficient c_d, where the median is indicated by the horizontal line.

For all points (d_k, q_i) , $k \in \{1, \dots K\}$, $i \in \{1, \dots \texttt{NMAX}\}$, of this diameter-flow grid, the results of Darcy-Weisbach and Hazen-Williams should coincide, i.e.

$$\frac{c_{hl}}{C_{HW}^{c_d}\, d_k^{4.87}}\, q_i^{c_d} \;=\; \frac{8\, c_{DW}}{g\, \pi^2\, d_k^5}\, u(q_i, d_k)\, q_i^2 \,.$$

For two different flow points q_i and q_j with $i, j \in \{1, \dots, \texttt{NMAX}\}$, $i \neq j$ and fixed diameter d_k one obtains

$$\left(\frac{q_i}{q_j}\right)^{c_d - 2} \;=\; \frac{u(q_i, d_k)}{u(q_j, d_k)}$$

which implies

$$c_d \;=\; 2 + \frac{\log(u(q_i, d_k)) - \log(u(q_j, d_k))}{\log q_i - \log q_j} \,. \tag{3.6.1}$$

Using a value of `NMAX = 100`, this results for each diameter in 4,900 pairs of different flow values. Therefore 4,900 values for c_d are obtained, and, to disregard outliers, in WaTerInt the median is chosen as value for the adjusted Hazen-Williams formula.

Analogously one obtains

$$C_{HW_k} = \left(\frac{c_{hl}\, g\, \pi^2}{8\, c_{DW}} d_k^{0.13}\, q^{c_d - 2}\; [u(q, d_k)]^{-1} \right)^{\frac{1}{c_d}} \tag{3.6.2}$$

for $k \in \{1, \dots, K\}$. For every diameter this results in 100 values, where again the median is taken in WaTerInt.

Figure 3.11 contains details of the data used during this determination. Obviously for small flow the spread increases, thus for illustration Figure 3.12 is based on a minimum flow of $\texttt{QMIN} = 100$ m/h or $\texttt{QMIN} = 100$ ft/s, respectively.

	QMIN = 1	QMIN = 100
Two-Loop Network	128.46 – 149.34	127.93 – 149.79
Hanoi	137.88 – 168.86	137.92 – 168.91
New York (12") or (4")	80.69 – 114.99	80.70 – 115.01

Table 3.2: Range of AHW-coefficients for standard test networks, obtained as median value from the grid spanned by available diameters and the linear flow discretization.

As already explained, small flow rates are feasible for the optimization problems used, and even part of the optimal solution obtained with Hazen-Williams formula. Thus, the minimum value 1 m/h or 1 ft/s is chosen. For the test networks and default parameters the ranges of the adjusted Hazen-Williams coefficients (AHW-coefficients) are summarized in Table 3.2, obviously the outliers do not significantly change the result.

Computational results show that a closer approximation of Darcy-Weisbach is obtained. For all standard test networks these are provided in Tables D.20 to D.24, and compared to those obtained with Hazen-Williams or Darcy-Weisbach formula in Section 5.3.1.

Chapter 4

Verification of Optimal Solution of NOP and DNOP

"Die traurige Wahrheit ist, dass Fehlerschätzungen von numerischen Simulationen realistischer Prozesse in der Praxis wenig verbreitet sind, da schwierig und zeitaufwendig."[1]

— Jens Lang, DMV-Jahrestagung Rostock, 16.09.03.

In this chapter, additional time needed for verification of a branch and bound algorithm is investigated on the water network design optimization problems.

Five parts of the branch and bound algorithm of Sherali and Subramanian have to be adjusted and the verification procedures used in WaTerInt are explained in detail. The verified version of the algorithm for the Darcy-Weisbach problem is described at the same time.

It occurs that the time needed for obtaining a verified solution is on average about fifteen times the time needed without rigorous calculations.

4.1 Incentives for Verification of NOP and DNOP

All known algorithms for the water distribution network design are based on floating point calculations. Even for the deterministic ones, no error estimates for the optimal solutions exist so far. However, even for early algorithms, which have been designed to start with initial values for the solution vector, numerical problems are discussed.

Fujiwara et al. report differences in the costs for the same initial flow distribution, as they calculated 486,562 whereas Alperovits and Shamir obtained 493,776. On the

[1] "The sad truth is that error estimates of numerical simulations of realistic processes are not very common in practice as they are both difficult and time-consuming."

one hand these differences may be due to the use of different head-loss or discard coefficients, but as a second reason Fujiwara points out that

> "the linear programming subroutine used in the present work might also differ from the one they have employed, thereby introducing different levels of accuracy at LP iterations. This point has to be emphasized because the coefficient matrix formed by the constraint set of a linear program is largely sparse and the *computational results could be sensitive to any perturbation in the data.* [F&a87]"

Neumaier emphasizes this objective of knowing about the correctness of the result. When summarizing challenges in global optimization, he points out that "ensuring reliability is perhaps the most pressing issue" [N03].

Even if there are many uncertainties in the raw data for the optimization problem as shown in Section 2.1, it is important to know about the reliability of the calculation. This is emphasized by the results of WaTerInt for the New York test network considering $12''$ increments for the diameter and using floating point arithmetic as suggested in [S01]. By exchanging only the linear programming solver, the following lower and upper bounds for the optimal value are obtained with WaTerInt (see Table D.3 and D.4):

`cplex`	$38,049,839.4065$	$38,049,864.8659$
`linprog`	$38,052,280.8570$	$38,052,280.8570$

The lower bound of calculations with `linprog` being greater than the upper bound of `cplex` shows that solutions must have been cut off during the branch and bound algorithm, or that a non-feasible point is considered as feasible.

To get a verified optimal solution in floating point arithmetic avoiding these effects, only five stages of the algorithm have to be altered:

(i) When calculating the relaxations, it has to be verified that no admissible points are cut off. Especially in the concave-convex case the determination of the points $\overline{\lambda}$ an $\widetilde{\lambda}$ has to be verified.

(ii) The adjustments of the flow bounds according to MSTR and the head constraint propagation have to be verified.

(iii) Pruning a subproblem as the lower bound is greater than the best known upper bound results in two verification steps: on one hand the lower bound has to be verified, and on the other the incumbent upper bounds.

(iv) Pruning a subproblem out of the reason of non-solvability of the lower bounding problem has to be verified.

(v) Finally the last lower bound has to be verified.

These topics are described in detail in the following section.

4.2 Verification of the Branch and Bound Algorithm

4.2.1 Verification of Relaxations

In order to obtain rigorous results, the relaxations part of LNOP must be implemented ensuring that no feasible solutions are cut off due to floating point arithmetic. Otherwise, the computed optimal value may not be a lower bound for NOP, and subproblems may be discarded which contain global optimal solutions. If all approximate floating point input data of LNOP is replaced by intervals containing the true value, a family of linear relaxations with real input data within these intervals is obtained.

This family of linear relaxations contains the lower bounding problem of NOP. Hence, a rigorous lower bound for all optimal values provides a rigorous lower bound for LNOP. The family can by generated by computing all input data of LNOP with interval arithmetic. This straightforward approach results in an overestimation that can be avoided by using monotonicity arguments.

For the verification of the relaxations, only the nonlinearity of NOP has to be considered, as it has to be ensured that the range defined by the inequalities contain the range of the nonlinear functions

$$q_{ij}(\lambda_{ij}) = \mathrm{sgn}\left[\lambda_{ij} v_{\min_{ij}} + (1-\lambda_{ij}) v_{\max_{ij}}\right] \left|\lambda_{ij} v_{\min_{ij}} + (1-\lambda_{ij}) v_{\max_{ij}}\right|^{\frac{1}{c_d}}$$

for all $(i,j) \in A$ and $0 \le \lambda_{ij} \le 1$.

When constructing the relaxations, the bounds of the flow interval $q_{\min_{ij}}$ and $q_{\max_{ij}}$ can be regarded as real numbers, whereas $v_{\min_{ij}}$ and $v_{\max_{ij}}$ cannot always be represented in floating point arithmetic. Hence, these are replaced by intervals

$$[v_{\min_{ij}}] = \left(\mathrm{sgn}\left(q_{\min_{ij}}\right) \left|q_{\min_{ij}}\right|^{[c_d]}\right)^I$$

and analogously $[v_{\max_{ij}}]$ and computed with natural interval extensions.

According to Section 2.3.1 the three cases, $q_{ij}(\lambda_{ij})$ being concave-convex, concave and convex are treated separately. In the first case, i.e. $q_{ij}(\lambda_{ij})$ is concave-convex, let λ^*_{ij} be defined with $q(\lambda^*_{ij}) = 0$, which implies

$$\lambda^*_{ij} = \frac{1}{1 - \frac{v_{\min_{ij}}}{v_{\max_{ij}}}} .$$

For calculating approximations for the points $\overline{\lambda}$ and $\widetilde{\lambda}$, the verified bisection described in Lemma 1.1.19 is used. A verified lower bound of the zero of the function $\overline{\varphi}_{ij}$ is determined with

$$\overline{\varphi}_{ij}(\lambda) = q_{\min_{ij}} - q_{ij}(\lambda) - (1-\lambda)\, q'_{ij}(\lambda) \quad \text{for} \quad \lambda \in [0\,,\ \lambda^*_{ij})\,,$$

and a verified upper bound of $\widetilde{\varphi}_{ij}$ with

$$\widetilde{\varphi}_{ij}(\lambda) = q_{\max_{ij}} - q_{ij}(\lambda) + \lambda\, q'_{ij}(\lambda) \quad \text{for} \quad \lambda \in (\lambda^*_{ij},\ 1]\,.$$

These functions are illustrated in Figure 4.1.

For $0 \leq \lambda < \lambda^*_{ij}$ the function $q_{ij}(\lambda)$ is strictly concave and for $\lambda^*_{ij} < \lambda \leq 1$ strictly convex. Therefore $\overline{\varphi}_{ij}(\lambda)$ and $\widetilde{\varphi}_{ij}(\lambda)$ are strictly monotone increasing. So no $\overline{\lambda}_{ij}$ exists if $\overline{\varphi}_{ij}(0) > 0$ and no $\widetilde{\lambda}_{ij}$ exists if $\widetilde{\varphi}_{ij}(1) < 0$.

The assumption

$$\overline{\varphi}_{ij}(0) > 0 \quad \text{and} \quad \widetilde{\varphi}_{ij}(1) < 0$$

implies a contradiction to the concave-convex structure of the function $q_{ij}(\lambda)$. Therefore a verified inclusion $\overline{\varphi}_{ij}(0)^I > 0$ guarantees the existence of $\widetilde{\lambda}_{ij}$ and analogously $\widetilde{\varphi}_{ij}(1)^I < 0$ the existence of $\overline{\lambda}_{ij}$.

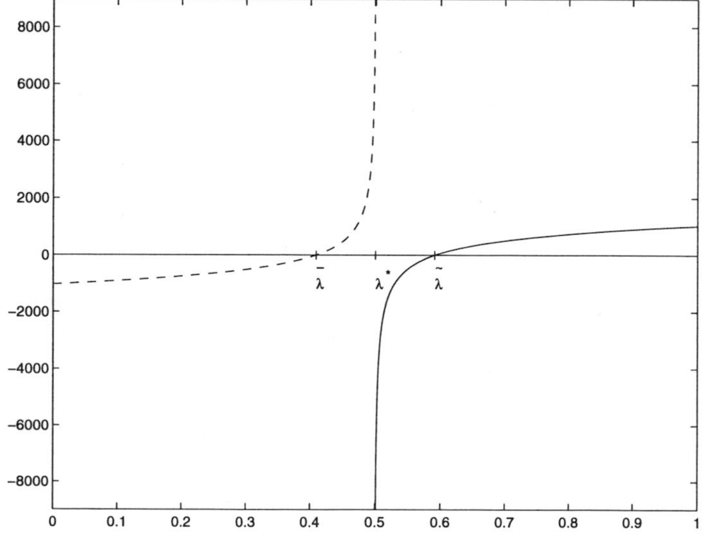

Figure 4.1: Illustration of $\overline{\varphi}_{ij}(\lambda)$ (dashed) and $\widetilde{\varphi}_{ij}(\lambda)$ (solid) with $q(\lambda^*_{ij}) = 0$ calculated for $q_{\min_{ij}} = -1120$ and $q_{\max_{ij}} = 1120$ and a discard coefficient of 1.852.

The inequalities can be verified by using interval calculations and the corresponding interval bounds in the inequalities, as λ is nonnegative and the coefficient of the flow is 1 or -1. So for all $(i,j) \in A$ the following verified relaxations are chosen according to case differentiation:

(i) $q_{\min_{ij}} < 0$ and $q_{\max_{ij}} > 0$ i.e. $q(\lambda_{ij})$ is concave-convex:

(a) Verified inclusions for $\overline{\lambda}_{ij}$ and $\widetilde{\lambda}_{ij}$ exist:

The concavity of $q_{ij}(\hat{\lambda})$ for $0 \leq \hat{\lambda} \leq \overline{\lambda}_{ij}$ and the construction of $\overline{\lambda}_{ij}$ imply

$$q_{ij} \leq q_{ij}(\hat{\lambda}) + (\lambda_{ij} - \hat{\lambda})\, q'_{ij}(\hat{\lambda}) \quad \text{for} \quad 0 \leq \hat{\lambda} \leq \overline{\lambda}_{ij}\,.$$

Therefore verified relaxations are obtained with

$$(-q'_{ij}(\hat{\lambda})^I)^L \lambda_{ij} + q_{ij} \quad \leq \quad (q_{ij}(\hat{\lambda})^I - \hat{\lambda}\, q'_{ij}(\hat{\lambda})^I)^U \text{ for } 0 \leq \hat{\lambda} \leq \overline{\lambda}_{ij}^L\,,$$

and in WaTerInt $\hat{\lambda} = 0\,,\ \frac{\overline{\lambda}_{ij}^L}{2}\,,\ \overline{\lambda}_{ij}^L$ respectively $\hat{\lambda} = 0\,,\ \overline{\lambda}_{ij}^L$ are selected. Analogously for $\widetilde{\lambda}_{ij}$ verified relaxations are

$$(q'_{ij}(\hat{\lambda})^I)^L \lambda_{ij} - q_{ij} \quad \leq \quad (\hat{\lambda}\, q'_{ij}(\hat{\lambda})^I - q_{ij}(\hat{\lambda})^I)^U \quad \text{for } \widetilde{\lambda}_{ij}^U \leq \hat{\lambda} \leq 1\,,$$

and again chosen for $\hat{\lambda} = \widetilde{\lambda}_{ij}^U\,,\ \frac{1+\widetilde{\lambda}_{ij}^U}{2}\,,\ 1$ respectively $\hat{\lambda} = \widetilde{\lambda}_{ij}^U\,,\ 1\,.$

(b) $\widetilde{\lambda}_{ij}$ does not exist:

For $\hat{\lambda} = 0\,,\ \frac{\overline{\lambda}_{ij}^L}{4}\,,\ \frac{\overline{\lambda}_{ij}^L}{2}\,,\ \frac{3\overline{\lambda}_{ij}^L}{4}\,,\ \overline{\lambda}_{ij}^L$ respectively $\hat{\lambda} = 0\,,\ \frac{\overline{\lambda}_{ij}^L}{2}\,,\ \overline{\lambda}_{ij}^L$

$$(-q'_{ij}(\hat{\lambda})^I)^L \lambda_{ij} + q_{ij} \quad \leq \quad (q_{ij}(\hat{\lambda})^I - \hat{\lambda}\, q'_{ij}(\hat{\lambda})^I)^U$$

and

$$(q_{\min_{ij}} - q_{\max_{ij}})^L \lambda_{ij} - q_{ij} \quad \leq \quad -q_{\max_{ij}}\,.$$

(c) $\overline{\lambda}_{ij}$ does not exist:

$$(q_{\max_{ij}} - q_{\min_{ij}})^L \lambda_{ij} + q_{ij} \quad \leq \quad q_{\max_{ij}}$$

and for $\hat{\lambda} = \widetilde{\lambda}_{ij}^U\,,\ \frac{1+\widetilde{\lambda}_{ij}^U}{4}\,,\ \frac{1+\widetilde{\lambda}_{ij}^U}{2}\,,\ \frac{3(1+\widetilde{\lambda}_{ij}^U)}{4}\,,\ 1$ respectively $\hat{\lambda} = \widetilde{\lambda}_{ij}^U\,,\ \frac{1+\widetilde{\lambda}_{ij}^U}{2}\,,\ 1$

$$q'_{ij}(\hat{\lambda})^L \lambda_{ij} - q_{ij} \quad \leq \quad (\hat{\lambda}\, q'_{ij}(\hat{\lambda})^I - q_{ij}(\hat{\lambda})^I)^U\,.$$

(ii) $q_{\min_{ij}} \geq 0$ i.e. $q(\lambda_{ij})$ is concave:

$$(q_{\min_{ij}} - q_{\max_{ij}})^L \lambda_{ij} - q_{ij} \quad \leq \quad -q_{\max_{ij}}$$

and for $\hat{\lambda} = 0\,,\ 0.25\,,\ 0.5\,,\ 0.8\,,\ 0.9$ respectively $\tilde{\lambda} = 0\,,\ 0.5\,,\ 0.9$

$$(-q'_{ij}(\hat{\lambda})^I)^L \lambda_{ij} + q_{ij} \quad \leq \quad (q_{ij}(\hat{\lambda})^I - \hat{\lambda}\, q'_{ij}(\hat{\lambda})^I)^U\,.$$

(iii) $q_{\max_{ij}} \leq 0$ i.e. $q(\lambda_{ij})$ is convex:

$$(q_{\max_{ij}} - q_{\min_{ij}})^L \lambda_{ij} + q_{ij} \quad \leq \quad q_{\max_{ij}}$$

and for $\hat{\lambda} = 0.1\,,\ 0.2\,,\ 0.5\,,\ 0.75\,,\ 1$ respectively $\hat{\lambda} = 0.1\,,\ 0.5\,,\ 1$

$$q'_{ij}(\hat{\lambda})^L \lambda_{ij} - q_{ij} \quad \leq \quad (\hat{\lambda}\, q'_{ij}(\hat{\lambda})^I - q_{ij}(\hat{\lambda})^I)^U\,.$$

Again, when using the Darcy-Weisbach equation, analogous relaxations can be used for all fixed arcs, i.e. λ_{ijk} with $(i,j) \in A \setminus P$ and $k \in \mathcal{K}$ with $\widetilde{x_{ijk}} \neq 0$. As in Section 3.5.4, it has to be ensured that no points are cut off due to the definition of p_k in the interval $[\lambda^{\triangleleft}_{ijk}, \lambda^{\triangleright}_{ijk}]$ for $(i,j,k) \in \mathcal{L}$. In this case the inequalities part of the relaxations are to be verified for intervals $[\lambda^{*}_{ijk}]$, $[\lambda^{\triangleleft}_{ijk}]$ and $[\lambda^{\triangleright}_{ijk}]$.

4.2.2 Adjustment of Flow Bounds

The procedure of adjusting the flow bounds of Section 2.3.2 can be simplified using intervals. Instead of regarding the flow bounds separately, they are directly treated as an interval $[q_{ij}] = [q_{\min_{ij}}, q_{\max_{ij}}]$. The supply respectively demand $[b_i]$ is regarded as floating point interval containing the real value b_i for all $i \in N$.

This implies the following adjustment $[\widetilde{q}_{ij}]$ of $[q_{ij}]$ for all tree-arcs (i,j), again ordered according to post-order tree traversal:

$$[\widetilde{q}_{ij}] = [q_{ij}] \cap \left([b_i] + \sum_{k:(k,i)\in A} [q_{ki}] - \sum_{\substack{k:(i,k)\in A \\ k\neq j}} [q_{ik}] \right) \cap \left(-[b_j] - \sum_{\substack{k:(k,j)\in A \\ k\neq i}} [q_{kj}] + \sum_{k:(j,k)\in A} [q_{jk}] \right)$$

Using INTLAB [R03] addition and subtraction of intervals guarantees that no admissible points are cut off during this procedure.

4.2.3 Head Constraint Propagation

The constraint propagation techniques of Sections 2.4 and 3.5.2 can analogously be verified. For the determination of the adjusted bounds, interval evaluations are necessary, i.e. in the first part bounds for the admissible head loss in the pipe (i,j) are determined for newly to be designed arcs by

$$[\widetilde{\phi}_{\min_{ij}}, \widetilde{\phi}_{\max_{ij}}] = \begin{cases} [\Phi_K([q_{\min_{ij}}],[L_{ij}])^L\,,\ \Phi_1([q_{\max_{ij}}],[L_{ij}])^U] & \text{for } q_{\min_{ij}} > 0 \\ [\Phi_1([q_{\min_{ij}}],[L_{ij}])^L\,,\ \Phi_K([q_{\max_{ij}}],[L_{ij}])^U] & \text{for } q_{\max_{ij}} < 0 \\ [\Phi_1([q_{\min_{ij}}],[L_{ij}])^L\,,\ \Phi_1([q_{\max_{ij}}],[L_{ij}])^U] & \text{else,} \end{cases}$$

and for fixed arcs instead by

$$[\widetilde{\phi}_{\min_{ij}}, \widetilde{\phi}_{\max_{ij}}] = \sum_{k=1}^{K} \left[\Phi_k([q_{\min_{ij}}],[\widetilde{x_{ijk}}])^L\,,\ \Phi_k([q_{\max_{ij}}],[\widetilde{x_{ijk}}])^U\right]\,.$$

The adjustments needed for the second part and for the Darcy-Weisbach problem are analogous.

For the Darcy-Weisbach problem, to avoid getting tiny absolute values for the flow bounds, again an obtained maximum flow out of the interval $(0, \texttt{HEPS})$ is set to `HEPS` and a minimum flow out of $(-\texttt{HEPS}, 0)$ to `-HEPS` in the first part, and analogously in the second. In WaTerInt, a value of 0.01 for the constant `HEPS`, called `FEPS` in the second part, is used.

No further considerations about pipes (i, j), where $q_{\min_{ij}} = q_{\max_{ij}}$ are necessary, as reliable results are obtained. Thus no branching strategy 5 or 6 exists for verified calculations.

4.2.4 Verification of Lower Bounds

The process of verifying the lower bounds is based on the bounds developed by Jansson and Neumaier. The lower bounds used in the branch and bound algorithm of NOP are obtained by solving linear problems.

An approximate lower bound calculated with a linear solver like `linprog` or `cplex` can easily be verified with Equation 1.2.1

$$\underline{f}^* = \min \left\{ a^\top y + [b]^\top z + \underline{x}^\top [d]^+ - \overline{x}^\top [d]^- \right\}$$

as shown in Lemma 1.2.3. For the determination of the approximate lower bound, the midpoint of the corresponding interval matrices are used.

From the underlying model of NOP, the input data of the linear lower bounding problem is considered as follows:

The cost vector is supposed to be given as vector $c \in I\!R^n$ or as result of a function of the diameter, e.g. $c_k = 1.1 \cdot d_k^{1.24}$ for $k = 1, \dots, K$. It is assumed to be a floating point vector $c \in I\!R^n$.

The matrices used in the linear optimization programs depend on the flow bounds and therefore differ in each step of the branch and bound algorithm. As shown in Section 4.2.1, the inequality constraints can be represented by a simple floating point matrix A and a vector a, whereas the equality constraints are described by interval matrix $[B]$ and interval vector $[b]$.

As suggested in [J02], the Lagrange multipliers y^* and z^* obtained with the linear solver are used as vectors y and z when calculating $\underline{f}^*$. The notation of `cplex` corresponds to [J02], whereas both vectors y^* and z^* obtained with `linprog` are used with opposite sign in WaTerInt.

As all lower bounds calculated during the branch and bound procedure are considered when pruning subproblems, they are verified directly after calculation. Therefore the last verification step, i.e. the verification of the last lower bound does not have to be treated separately.

4.2.5 Verification of Upper Bound

"Correct to within an order of magnitude = wrong."[2]

In the branch and bound algorithm proposed by Sherali, Subramanian and Loganathan ([S&a01], [S01]), a subproblem P is pruned if for the lower bound $\text{LB}(P)$ and for the best known upper bound UB

$$\text{LB}(P) \geq (1-\varepsilon)\,\text{UB}$$

holds true, where ε is the accuracy, per default in WaTerInt a value of $\varepsilon = 10^{-6}$ is used.

For verified calculations, a subproblem P can only be pruned if

$$\text{LB}(P) \geq \text{UB}\,.$$

This implies a significant increase in the number of subproblems to be evaluated during the branch and bound algorithm as shown in Tables D.31 to D.35, when comparing the computational time needed for different accuracies.

For keeping the amount of additional computation low, the upper bound is verified a posteriori, i.e. only the last upper bound is verified as VUB. During the branch and bound algorithm the incumbent upper bounds used for pruning lower bounding problems are increased by a factor η. Therefore the solution has to be recalculated, if for the best known upper bound

$$\text{UB} + \eta < \text{VUB}\,.$$

To avoid this, additionally the possibility is provided to verify every upper bound obtained directly, as part of verification strategy 3. Obviously, this strategy incurs longer computational times.

In mathematical literature, a tendency to publish optimal values that are slightly better than the best known solutions (see Table 5.8, 5.9 and 5.11) can be observed. The conflict between calculation time and the accuracy of the calculated optimal value is obvious. In verified calculation the accuracy is measured with the diameter of the interval that contains the optimal value.

[2] *Glossary for Mathematical Writing,* H. Pètard. A brief dictionary of phrases used in mathematical writing. Amer. Math. Monthly, 73:196 – 197, 1966. — Adapted by Nicholas J. Higham. *Handbook of Writing for the Mathematical Sciences.* SIAM, ISBN 0-89871-420-6, 1998.

From the point of engineering application, no small diameter of the interval for the optimal costs is needed, in agreement with the statement of Carl Friedrich Gauß:

> "Der Mangel an mathematischer Bildung gibt sich durch nichts so auffallend zu erkennen, wie durch maßlose Schärfe im Zahlenrechnen."[3]

In response, in WaTerInt the accuracy level ε can be provided as parameter for the verified calculations as well. Therefore the branch and bound procedure is stopped when

$$\min_P\{\mathrm{LB}(P)\} > (1-\varepsilon)\,\mathrm{UB}\,.$$

In this case the global lower bound GLB is set to

$$\mathrm{GLB} = \min_P\{\mathrm{LB}(P)\}\,.$$

The main issue for verifying the upper bound is to guarantee that the **solution vector $\boldsymbol{x}^*$ is admissible.** An algorithm for this part of the verification is developed in the rest of this section and summarized in Table 4.1.

In addition to the usual symbols for NOP and DNOP, throughout this section the following notation is used: Let a be defined as the number of arcs in the network, which is the same as the cardinality of A, let p be the number of newly to be designed arcs, let n be the number of nodes and s the number of segments, i.e.

$$a := |A|\,, \qquad p := |P|\,, \qquad n := |N|\,, \qquad s := a \cdot K\,.$$

Let $x = (x_{ijk})_{(i,j)\in A,\ k=1,\dots,K} \in I\!R^s$ be the vector of the length of the segments. Let the flow vector be denoted by $q = (q_{ij})_{(i,j)\in A} \in I\!R^a$ and let the head vector be defined by $H = (H_i)_{i\in N} \in I\!R^n$.

So it is the objective to calculate an inclusion $[y] \in I\!I\!R^{s+a+n}$ "in the near" of the approximate solution vector $(\widetilde{x}\,,\ \widetilde{q}\,,\ \widetilde{H})^\top$, which contains an admissible vector y^*, i.e.

$$y^* = \begin{pmatrix} x^* \\ q^* \\ H^* \end{pmatrix} \in \begin{pmatrix} [x] \\ [q] \\ [H] \end{pmatrix} = [y]\,.$$

[3] A lack of mathematical education is chiefly recognizable by an excessive precision in calculation with numbers. – Citation taken from [Ku96].

Consequently, the following five groups of constraints of the original problem introduced in Section 2.2.1 have to be regarded:

(i) $\Phi_{ij}(q_{ij}, x_{ij}) = (H_i + E_i) - (H_j + E_j)$ for all $(i,j) \in A$

(ii) $\sum_{k=1}^{K} x_{ijk} = L_{ij}$ for all $(i,j) \in P$

(iii) $\sum_{j:(1,j)\in A} q_{1j} - \sum_{j:(j,1)\in A} q_{j1} \leq b_1$

$\sum_{j:(i,j)\in A} q_{ij} - \sum_{j:(j,i)\in A} q_{ji} = b_i$ for all $i \in D$

(iv) $q_{\min_{ij}} \leq q_{ij} \leq q_{\max_{ij}}$ for all $(i,j) \in A$

$H_{iL} \leq H_i + E_i \leq H_{iU}$ for all $i \in N$

$x_{ijk} \geq 0$ for all $(i,j) \in P\,,\quad k = 1, \dots, K$

(v) $x_{ijk} = \widetilde{x_{ijk}}$ for all $(i,j) \in A \setminus P\,,\ k = 1, \dots, K\,.$

A common way to determine an inclusion that guarantees to contain an admissible vector for a linear optimization problem is described in [J02]. In this algorithm, an inclusion of the equalities is determined for an approximate solution. If this inclusion is not inside the bounds or violates the inequalities, Jansson iteratively solves a perturbed problem to improve the approximate solution. For the perturbed problem, he tightens the simple bounds and the right hand side of the inequalities.

One of the difficulties of NOP is to find an admissible solution fulfilling the nonlinear and linear equality constraints. So solving a perturbed problem would in this case imply to rerun the branch and bound procedure, which is extremely time consuming. Therefore in this verification process, the structure of the problem is utilized where possible.

First of all, the amount of equalities and inequalities to be considered is reduced as follows:

The equality **constraints (v)** are fulfilled according to the construction in the branch and bound algorithm of setting a priori $x_{ijk} = [\widetilde{x_{ijk}}]$ for all $(i,j) \in A \setminus P$, $k = 1, \dots, K$. So they are not to be considered when verifying admissibility.

The flow bounds part of **constraints (iv)** are added by Sherali and Subramanian for branch and bound purposes, as their algorithm is based on bisecting the hyperrectangle spanned by these flow bounds. These constraints do not imply any further restrictions as they are deduced from the conservation of flow according to the network structure. So they are neither to be considered when verifying the admissibility of a solution.

For **constraint (iii)**, it can be assumed without loss of generality that the equation

$$\sum_{i=2}^{n} b_i = -b_1$$

holds true, which is fulfilled for all test networks calculated, see Appendix A. Therefore according to Lemma 2.5.10, the first inequality can be regarded as equality.

Further on, Lemma 2.5.10 implies that the first row of the augmented matrix of the inhomogeneous linear equality system is linearly dependent on the other rows. Therefore, this inequality always holds true for a flow fulfilling the rest of the equalities and can be omitted in the verification process.

For **constraints (ii)** and the bounds $x_{ijk} \geq 0$ for all $(i,j) \in P$, $k = 1, \dots, K$ the values obtained for the length segments are improved with the adjacency property. To ensure this property, the monotonicity and convexity conditions of Lemma 2.5.7 and Lemma 3.4.15 have to be checked.

4.2.1 Definition and Lemma.

(i) Theorem 2.5.8 and Corollary 3.4.16 imply for $k = 1 \dots, K$ only one or two adjacent x_{ijk} to be non-zero for all arcs $(i,j) \in A$. Therefore, for fixed (i,j) all elements of the vector $(x_{ij1}, \dots, x_{ijK})^\top$ except the maximum and the larger one of its adjacent elements are directly set to zero.

Let the remaining non-zero elements of the vector (x_{ijk}) be x_{ijk_1} and x_{ijk_2} with $k_1 < k_2$ for all $(i,j) \in P$. Then constraints (ii) imply

$$x_{ijk_1} = L_{ij} - x_{ijk_2} \qquad \text{for all } (i,j) \in P\,.$$

The vectors $x^1, x^2 \in I\!R^p$ are defined as

$$x^1 := (x_{ijk_1})_{(i,j)\in P} \quad \text{and} \quad x^2 := (x_{ijk_2})_{(i,j)\in P}\,.$$

(ii) Analogously let the cost vectors $c_1, c_2 \in I\!R^p$ be defined as

$$c_1 := \left(c_{(i,j)k_1}\right)_{(i,j)\in P} \quad \text{and} \quad c_2 := \left(c_{(i,j)k_2}\right)_{(i,j)\in P}\,.$$

Then the costs of NOP can be calculated as

$$c_2^\top x^2 + c_1^\top x^1 = (c_2^\top - c_1^\top)\, x^2 + c_1^\top L_{ij}\,,$$

so furthermore let the adjusted pipe costs be defined as $\widetilde{c} \in I\!R^p$ with

$$\widetilde{c} := c_2 - c_1\,.$$

(iii) Let $[x^2]$ be an inclusion for x^2. Then the vector $[x]$ is defined by the components

$$[x_{ijk}] := \begin{cases} [\widetilde{x_{ijk}}] & \text{for all } (i,j) \in A \setminus P \\ [x^2_{ijk}] & \text{for all } (i,j) \in P \text{ and } k = k_2 \\ [L_{ij}] - [x^2_{ijk}] & \text{for all } (i,j) \in P \text{ and } k = k_1 \\ [0] & \text{else}. \end{cases}$$

The next definition contains a nonlinear function F as result of the reduction of the constraints. The rest of this section focuses on determining an inclusion for the zero of the function F.

4.2.2 Definition. Let the nonlinear function $F : \mathbb{R}^{p+a+n} \to \mathbb{R}^{a+n-1}$ be defined for NOP as $F = F_{HW}$, with

$$F_{HW}\left(\begin{pmatrix} x^2 \\ q \\ H \end{pmatrix}\right) = \begin{pmatrix} \left(\operatorname{sgn}(q_{ij})\,|q_{ij}|^{c_d}\,((\alpha_{k_2}-\alpha_{k_1})x_{ijk_2} + \alpha_{k_1}L_{ij}) - H_i + H_j - E_i + E_j\right)_{(i,j)\in P} \\ \left(\operatorname{sgn}(q_{ij})\,|q_{ij}|^{c_d}\sum_{k=1}^{K}\alpha_k\,\widetilde{x_{ijk}} - H_i + H_j - E_i + E_j\right)_{(i,j)\in A\setminus P} \\ \left(\sum_{j:(i,j)\in A} q_{ij} - \sum_{j:(j,i)\in A} q_{ji} - b_i\right)_{i\in N,\, i\neq 1} \end{pmatrix}$$

or respectively for DNOP as $F = F_{DW}$ with

$$F_{DW}\left(\begin{pmatrix} x^2 \\ q \\ H \end{pmatrix}\right) = \begin{pmatrix} \left(\beta_{k_2}w_{k_2}(q_{ij})x_{ijk_2} + \beta_{k_1}w_{k_1}(q_{ij})(L_{ij}-x_{ijk_2}) - H_i + H_j - E_i + E_j\right)_{(i,j)\in P} \\ \left(\sum_{k=1}^{K} w_k(q_{ij})\,\beta_k\,\widetilde{x_{ijk}} - H_i + H_j - E_i + E_j\right)_{(i,j)\in A\setminus P} \\ \left(\sum_{j:(i,j)\in A} q_{ij} - \sum_{j:(j,i)\in A} q_{ji} - b_i\right)_{i\in N,\, i\neq 1} \end{pmatrix}.$$

The following theorem contains, as result of these reductions of the constraints, a smaller, equivalent problem.

4.2.3 Theorem. Let $[y_F] = ([x^2], [q], [H])^\top$ be an interval vector within the bounds, i.e.

$$H_{iL} \leq [H_i] + E_i \leq H_{iU} \quad \text{for all } i \in N \qquad \text{and} \qquad 0^p \leq [x^2] \leq (L_{ij})_{(i,j)\in P}$$

such that $[y_F]$ is an inclusion for the zero of F, i.e. it contains a vector

$$y_F^* \in I\!R^{p+a+n} \quad \text{with} \quad F(y_F^*) = 0^{a+n} .$$

According to Definition 4.2.1 (iii) suppose $[y] = ([x], [q], [H])^\top$.

Then there exists at least one $y^* \in [y]$, which is an admissible point of NOP.

Proof. Constraint (i) and (iii) are fulfilled for any zero of F, constraint (ii) and (v) are fulfilled for at least one point $x^* \in [x]$ according to Lemma 4.2.1. The head and length bounds are fulfilled according to the premises of this theorem. The flow bounds are deduced from the flow conservation conditions and therefore fulfilled for all flows fulfilling constrains (iii). □

It is common practice to determine an inclusion of the zero of a nonlinear function F by using Theorem 1.1.27. Therefore, the Jacobian matrix of F has to be determined, which is provided in WaTerInt using symbolic differentiation.

4.2.4 Lemma. Regarding a node, the index i is used for all $i \in N$, and suppose (k_j, l_j) denote arc j for all $j \in A$. Then, let the node-edge-adjacency-matrix $Q = (q_{ij}) \in \{-1, 0, 1\}^{n\times a}$ be defined with the Kronecker delta as

$$q_{ij} = \delta_{i\,k_j} - \delta_{i\,l_j}$$

and let Q_1 denote the matrix containing all columns of Q whose arcs belong to P, and Q_2 contain the rest of the columns of Q. Let $Q_3 \in \{-1, 0, 1\}^{n-1\times a}$ consist of all rows of Q except the first one.

Furthermore let the matrix $D \in I\!R^{p\times p}$ be defined for NOP as $D = D_{HW}$ by

$$D_{HW} = \text{diag}\left\langle (\text{sgn}\,(q_{ij})\,|q_{ij}|^{c_d}(\alpha_{k_2} - \alpha_{k_1}))_{(i,j)\in P}\right\rangle$$

and for DNOP as $D = D_{DW}$ with

$$D_{DW} = \text{diag}\left\langle (\beta_{k_2} w_{k_2}(q_{ij}) - \beta_{k_1} w_{k_1}(q_{ij}))_{(i,j)\in P}\right\rangle$$

Suppose $v = (v_{ij})_{(i,j)\in A} \in I\!R^a$ is defined for NOP by

$$v_{HW_{ij}} = \begin{cases} c_d\,|q_{ij}|^{c_d-1}\,((\alpha_{k_2} - \alpha_{k_1})x_{ijk_2} + \alpha_{k_1} L_{ij}) & \text{for } (i,j) \in P \\ c_d\,|q_{ij}|^{c_d-1}\,\displaystyle\sum_{k=1}^{K} \alpha_k\,\widetilde{x_{ijk}} & \text{for } (i,j) \in A \setminus P \end{cases}$$

and for DNOP for $q_{ij} \neq 0$ as

$$v_{DW_{ij}} = \begin{cases} \beta_{k_2} x_{ijk_2} \frac{\partial w_{k_2}}{\partial q_{ij}}(q_{ij}) + \beta_{k_1}(L_{ij} - x_{ijk_2}) \frac{\partial w_{k_1}}{\partial q_{ij}}(q_{ij}) & \text{for } (i,j) \in P \\ \sum_{k=1}^{K} \beta_k \widetilde{x_{ijk}} \frac{\partial w_k}{\partial q_{ij}}(q_{ij}) & \text{for } (i,j) \in A \setminus P\,. \end{cases}$$

Let $V_1 \in I\!R^{p \times a}$ denote the p rows of diag $\langle v \rangle$ belonging to P and $V_2 \in I\!R^{a-p \times a}$ the rest of the rows of diag $\langle v \rangle$.

Then the Jacobian of the function F as defined in Definition 4.2.2 is

$$\mathcal{D}F(\begin{pmatrix} x^2 \\ q \\ H \end{pmatrix}) = \begin{pmatrix} D & V_1 & -Q_1^\top \\ 0^{(a-p)\times p} & V_2 & -Q_2^\top \\ 0^{(n-1)\times p} & Q_3 & 0^{(n-1)\times n} \end{pmatrix},$$

where for the Darcy-Weisbach problem $q_{ij} \neq 0$ has to be assumed.

Proof. Let $c \in I\!R_+ \setminus \{0\}$. The function $\varphi : I\!R \to I\!R$ defined with $\varphi(q) = \operatorname{sgn}(q)\ |q|^c$ is differentiable in $I\!R$ with $\varphi'(q) = c\ |q|^{c-1}$. Therefore, for NOP the Jacobian follows as specified, the Jacobian for DNOP can be determined by direct differentiation. □

Attempts to use F directly to obtain a verified inclusion for the solution of NOP cause the problem of verifying an inclusion for the zero of a nonlinear function $f : I\!R^{p+a+n} \to I\!R^{a+n-1}$ within simple bounds.

As verification algorithms are based on the Brouwer fixed-point theorem for functions $F : I\!R^{a+n-1} \to I\!R^{a+n-1}$, one major decision is the determination of the $p+1$ indices for which the approximate solution is taken as a degenerate interval.

For example, for the selection of these indices for the New-York City network problem, there are $\binom{65}{7} = 696,190,560$ possibilities, some of which numerically seem to have no full rank of the Jacobian. Standard procedures would focus on getting a full rank of the corresponding Jacobian matrix using the permutation matrix obtained with an LU decomposition, or on fixing those where the approximate solution is on or close to the simple bounds or a mixture of these. Even tough, especially for the New York network, only zeros of F verified outside the simple bounds for one or two dimensions could be found.

Therefore, the problem is again reduced, and two cases are distinguished according to the network structure: first, networks where the number of newly to be designed arcs equals the number of arcs, and second, expansion networks.

In the **first case, i.e. $p = a$**, the problem can be split into two linear subproblems. First an inclusion for the flow is calculated, and then, for all flows element of this inclusion, an inclusion for the length and the head is determined. The latter is possible, as it is guaranteed that $a + n > a$.

Thus, an inclusion for the flow of the restriction $F_1 := F|_q : I\!R^a \to I\!R^{n-1}$ with

$$F_1(q) = \left(\sum_{j:(i,j)\in A} q_{ij} - \sum_{j:(j,i)\in A} q_{ji} - b_i \right)_{i\in N\,,\, i\neq 1}$$

is to be determined.

As the number of arcs of the network can be assumed to be greater than the number of nodes, this results in the problem of finding a verified solution of the underdetermined linear system $F_1(q) = 0$, where an inclusion $[q]$ can be determined as described in Remark 1.1.32.

Let then $[\varrho] = ([\varrho_{ij}])_{(i,j)\in A}$ be defined for NOP as

$$[\varrho_{HW_{ij}}] = \operatorname{sgn}([q_{ij}])\,|[q_{ij}]|^{c_d}\,([\alpha_{k_2}] - [\alpha_{k_1}]) \quad \text{for } (i,j)\in A$$

and $[b] = ([b_{ij}])_{(i,j)\in A}$ be defined as

$$[b_{HW_{ij}}] = E_i - E_j - \operatorname{sgn}([q_{ij}])\,|[q_{ij}]|^{c_d}\,[\alpha_{k_1}]\,[L_{ij}] \quad \text{for } (i,j)\in A$$

and for DNOP respectively

$$[\varrho_{DW_{ij}}] = [\beta_{k_2}]\,[w_{k_2}([q_{ij}])] - [\beta_{k_1}]\,[w_{k_1}([q_{ij}])] \quad \text{for } (i,j)\in A$$

and $[b] = ([b_{ij}])_{(i,j)\in A}$ be defined as

$$[b_{DW_{ij}}] = E_i - E_j - [\beta_{k_1}]\,[L_{ij}]\,[w_{k_1}([q_{ij}])] \quad \text{for } (i,j)\in A\,.$$

Let simple bounds $l, u \in I\!R^{a+n}$ be denoted with

$$l = \begin{pmatrix} 0^p \\ (H_{iL} - E_i)_{i\in N} \end{pmatrix} \quad \text{and} \quad u = \begin{pmatrix} (L_{ij})_{(i,j)\in A} \\ (H_{iU} - E_i)_{i\in N} \end{pmatrix}.$$

Then an inclusion for the linear interval function $F_2 : I\!R^{a+n} \to I\!R^a$,

$$F_2(\begin{pmatrix} x^2 \\ H \end{pmatrix}) = \left(([\varrho_{ij}]\,x_{ijk_2} - H_i + H_j - [b_{ij}])_{(i,j)\in A} \right)$$

is calculated. For this, the approximate solution determined in the branch and bound algorithm is used. This solution is adjusted to satisfy the simple bounds and the last n indices are chosen to be fixed.

If this inclusion is not within the bounds, i.e the inequalities $l \leq ([x^2], [H])^\top \leq u$ do not hold true, the components of the bounds violated are adjusted iteratively as follows: If for $\nu \in A$ the component of the lower bound l_ν is not lower than the determined inclusion, it is substituted by $\widetilde{l_\nu}$ with

$$\widetilde{l_\nu} = l_\nu(1+\eta) + \eta$$

and accordingly for an upper bound violated $\widetilde{u_\nu} = u_\nu(1-\eta) - \eta$, where a parameter $\eta = 10^{-10}$ is used in WaTerInt. Then a solution of the auxiliary linear optimization problem

$$\begin{cases} \min \widetilde{c}^\top x^2 \\ \text{s.t.} \\ F_2((x^2, H)^\top) = 0^a \\ \widetilde{l_\nu} \leq \left(x^2, H\right)^\top \leq \widetilde{u_\nu} \end{cases}$$

is used as new approximate solution for determining an inclusion of F_2.

In the **second case, i.e. $\boldsymbol{p \neq a}$** an analogous algorithm could again separate F into two linear systems. But it cannot be guaranteed that $p + n \geq a$, i.e. the second verification of the first problem would not always be possible. In fact, for the New-York network problem the dimension is $p + n = 6 + 26 < 33 = a$.

Therefore, in the first step the following nonlinear restriction $F_3 : \mathbb{R}^{a+n} \longrightarrow \mathbb{R}^{a-p+n-1}$ of $F : \mathbb{R}^{p+a+n} \longrightarrow \mathbb{R}^{a+n-1}$ is regarded, i.e. for NOP $F_3 = F_{3_{HW}}$ with

$$F_{3_{HW}}\left(\begin{pmatrix} q \\ H \end{pmatrix}\right) = \begin{pmatrix} \left(\operatorname{sgn}(q_{ij})\, |q_{ij}|^{c_d} \sum_{k=1}^{K} \alpha_k\, \widetilde{x_{ijk}} - H_i + H_j - E_i + E_j\right)_{(i,j)\in A\setminus P} \\ \\ \left(\sum_{j:(i,j)\in A} q_{ij} - \sum_{j:(j,i)\in A} q_{ji} - b_i\right)_{i\in N,\, i\neq 1} \end{pmatrix}$$

and for DNOP $F_3 = F_{3_{DW}}$ with

$$F_{3_{DW}}\left(\begin{pmatrix} q \\ H \end{pmatrix}\right) = \begin{pmatrix} \left(\sum_{k=1}^{K} w_k(q_{ij})\, \beta_k\, \widetilde{x_{ijk}} - H_i + H_j - E_i + E_j\right)_{(i,j)\in A\setminus P} \\ \\ \left(\sum_{j:(i,j)\in A} q_{ij} - \sum_{j:(j,i)\in A} q_{ji} - b_i\right)_{i\in N,\, i\neq 1} \end{pmatrix}.$$

Verification of Upper Bound

Determination of approximate solution $(\widetilde{x}^2, \widetilde{q}, \widetilde{H})^\top$

if (`No_narcs` $=$ `No_arcs`)

    Determine inclusion $[q]$ of F_1

    for 1 to `max_iterations`

        Calculate inclusion $([x^2],[H])^\top$ of F_2

        if $([x^2],[H])^\top$ is within described bounds
            Stop: $[y] = ([x],[q],[H])^\top$ contains admissible point
        else
            Adjust lower and upper bounds for auxiliary linear problem
            Replace $\widetilde{x}^2$ and $\widetilde{H}$ by solution of auxiliary linear problem
        fi
    rof

else (`No_narcs` $\neq$ `No_arcs`)

    for 1 to `max_iterations`
        Determine inclusion $([q],[H])^\top$ of F_3 using `verifyudnlss`

        if $[H] \in [H_{iL}\,,\, H_{iU}]$
            Break: $([q],[H])^\top$ can be fixed for determination of $[x^2]$
        else
            Adjust indices which are fixed when calling `verifyudnlss`
        fi
    rof
    if $[H] \notin [H_{iL}\,,\, H_{iU}]$
        Stop: no inclusion found within bounds
    fi

    Determine the interval $[x^2]$
    for all $(i,j) \in P$ with $[x_{ijk_2}] \notin [0\,,\, L_{ijk}]$
        Redefine the index vector and accordingly c_1, c_2, $[x^2]$
    rof
    if $[x^2_{ij}] \in [0\,,\, L_{ijk_2}]$ for all $(i,j) \in P$
            Stop: $[y] = ([x],[q],[H])^\top$ contains admissible point

    fi
fi

Table 4.1: Algorithm for verifying the admissibility of the solution vector for the upper bound as implemented in WaTerInt.

An inclusion within the head bounds is determined as follows: For the fixed indices H_1 is used, as according to the test examples $H_{1L} = H_{1U}$, and iteratively all indices are added where an inclusion violates the bounds. For the approximate solution, always the one obtained with the branch and bound algorithm is used. Only for the indices where bounds are violated, the values are set to the ones of the corresponding simple bounds.

Afterwards the length of the newly to be designed arcs $(i, j) \in P$ are determined for NOP as

$$[x_{ijk_2}] = \frac{[H_i] - [H_j] + [E_i] - [E_j] - \operatorname{sgn}([q_{ij}])\,|[q_{ij}]|^{c_d}\,[\alpha_{k_1}]\,[L_{ij}]}{\operatorname{sgn}([q_{ij}])\,|[q_{ij}]|^{c_d}\,([\alpha_{k_2}] - [\alpha_{k_1}])}$$

and for DNOP as

$$[x_{ijk_2}] = \frac{[H_i] - [H_j] + [E_i] - [E_j] - [\beta_{k_1}]\,[L_{ij}]\,[w_{k_1}([q_{ij}])]}{[\beta_{k_2}]\,[w_{k_2}([q_{ij}])] - [\beta_{k_1}]\,[w_{k_1}([q_{ij}])]} .$$

If the vector $([x_{ijk_2}])_{(i,j)\in P}$ is not within the simple bounds, the index vector defining x^2 is adjusted for the indices violating the bounds. Then $[x_{ijk_2}]$ is calculated again and checked to be within the bounds. According to the experience with the test examples provided in Section 5.2, an inclusion within the simple bounds can be obtained with this procedure.

4.2.6 Non-solvability of Lower Bounding Problem

During the branch and bound algorithm subproblems are pruned out of the reason of non-solvability of the lower bounding problem. Three possibilities are considered in WaTerInt, where it is possible to prove that the subproblem cannot contain an optimal solution:

(i) No feasible point can exist according to the adjustment of flow bounds.

(ii) The lower bound is larger than best known upper bound.

(iii) The linear lower bounding problem has no feasible point.

In the **first case,** the flow bounds determined after the maximum spanning tree based reduction procedure are investigated. As these bounds are reliable, if in one dimension a lower bound greater than the corresponding upper bound is obtained, the subproblem can be pruned. In Figure 4.2, a simple example illustrates that this case may occur during the branch and bound algorithm.

The **second case** is analogous to Section 4.2.4. The return values of the linear solver for the Lagrange parameters can be used to calculate the verified lower bound. Even if the linear lower bounding problem is infeasible, with these values a large lower bound is often obtained. This holds true as the infimum of the empty set is per definition infinity, and therefore the subproblem can be pruned if its lower bound is larger than the best known upper bound. For getting return values of Lagrange parameters even if the solver `cplex` identifies the problem to be infeasible, the preprocessing has to be turned off, i.e. the parameter `CPX_PARAM_PREIND` has to be set to a value of 0.

In a number of cases, the value of this bound is not large enough to prune the subproblem. Then as **third possibility,** the algorithm presented in Table 1.3 is used. As far as observed with WaTerInt, this procedure meets the expected results.

There may remain a few cases where these three procedures do not identify infeasibility in contrast to the result of the linear solver. This possibly could generate the problem of perpetual splitting of one subproblem:

If the linear solver does not find a solution of the lower bounding problem, no upper bound is searched, as no flow could be determined. Therefore, for subsequent subproblems the lower bounds remain constant, and no feasible points can be found.

To avoid this for the verified calculations in WaTerInt, one further flag is added to every subproblem, indicating if the linear solver failed to find a solution of the corresponding lower bounding problem. Then, in the branch and bound step, the subproblem with the least lower bound is selected only from those problems not having this flag set. When no problems without this flag remain, or when the a priori determined accuracy for the subproblems without this flag is reached, the problems with this flag set are again considered for selection.

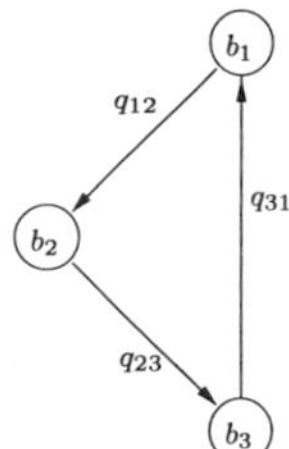

Figure 4.2: Example for Illustration of MSTR: Suppose the supply is $b_1 = 10$ and the demand $b_2 = -4$ and $b_3 = -6$. Assume flow bounds of $[-10, 10]$ for all arcs, q_{31} to be designated as non-tree arc and q_{23}, q_{12} be the order for post-tree traversal. Suppose the interval $[q_{31}]$ is branched and at some step the part $[q_{31}] = [5, 10]$ is active. Then independent of the actual values of $[q_{12}]$ and $[q_{23}]$ applying MSTR results in $\widetilde{q}_{23} = q_{23} \cap (b_2 + [q_{12}]) \cap (-b_3 + [q_{31}]) = \emptyset$. As the calculation of these bounds are verified, the subproblem can be pruned.

4.3 Two Phase Branch and Bound Algorithm

Applying the verification algorithms described in this chapter to the standard test problems, two difficulties remain unsolved:

(i) The possibility of verifying an upper bound depends on the approximate solution. The results with the procedure described in Section 4.2.5 are summarized in Tables D.31 to D.35.

(ii) In the branch and bound procedure for the New York network, an admissible vector is found quite late, which implies a lot of subproblems are to be evaluated and stored, before they can be pruned for their verified lower bound being greater that the incumbent upper bound.

To address these, a two phase branch and bound algorithm is added to WaTerInt, specified with verification strategy 2. In the first phase, a floating point solution is determined without verification. In this case, the parameter `accuracy` is fixed to a value of 10^{-6} . This solution is verified with the procedure described in Section 4.2.5, and this verified upper bound is fixed for the rest of the algorithm.

In the second phase, the branch and bound algorithms is executed again. This time, calculations and incumbent lower bounds are verified and the maximum number of iterations is reduced to $\lceil$`max_iter / 2`$\rceil$. For pruning subproblems, the already verified upper bound is used and no further work is done to check for incumbent upper bounds. Consequently, the subproblem with the least lower bound is always considered first, and no flag for infeasibility is needed. As soon as the required accuracy is reached, the minimum lower bound of all remaining subproblems is taken as global lower bound.

Thus, the verification is done as late as possible according to the objective emphasized by Jansson [J94] with the following citation of Wilkinson:

> "In general it is the best in algebraic computations to leave the use of interval arithmetic as late as possible so that it effectively becomes an a posteriori weapon."

The results of this two phase branch and bound algorithm, implemented as verification strategy 2 in WaTerInt, are provided in Appendix D.1.2 for the New York networks. Two further verification strategies have also been added. In some cases it seems reasonable to verify any upper bound obtained during the branch and bound algorithm, especially if only few upper bounds are obtained at all. This is implemented as verification strategy 3 in WaTerInt.

Finally, when using the two phase branch and bound algorithm, it can be preferable in the second stage to change to branching strategy 3 after a number of iterations, i.e. to implement the possibility to improve the upper bound obtained after the first phase. This is implemented by strategy 4 and improves the calculation of the two-loop example, where a maximum number of iterations is reached for an accuracy of 10^{-6} for the two phase branch and bound, but not when using verification strategy 4. A detailed comparison of these strategies is part of Tables D.29 and D.30.

4.4 Results of Verification

The results of the calculated test networks show that it is possible to verify the optimal value of the nonlinear water distribution network problem with known verification techniques. The a posteriori verification of the lower bound of a linear minimization problem forms a central part of the verification of the branch and bound algorithm.

The major challenges during the verification process still remain in the verification of the upper bound, i.e. the verification of the admissibility of a solution vector. This part of the verification process depends essentially on the approximate solution. As the floating point results obtained with `cplex` are closer to the optimum than those obtained with `linprog`, verified calculations are only provided for the linear solver `cplex`.

For the verification part of the algorithm introduced in Section 4.2.5, a number of special properties of the constraints are used. Further research could focus on more general methods at this point.

The two-loop expansion network is introduced as an example, where simple use of floating point calculations does not guarantee a solution to be found, for details see Section 5.2.2. Essentially, the solution of the two-loop test example is used to create a small expansion problem. For this network, no solution with the linear solver `cplex` and branching strategy 1 can be found on `sdome`, see Table D.2. Due to the standard `cplex`-preprocessing, subproblems are identified to be infeasible even if they contain the optimal solution vector. Therefore, no admissible point has been found at the end of the branch and bound algorithm, when no subproblems remain. Of course, in this case the two phase branch and bound algorithm will not terminate with a solution either. But with verified calculations, the optimal solution can be determined.

Consequently, this two-loop example shows that the calculated results agree with the following quotation of Hamming:

> "The beginner in the fields of computing always has the feeling that if he is lucky he will be able to ignore roundoff effects, and thus he tends to ignore roundoff theories. And if he is lucky, he can;" [H62]

Other numerical artifacts are described in Section 5.2. These as well as the problem that for the New York network and different linear solvers the lower bound may be larger than the upper bound, as mentioned at the beginning of this chapter, are avoided using verified calculations.

The main objective has been to investigate the additional time needed for the verification of this branch and bound algorithm. For the Hazen-Williams problem, extracts of Tables D.13 and D.14 are provided in Table 4.2 to compare the approximate CPU-times in seconds for finding a solution with an accuracy of 10^{-6}.

	HW (`lambda`)	HW (`sdome`)
Two-Loop Test Example	4 / 52	6 / 77
New York Network (12")	420 / 11,779	1,005 / 17,640
New York Network (4")	1,173 / 20,993	2,691 / 36,978
Hanoi Network	185 / 2,730	430 / 4,225

Table 4.2: Comparison of additional times for verification, for each example the times in seconds are provided in pairs, i.e. floating point / verified.

Thus, roughly spoken for the current implementation of WaTerInt 2.1, the time needed for determination of a verified optimum is on average about fifteen (ranging from about 10 to 28) times the time needed for floating point calculation. This factor is approximately the same for the Darcy-Weisbach problem, as summarized in Table D.19 and for the adjusted Hazen-Williams calculations as part of Table D.26, except an increased number of subproblems for the New York problems in the latter case as for a number of incumbent upper bounds no inclusion could be determined.

4.5 Evaluation of Computational Time

"Life is too short to spend writing for loops."[4]

Using the MATLAB Profiler utility, it can be measured what parts of WaTerInt consume most computation time. For understanding the additional time for the verification, the MATLAB profile summary report is provided in extracts for the New York (12") network in Appendix C. According to the MATLAB documentation, these include a list of all the functions and subfunctions called by the main routine `wds`, the number of times the function was called, the total time spent in a function, including all child functions called, in seconds, and the total time spent in a function, not including time for any child functions called, in seconds. The overall computational time is longer than obtained within Tables D.3 and D.10, as the Profiler itself uses some time.

When using floating point calculation, essentially two functions can be identified where a significant amount of time is spent: the determination of the relaxation `'relax'` and the calls of the linear solver `'lp_cplex_mex'`. As shown in Table C.3, most of the time is spent for the relaxation in the verified version, due to the amount of calls of the function

$$\texttt{sgn\_pow (intval(q))} \; = \operatorname{sgn}([q])|[q]|^{[c_d]} \, ,$$

for the Hazen-Williams problem or `verify_psi` for Darcy-Weisbach, see local subfunctions to the module `vnop` in Appendix B.5.

[4] *Getting Started with MATLAB Version 5*, 1998, The MathWorks, Inc.

About 60% of the overall computation time is needed by the private INTLAB function `/intval/power`, which is called by the WaTerInt functions `sgn_pow` and `q_dt`. Therefore, further research for improving the additional time needed, may focus on these procedures.

In the current implementation, the computational times for basic binary interval operations are significant, as e.g. in the New York (12") example on `sdome` the overall selftime of the interval functions `times`, `minus`, `rdivide` and `plus` is 29 % and on `lambda` 28 %. The total time of the INTLAB function `power` is 49 % on `sdome` and 48 % on `lambda`. This may be reduced significantly when not depending on interpretation overhead of MATLAB, especially for object orientation.

In addition, the amount of calls of the INTLAB function `setround` is interesting. The self time is 2 % on `sdome` and 11 % to 12 % on `lambda` for all examples part of Table C.3.

The introductory citation of this subsection concerning for loops refers to MATLAB version 5, where loops are extremely slow due to interpretation overhead. So for this version, it is advisable to use scalar products instead as far as possible. As known, this does no longer holds for MATLAB version 6 using floating point data, but for interval data this is not the case.

So in summary, a major factor of computational time results from the interpretation overhead, and expecially the operator concept of MATLAB. The additional time needed for verification may be significantly reduced if the interval data type could be directly integrated.

Chapter 5

The Software Package WaTerInt

The algorithms described in chapter two to four are implemented in the software package WaTerInt. A short documentation and a description of the modular structure are part of this chapter.

Details about the test networks and the input data used are provided. The solutions obtained for the Hazen-Williams problem with WaTerInt are compared to those known from literature and discussed in detail for the standard test networks as well as for a newly added small two-loop expansion network.

The results are compared to the solutions obtained with the Darcy-Weisbach and adjusted Hazen-Williams formulae and finally the solutions for different hydraulic parameters are discussed.

5.1 Implementation Details of WaTerInt

WaTerInt implements the branch and bound algorithm introduced by Sherali, Subramanian and Loganathan [S&a01] based on floating point calculations, as well as the verified version, and the branch and bound algorithm for the Darcy-Weisbach problem. It is based on MATLAB, ILOG CPLEX and on INTLAB. The interface from MATLAB to CPLEX is based on mex-files[1] from Musicant.

WaTerInt 2.1 can be executed as a MATLAB function with the syntax

```
[opt, acc]
  = wds ( town, [head_loss_formula], [k_value], [viscosity], [source_head],...
                [head_cost], [lp_solver], [verification], [accuracy],...
                [no_relax], [branching], [max_iter], [verbose] )
```

and is provided in Appendix B in extracts. Table 5.1 contains an overview of the input parameters, the default values are summarized in Table 5.2.

[1] Dave R. Musicant. *MATLAB/CPLEX MEX-Files.* Computer Sciences Department, University of Wisconsin, Madison, 2000. http://www.mathcs.carleton.edu/faculty/dmusican/cplex/

`town`	Specification of test-example, possible values: `'twoloop'`, `twoloop_exp`, `'newyork_12'`, `'newyork_4'` or `'hanoi'`
`head_loss_formula`	Specification of head-loss formula used for calculations, possible values: `'HW'` (Hazen-Williams), `'DW'` (Darcy-Weisbach) or `'AHW'` (adjusted Hazen-Williams)
`k_value`	Specification of Hazen-Williams coefficient or equivalent sand roughness in [cm] or [inch], depending on town
`viscosity`	Specification of kinematic viscosity (only for Darcy-Weisbach and for adjusted Hazen-Williams) in [m^2/s] or [ft^2/s]
`source_head`	Specification of fixed value for head at source node; if empty, default bounds according to network data are used
`head_cost`	Energy cost to raise head at source node in [US$/m]; has to be zero if `source_head` is specified
`lp_solver`	Linear programming solver used, possible values: `'linprog'` or `'cplex'`
`verification`	`(0)` for use of floating point calculations, `(1)` for verified calculations, `(2)` for two phase branch and bound algorithm, `(3)` for verification of upper bound in every step or `(4)` for two phase branch and bound with adjustments after $\lfloor \frac{\texttt{max\_iter}}{2} \rfloor$ steps
`accuracy`	Accuracy used for pruning subproblems, floating point number out of the interval [0,1]; when calculations are verified, accuracy for termination of branch and bound algorithm
`no_relax`	Number of tangents for relaxation, possible values: `(4)`, `(6)`, or `(0)` for calculation without relaxation
`branching`	Branching strategy used, possible values: `(1)` for the default strategy, `(2)` for the strategy based on distance to feasibility, `(3)` or `(4)` for strategy `(1)` or `(2)` with head constraint propagation, and `(5)` or `(6)` in case of floating point calculations with adjusted head constraint propagation
`max_iter`	Maximum number of iterations during branch and bound algorithm
`verbose`	Specification of amount of output during calculations, possible values: `(0)` minimal output, `(1)` limited output and `(2)` full output including information about relaxations

Table 5.1: Input parameters of WaTerInt. Except of the parameter `'town'`, all others are optional. If empty, the default value according to the selected network is used as summarized in Table 5.2.

Parameter	Default value	Condition
`head_loss_formula`	'HW'	
`k_value`	100	for HW and New York
	130	for HW and two-loop, two-loop expansion or Hanoi
	0.24	for DW or AHW and New York
	0.0003	for DW or AHW and two-loop or two-loop expansion
	0.025	for DW or AHW and Hanoi
`viscosity`	[]	for HW
	$1.407 \cdot 10^{-5}$	for DW or AHW and New York
	$1.306 \cdot 10^{-6}$	for DW or AHW and two-loop, two-loop exp. or Hanoi
`source_head`	[]	
`head_cost`	0	
`verification`	0	
`lp_solver`	'cplex'	
`accuracy`	10^{-6}	
`no_relax`	0	for DW and two-loop or Hanoi
	6	in all other cases
`branching`	5	for two-loop expansion or New York without verification
	3	for two-loop expansion or New York with verification
	1	for all other networks
`max_iter`	2,000	for two-loop
	1,000	for two-loop expansion
	40,000	for New York (12")
	80,000	for New York (4")
	20,000	for Hanoi
`verbose`	0	

Table 5.2: Default values for the input parameters used in WaTerInt 2.1. The head-loss formula is abbreviated as follows: Hazen-Williams (`HW`), Darcy-Weisbach (`DW`) and adjusted Hazen-Williams (`AHW`).

As the head constraint propagation reduces the time for expansion networks, for these the branching strategy includes this procedure per default, and for floating point calculation the additional improvement is also selected.

In WaTerInt, the energy costs and initial head can be determined by the input parameters `head_cost` and `source_head`. When energy costs are specified, the minimum head is taken from the bounds defined for the network and the maximum head is set to a value specified by the constant `HMAX` above this, i.e. per default 100 m or 100 ft, respectively. If the parameter `head_cost` is empty or zero, either the head specified by `source_head` is taken as a fixed value or, if not specified, the originally defined head bounds are used and the energy costs are considered to be zero.

According to the test results published in [S01], one parameter determines the number of equations for the relaxations. In addition to the six equations presented in Sections 2.3.1, 3.5.4 and 4.2.1, four equations are implemented analogously.

One main objective of this implementation is to gain practical results for the additional time needed for verification. The results are to be checked on the well-known test networks of New York City and Hanoi and the standard two-loop example from literature. All these networks contain just one source node, so the supply is assumed to be given as first element of the demand vector. For conversion of floating point input data to intervals, four places after the decimal point are considered.

The return value `'opt'` is the best known upper bound for the optimal value, and `'acc'` the accuracy, i.e. the best known upper bound minus best known lower bound.

The output of the optimal solution is passed to files in local subdirectories `'results'` and `'calculated'`. In the first, the files are named according to the actual time, and for every calculation the following four files are written:

(i) `<date>.mat` — a copy of the MATLAB workspace including the results,

(ii) `<date>.txt` — output of optimal solution in ASCII format,

(iii) `<date>.wds` — output of optimal solution for hydraulic comparisons, i.e. instead of the pipeline flow rate q_{ij} the average fluid velocity v_{ijk} is provided in [m/s] for all arcs (i, j) and all selected diameters d_k, $k \in \mathcal{K}$, and SI-units are used

(iv) `<date>.tex` — output tables of the optimal solution for LaTeX purposes.

The second subdirectory contains files named according to the test network and the parameters used. These contain the optimal solutions or the error messages in binary MATLAB format, and are read if an already calculated example is executed again. These files are not written or read when using verbose output, as in this case the computational time can differ.

The software package WaTerInt includes the modules listed in Figure 5.1. The main module `'wds'` contains the overall algorithm, especially input and error handling, i.e. the setting of parameters and control of files in the directory `'calculated'`.

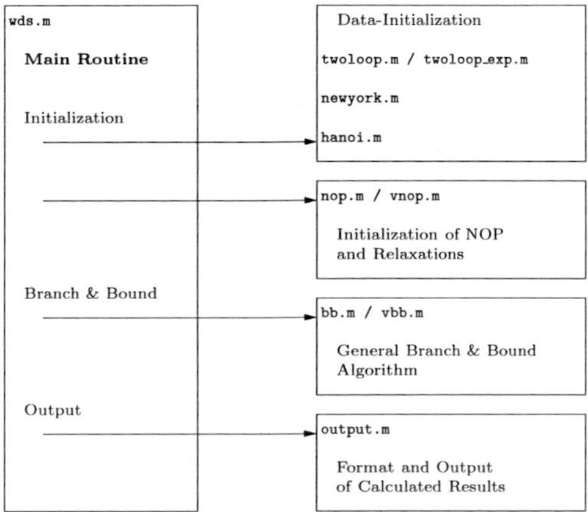

Figure 5.1: Modules and function calls of WaTerInt

In WaTerInt, global variables are defined for the dimensions of the test problem, as summarized in Table 5.3. Table 5.4 contains the variables describing the network structure and the main variables used for the optimization problem are part of Table 5.5.

The module `'nop'`, and for the verified part `'vnop'` respectively, contains the initialization of the lower and upper bounding problems as far as not adjusted during the branch and bound algorithm. Furthermore, `'nop'` contains Kruskal's algorithm for identifying a spanning tree and in particular the non-tree arcs, whose indices are stored in the vector `non_tree`. The resulting index vector `post_tree` contains the tree-arcs ordered as needed during the branch and bound procedure for post order tree-traversal. In case of expansion networks and the Darcy-Weisbach formula, first the fixed arcs are chosen instead of ordering the arcs according to the diameter of the interval given by the flow bounds, to preferentially select fixed arcs as non-tree arcs.

The main functions defined in `'nop'`, and for the verification in `'vnop'` respectively, include the adjustments needed for the lower and upper bounding problem, `adjust_lbeq` and `upper_bounding`, a function `relax` for generating the relaxations of Sections 2.3.1, 3.5.4 and 4.2.1, and a procedure `select_idx` for selecting the branching index. When determining the head-loss according to Darcy-Weisbach, in this procedure flow with small absolute value is regarded as zero.

For improving the computational time, an equivalent formulation reducing the dimension of the **lower bounding problem** described in Section 2.3.1 is part of WaTerInt. The variables x^1_{ijk} and x^2_{ijk} for $(i,j) \in A \setminus P$, $k = 1, \dots, K$ are not considered as decision variables.

No_arcs	Number of arcs A in the network
No_nodes	Number of nodes N in the network, in all given examples just one source node exists
No_dia	Number of available diameters
No_seg	No_arcs * No_dia
No_relax	Number of tangents for relaxation
No_lambda	Number of nonlinearities for which relaxations are to be determined
No_narcs	Cardinality of P, i.e. amount of newly to be designed arcs
No_nseg	No_narcs * No_dia

Table 5.3: List of global variables used in WaTerInt to describe the dimension of the problem. In addition, three global variables are used: one string containing all parameters selected, the function handle wds_error, and wds_filename for archiving the results in the directory calculated.

arcs	Matrix of dimension No_arcs $\times$ 2 containing in row i the nodes connected by the directed link i, $1 \leq i \leq$ No_arcs
arc_length	Vector of length No_arcs containing length L_{ij} for arc (i,j)
arcs_fixed	Vector of length No_arcs - No_narcs containing the indices of the arcs element of $A \setminus P$
length_fixed	Vector of length No_arcs * No_dia containing the fixed length $\widetilde{x_{ijk}}$ for the fixed arcs $(i,j) \in A \setminus P$ and zero for all diameters belonging to the arcs $(i,j) \in P$
seg_fixed	Vector of length No_seg - No_nseg containing the indices of fixed segments
node_edge	Matrix of size No_nodes $\times$ No_arcs as part of Lemma 4.2.4: node_edge(i,:) = (arcs(:,1)==i)'-(arcs(:,2)==i)'

Table 5.4: Variables of WaTerInt describing the network structure.

d_k	`avail_diams`	Vector containing the available diameters
c_k	`pipe_cost`	Vector of the same size as `avail_diams`, cost for a pipe segment of diameter d_k is given by `pipe_cost(avail_diams==`d_k`)`
$\boldsymbol{x_{ijk}}$	`length`	**Length of segment of diameter k to be determined in arc (i,j), vector of length `No_seg`**
$\boldsymbol{q_{ij}}$	`flow`	**Flow, vector of length `No_arcs`**
$q_{\min_{ij}}$, $q_{\max_{ij}}$	`flow_min, flow_max`	Flow bounds, vectors of length `No_arcs`
b_i	`demand`	Vector of length `No_nodes`, `demand(1)` contains water supply, the rest of the vector the demand
E_i	`elevation`	Vector of length `No_nodes` containing ground elevation
$\boldsymbol{H_i}$	`head`	**Head, vector of length `No_nodes`**
H_{iL}, H_{iU}	`head_min, head_max`	head bounds including elevation, vectors of length `No_nodes`
F_1	`head_max(1)`	Stored in the upper head bound as the supply node has index 1 in all test networks
c_{s1}	`head_cost`	Present value of costs per unit energy head for raising head above E_1
$\boldsymbol{H_{s1}}$	`head(1)`	**Additional head to be developed at each source node $i \in S$**
C_{HW}, ϵ	`k_value`	Hazen-Williams coefficients, i.e. double (HW), or vector of length `No_dia` (AHW), and double containing equivalent sand roughness for Darcy-Weisbach

Table 5.5: Names of main variables of WaTerInt analogous to Table 2.1, decision variables are marked bold.

This is possible, as they have no influence on the value of the cost function and can be updated with the values of $x^1_{ijk} = \lambda_{ij}\widetilde{x_{ijk}}$ and $x^2_{ijk} = (1-\lambda_{ij})\widetilde{x_{ijk}}$ after solving the linear program. The significance of this reduction can be seen for example regarding the New York City network, where dimension 1082 is reduced to 272.

Then, this reduction implies the vector of decision variables

$$x = \begin{pmatrix} x^1_{ijk} \\ x^2_{ijk} \\ \lambda_{ij} \\ q_{ij} \\ H_i \end{pmatrix},$$

with the dimension `2*No_nseg + No_lambda + No_arcs + No_nodes`. In case of Darcy-Weisbach, the elements λ_{ij} are substituted by λ_{ijk} for all $(i,j) \in A \setminus P$ with $\widetilde{x_{ijk}} \neq 0$.

Thus, for calculations with the Hazen-Williams formula, one obtains `No_lambda = No_arcs`, while for Darcy-Weisbach `No_lambda` is the amount of fixed segments with a length greater than zero.

According to the simplifications the objective function shrinks to

$$\sum_{(i,j)\in P} \sum_{k=1}^{K} c_k(x^1_{ijk} + x^2_{ijk}) \; + \; c_{s1}\, H_{s1} \; .$$

The constraints of the original problem are separated according to CPLEX and MATLAB optimization toolbox standards. The linear equalities

$$\begin{cases} \displaystyle\sum_{k=1}^{K} \alpha_k (v_{\min_{ij}} x^1_{ijk} + v_{\max_{ij}} x^2_{ijk}) & \\ \qquad\qquad + (H_j - H_i) = (E_i - E_j) & \text{for all } (i,j) \in P \\ \displaystyle\lambda_{ij}(v_{\min_{ij}} - v_{\max_{ij}}) \sum_{k=1}^{K} \alpha_k \, \widetilde{x_{ijk}} \; + (H_j - H_i) & \\ \qquad\qquad \displaystyle = (E_i - E_j) - v_{\max_{ij}} \sum_{k=1}^{K} \alpha_k \, \widetilde{x_{ijk}} & \text{for all } (i,j) \in A \setminus P \\ \displaystyle\sum_{k=1}^{K} x^1_{ijk} + \sum_{k=1}^{K} x^2_{ijk} = L_{ij} & \text{for all } (i,j) \in P \\ \displaystyle\sum_{k=1}^{K} x^1_{ijk} = \lambda_{ij} L_{ij} & \text{for all } (i,j) \in P \\ \displaystyle\sum_{j:(i,j)\in A} q_{ij} - \sum_{j:(j,i)\in A} q_{ji} = b_i & \text{for all} i \in N \setminus \{1\} \end{cases}$$

or, for Darcy-Weisbach

$$\begin{cases} \sum_{k=1}^{K} \beta_k (w_{\min_{ijk}} x_{ijk}^1 + w_{\max_{ijk}} x_{ijk}^2) & \\ \qquad +(H_j - H_i) = (E_i - E_j) & \text{for all } (i,j) \in P \\ \sum_{k=1}^{K} \beta_k (w_{\min_{ijk}} - w_{\max_{ijk}}) \widetilde{x_{ijk}} \, \lambda_{ijk} + (H_j - H_i) & \\ \qquad = (E_i - E_j) - \sum_{k=1}^{K} \beta_k \, w_{\max_{ijk}} \widetilde{x_{ijk}} & \text{for all } (i,j) \in A \setminus P \\ \sum_{k=1}^{K} x_{ijk}^1 + \sum_{k=1}^{K} x_{ijk}^2 = L_{ij} & \text{for all } (i,j) \in P \\ \sum_{j:(i,j)\in A} q_{ij} - \sum_{j:(j,i)\in A} q_{ji} = b_i & \text{for all } i \in N \setminus \{1\} \end{cases}$$

are stored in the matrix `Aeq` and the vector `beq`. The first `No_arcs` equalities are adjusted at the beginning of each branch and bound step.

According to Lemma 2.5.10, the linear inequality $\sum_{j:(1,j)\in A} q_{1j} - \sum_{j:(j,1)\in A} q_{j1} \leq b_1$ is omitted in WaTerInt 2.1. Out of memory reasons, the matrix `A` and the vector `b` containing the inequalities defined by the relaxations are calculated for all arcs at the beginning of each branch and bound step. The lower and upper bounds are represented by the vectors `lb` and `ub`.

The function `upper_bounding` for constructing the corresponding matrix and vectors for the **linear upper bounding program** is part of the module `'nop'`. As described in Section 2.3.3, the flow is fixed according to the solution obtained by the relaxed program. Analogous to the lower bounding problem, the dimension of the decision variables is reduced. This results for instance in a reduction of dimension 521 to 116 for the New York City network. So the decision vector $x = (x_{ijk},\ H_i)^\top$ is of dimension `No_nseg + No_nodes` and the objective function is

$$\sum_{(i,j)\in P} \sum_{k=1}^{K} c_k \, x_{ijk} + c_{s1} H_1 \ .$$

Restrictions on the flow q_{ij} can be omitted, so no further linear inequalities exist.

The linear equalities are

$$\begin{cases} H_j - H_i + \operatorname{sgn}(q_{ij}) \, |q_{ij}|^{c_d} \sum_{k=1}^{K} \alpha_k \, x_{ijk} = E_i - E_j & \forall (i,j) \in P \\ \sum_{k=1}^{K} x_{ijk} = L_{ij} & \forall (i,j) \in P \\ H_j - H_i = E_i - E_j - \operatorname{sgn}(q_{ij}) \, |q_{ij}|^{c_d} \sum_{k=1}^{K} \alpha_k \, \widetilde{x_{ijk}} & \forall (i,j) \in A \setminus P \ , \end{cases}$$

or for Darcy-Weisbach

$$\begin{cases} H_j - H_i + \sum_{k=1}^{K} \beta_k \, x_{ijk} \, w_k(q_{ij}) = E_i - E_j & \forall (i,j) \in P \\ \sum_{k=1}^{K} x_{ijk} = L_{ij} & \forall (i,j) \in P \\ H_j - H_i = E_i - E_j - \sum_{k=1}^{K} \beta_k \, \widetilde{x_{ijk}} \, w_k(q_{ij}) & \forall (i,j) \in A \setminus P \, . \end{cases}$$

Whereas the flow constraint propagation is directly part of the module `bb`, the head constraint propagation is contained in the functions `adjust_head` and `adjust_flow`. The function `adjust_hazen_williams` contains the procedure described in Section 3.6. To directly compare verified and unverified solutions of the nonlinear optimization problem, for the verified calculations the vector containing the adjusted Hazen-Williams coefficients is determined with the same floating point based procedure and afterwards interpreted as an interval vector with zero diameter.

The procedure `test_adjacency` contains a check of the monotonicity and convexity assumptions of Lemma 2.5.7 and Lemma 3.4.15, to guarantee the adjacency property which is used for verifying the upper bound. The a posteriori verification of the optimal solution of a linear solver is part of the procedure `verify_lb` and the verification of infeasibility part of `verify_infeasibility`. The algorithm described in Section 4.2.5 is part of `verify_ub`.

Further on, `nop` contains Hazen-Williams and Darcy-Weisbach specific functions. In the latter case, a verified inclusion for the zero of the function $\psi_k(f)$ as part of Definition 3.4.1 is needed. As the derivative can easily be determined symbolically, the function `verify_psi` is added following the procedure `verify_nlss` of INTLAB.

The functions u_k of Definition 3.4.3, `u_fkt`, are only called in the context of w_k (`w_fkt`), therefore it is possible to set $u(0) = 0$ to simplify calculations. This function is needed for every arc diameter combination as well, therefore a long version `u_fkt_l` is provided in addition. For the verification of the upper bound, existence of the Jacobian of the function F as defined in Definition 4.2.2 has to be guaranteed. Therefore, in case of Darcy-Weisbach the procedure `verify_ub` is only called when $|q_{ij}| >$ `WEPS` for all arcs (i,j). Any attempt to determine the derivative of w_k (`w_fkt_dq`) for an interval containing zero results in an error.

During the **branch and bound procedure** part of the module `'bb'`, several variables are stored for each subproblem: the index selected for partitioning the hyperrectangle of the flow `d_idx`, the corresponding flow `d_q`, the flow bounds and the lower bound of the relaxed problem. In the verified version, `vbb`, the flow bounds are as well stored as floating point vectors instead of intervals for performance reasons. An additional `status` parameter indicates for every subproblem, if this was identified to be infeasible by the linear solver. For better performance, the struct array containing the list of subproblems to be worked on is preallocated to 1,000 subproblems. The constants used for internal configuration are part of Table 5.7.

The interface from MATLAB to cplex, called by `lp_cplex_mex`, has been extended to select whether the preprocessing should be turned off or not. Naturally, for verified calculations no preprocessing is applied, indicated by setting the argument following the simple bounds to 0, which is by default set to 1. During output, floating point solutions lower than `LEPS` are not printed as they are considered to be zero. To ensure outward rounding when printing inclusions for the optimal solutions up to the desired accuracy, local subfunctions `s_inf` and `s_sup` are defined in `'output'`.

One part of the calculations presented in this thesis has been executed on `'lambda'`, an IBM ThinkPad T43 with an Intel Pentium M Processor 750, 1860 MHz, 1024 MB SDRAM and operating system SUSE Linux 10.0, with the software versions:

(i) MATLAB 7.1.0.183 Release 14 (Service Pack 3)
with Optimization Toolbox 3.03

(ii) ILOG CPLEX 9.0

(iii) INTLAB 5.2

The other part has been calculated on an HP Superdome server, called `'sdome'`, with 64 GB RAM, 64 processors PA-RISC 8700 (750 MHz) and operating system HP-UX B.11.11 U 9000/800. They are based on the following versions:

(i) MATLAB 6.5.1.199709 Release 13 (Service Pack 1)
with Optimization Toolbox 2.3

(ii) ILOG CPLEX 8.1

(iii) INTLAB 5.2

Table 5.6 compares the computational times for two small standard test calculations with WaTerInt 2.1 for different MATLAB versions. Whereas it has been preferable to use the older one on `sdome`, the computational time decreases on `lambda` when using the latest MATLAB version. All results presented in this thesis are calculated with WaTerInt Version 2.1.

	`sdome`					`lambda`	
	6.5.1.199709 R13 SP1	7.0.0.19901 R14	7.0.1.24704 R14 SP1	7.0.4.352 R14 SP2	7.1.0.183 R14 SP3	6.5.1.199709 R13 SP1	7.1.0.183 R14 SP3
HW	76	158	158	163	156	45	52
DW	493	817	727	649	636	249	182

Table 5.6: CPU time in seconds needed for the calculation of the two-loop test example with Hazen-Williams formula (`verification = 1`) and Darcy-Weisbach (`verification = 0`) with default parameters for different MATLAB versions.

Constant	Value	Description
BBUB	10^{-2}	Factor for a priori pruning of subproblems for `verification = 1`
BEPS	10^{-10}	Accuracy around λ^* for searching zero during determination of relaxations when using Hazen-Williams formula, for Darcy-Weisbach a value of 10^{-5} is used, respectively
BS_P	0.9	Selection percentage used for branching strategy 1 and 2
DELTA_ACC	10^{-6}	Maximal deviation of nonlinear constraints for a solution to be feasible, used by floating point version of the branch and bound algorithm
D_EXP	4.87	Hazen-Williams exponent for diameters, for verified calculations `intval('4.87')`, respectively
ETA	10^{-10}	Factor for adjusting auxiliary bounds for finding a verified solution within original bounds
FEPS	0.01	If possible, flow bounds below this value are set to zero in `adjust_flow`
FMIN	10^{-4}	Lower starting values for bisection of `psi_fkt`
FMAX	10^{4}	Upper starting values for bisection of `psi_fkt`
HEPS	0.01	Absolute minimum flow to calculate head-loss in `adjust_head`
HMAX	100	Maximum head in meter or foot above specified minimum head allowed when using non-zero energy costs
LEPS	10^{-4}	For output of the floating point solution, length lower than LEPS is regarded as zero
MAX_DIFF	100	Absolute bound for the maximum flow difference used to select the branching variable
MAX_ITER	20	Maximum number of iteration when determining a decimal string from interval x
MAX_SUBS	1,000	Preallocation size of list of subproblems
MAX_UB_ITER	10	Maximum number of iteration to get a verified result within simple bounds
MEPS	10^{-6}	Head loss lower than this value is regarded as zero in `adjust_flow`
NMAX	100	Number of data points within flow bounds in `adjust_hazen_williams`
OPT_MAX	10^{6}	Maximum value to restrict region for linear solver
PEPS	10^{-10}	Minimum value to accept gradient to be non-zero, used in function `verify_psi`
PREC	`'%.4f'`	Precision string for transferring floating point input data to intervals `x = intval( sprintf(PREC, x))`
QMIN	1	Minimal flow to be regarded when determining coefficients for adjusted Hazen-Williams
REPS	0.01	Accuracy used for bisection
UEPS	10^{-10}	Points lower than UEPS are assumed to be zero in `u_fkt` and `u_fkt_1`
VEPS	10^{-12}	Factor for inflation of interval, used in `verify_udnlss`
V3_P	0.5	Percentage of `max_iter` where verification (4) changes from (2) to (3)
WEPS	0.1	Minimum absolute flow to start verification of upper bound
ZMAX	10^{-4}	Flow lower than ZMAX is assumed to be zero in `select_idx` for Darcy-Weisbach

Table 5.7: Internal constants of WaTerInt and their default values used for parametrisation of the algorithm.

5.2 Computational Results for the Test Networks

In the following, computational results for the standard networks are summarized. The input data for these networks are provided in Appendix A, and detailed data for the optimal layout of the test networks are presented in Appendix D.

For conversion of units the following standard relations are used: 1 yard = 0.9144 meter, 1 foot = 0.3048 meter and 1 inch = 25.4 millimeter.

5.2.1 Two-Loop Network

The two-loop test network was first introduced by Alperovits and Shamir as example for their linear programming gradient method presented in [A&S77]. The network topology is illustrated in Figure 5.2.

The optimization methods used tend to find a branched optimal solution instead of a looped one. As the layout of the network contains loops out of reliability considerations, some authors impose a minimum diameter of 1 inch for all pipes (see [L&a95]), as suggested already by Alperovits and Shamir [A&S77]. Obviously, this does not necessarily imply a practicable design for engineers as illustrated in Figures D.4 and D.5.

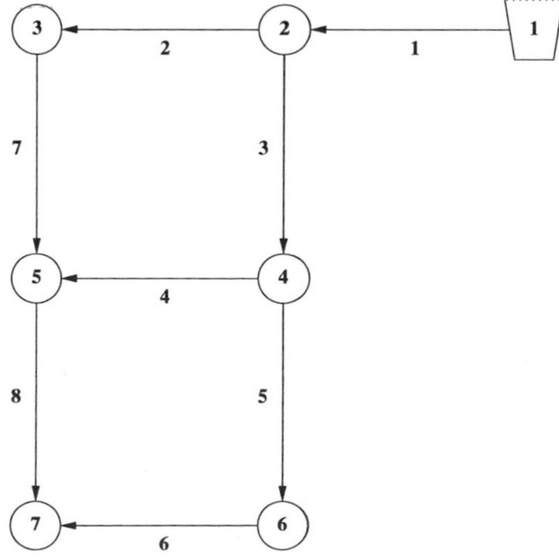

Figure 5.2: Layout of the two-loop test network

Computational results of previous publications are listed in Table 5.8 in extracts. Some confusions can be noticed in the input data when comparing these papers. Even if these inconsistencies are explained as far as they can obviously be seen, Table 5.8 and analogously Tables 5.9 and 5.11 should be read with caution.

Sherali, Subramanian and Loganathan [S&a01] permit "the selection of all commercially available pipe diameters". In [S&S97] it is summarized that "there exists some confusion in the literature in reference to this test problem, since the original paper ([A&S77]) considers only a restricted set of pipe diameters in stating the test problem, but the solution given includes a $4''$ pipe diameter for one of the links $(5, 7)$ that is not part of the printed data."

Interesting is the following explanation of Fujiwara and Dey [F&D87]: "Alperovits and Shamir [A&S77] have proposed an algorithm that takes advantage of this adjacency property in such a way that at each iteration they add one smaller and one larger pipe diameter to the given feasible diameter set and consider only these pipe diameters in setting up the model for the next iteration."

Probably a starting feasible set of diameters has been interpreted as a permanently restricted one. Therefore, WaTerInt contains no restrictions in the admissible diameters as well, see Table A.1.

1977	Alperovits and Shamir	[A&S77]		479,525
1979	Quindry et al.	[Q&a79]		441,522
1989	Kessler and Shamir	[K&S89]		417,500
1993	Sherali and Smith	[S&S93]	441,674	407,109$^{\natural}$
1994	Eiger et al.	[E&a94]		402,352†
1995	Loganathan et al.	[L&a95]		403,657
1997	Sherali and Smith	[S&S97]	436,684	403,386
1998	Sherali et al.	[S&a98]	436,915	403,390
2000	Geem	[G00]		419,000‡
2001	Subramanian	[S01]		403,386
	present solution			403,386

Table 5.8: Development of computational results for the minimum costs of the two-loop test network.
Next to last column with restricted set of candidate diameters, ($\natural$) indicates the use of a minimum head of 175 instead of 180 for node 2, ($\dagger$) indicates violations in the flow constraints, ($\ddagger$) indicates a solution considering only discrete diameters for the pipe segments and with different values in the Hazen-Williams equation.

The commercially available pipe diameters used are given in inch, for calculations they are transformed to centimeters. The corresponding costs are given in US$ per meter.

The maximum bounds for the head are chosen according to [S01] or [S&S93], the original problem contains only minimum bounds. Originally, Alperovits and Shamir fixed the head of the supply node to a value of $H_1 = 210$, the solution of Subramanian contains this value and he indicates the bounds for H_1 with n/a. In WaTerInt this value is taken as lower and upper bound for H_1 , see Table A.2.

The flow bounds are taken from [S&a98], where they are presented as results from determination of supply and demand data along with the continuity of flow considerations. A Hazen-Williams coefficient of 130 is used for all arcs. Kruskal's algorithm identifies the arcs seven and eight as non-tree arcs.

Verified calculations with WaTerInt show that the optimal value is part of the interval $[403,385.27\ ,\ 403,385.53]$, see Table D.8. With floating point calculations an optimal value of 403,385.5249 US$ is obtained after 4 seconds. This corresponds to the floating point calculation of Subramanian [S01] providing a solution of 403,385.418 US$ after a CPU time of 12 seconds on a SUN SPARC 10 Workstation.

A comparison of the results obtained on `lambda` and `sdome` is part of Tables D.13 and D.14 and for different linear solvers part of Tables D.27 and D.28. Especially two points may be of interest:

First with floating point calculations it is possible to obtain a lower bound being greater than the upper bound. This can be observed for example when using the linear solver `linprog` which results on `sdome` in the bounds for the optimum value of $[403,473.9334\ ,\ 403,406.1124]$.

The reason is the unreliability in the identification of infeasibility by the linear solvers. The lower bound is determined as minimum of all subproblems pruned out of the reason that their lower bound is greater than the best known upper bound reduced about a percentage of $\frac{\varepsilon}{100}$. When a subproblem is identified to be infeasible by the linear solver, the floating point algorithm trusts this result and prunes it. When an admissible point is determined in one subproblem and after splitting all subsequent subproblems are identified to be infeasible, no lower bound of this subproblem is considered for the calculation of the global lower bound. Therefore the lower bound may be larger than the best known upper bound.

And finally, when calculating on `lambda` an upper bound of 403,385.5249 US$ is obtained, whereas on `sdome` a greater value for the lower bound of 403,385.6092 US$ is obtained. Even if the difference is small and not significant from the engineering point of view, this emphasizes that using floating point calculations a lower or upper bound cannot be assumed to be reliable.

5.2.2 Two-Loop Expansion Network

"It's a dirty secret. Floating-point arithmetic is wrong."
— John Gustafson[2]

This academic example underlines the statement of Hansen and Walster that "numerical instability can remain hidden in the result of evaluating a floating-point expression, but not in an interval expression result". It is constructed from the optimal solution of the two-loop example while fixing the value obtained for the first five arcs according to a rounded solution calculated with WaTerInt. Detailed data is contained in A.3.

As this simple test network is derived from the standard two-loop example, engineering problems like the minimum diameter of one inch remain unsolved. But for illustrating the problems which can occur when relying on floating point arithmetic a simple structure and lower dimensions are preferable.

By trying to solve this problem with the floating point version of the branch and bound algorithm with an accuracy of 10^{-6}, no solution can be obtained on `sdome` when using the branching strategies suggested by Sherali, independently of the use of 4 or 6 tangents for relaxation and the linear solver. One exception is the use of `cplex`, 6 tangents for relaxation and branching strategy 2, see Table D.2. Using the head constraint propagation a solution can be obtained when using the floating point improvement as described in Section 2.4.

On `lambda` one obtains optimal solutions only when using the linear solver `cplex`, with `linprog` the problem is still identified to be infeasible.

5.2.3 New York Network

The test problem of the New York City water supply system is introduced in [S&L69]. Figure 5.3 illustrates the network of the New York water tunnels. It is an expansion of this network. Originally, Schaake and Lai wanted to identify those arcs which are to be enlarged to be able to meet an increasing demand. Therefore, they considered the optimization problem of designing parallel pipes for all arcs.

For comparability reasons, WaTerInt contains the same expansion problem as solved by Sherali, Subramanian and Loganathan [S&a01]. This is based on the original network which contains 20 nodes and 21 arcs. Parallel pipes are to be designed

[2] Citation of "Robert McMillan. *Sun researchers: Computers do bad math.* DIG News Service, 17.12.03. `http://www.infoworld.com/article/03/12/17/HNbadmath_1.html`"

for arcs 7, 16, 17, 18, 19 and 21. Therefore six dummy nodes (indices 21 - 26) are introduced with the corresponding arcs:

22:	(7, 21)	23:	(21, 8)
24:	(10, 22)	25:	(22, 17)
26:	(9, 23)	27:	(23, 16)
28:	(20, 24)	29:	(24, 11)
30:	(12, 25)	31:	(25, 18)
32:	(18, 26)	33:	(26, 19)

The following arcs are identified as non-tree arcs with Kruskal's algorithm in WaTerInt: 4, 12, 15, 21, 23, 25, 27, 29, 31, 33.

Some solutions published to this optimization problem are summarized in Table 5.9, for the sake of comparability only solutions using 12 inch increments are presented.

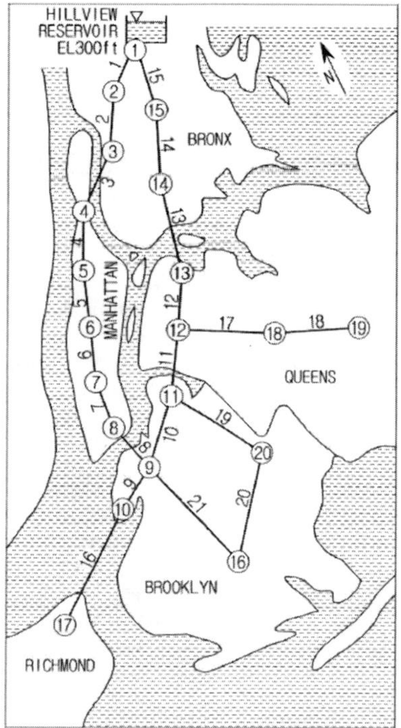

Figure 5.3: Water supply network of New York City[3]

[3]Courtesy of Zong Woo Geem (Reprint of [G00])

1969	Schaake and Lai	[S&L69]	78,084,928[†]
1985	Morgan and Goulter	[M&G85]	38,900,000[‡⃥]
1995	Loganathan et al.	[L&a95]	38,041,384
1996	Dandy et al.	[D&a96]	38,800,000[‡]
2000	Geem	[G00]	36,660,000[‡̸]
2001	Subramanian	[S01]	38,067,935
2005	Awad and von Poser	[A&P05]	38,500,000[‡]
	present solution		38,049,865

Table 5.9: Development of computational results for the minimum costs of the New York City network using 12" increments, different feasibility tolerance for the head-loss values have been used.
(†) Indicates a different set of newly to be designed arcs as parallel pipes are considered for all arcs, (‡⃥) no consistent rounding mode can be found in the cost data, (‡) solution considering only discrete pipe-diameters per segment and different set of newly to be designed arcs, (‡̸) solution considering only discrete pipe-diameters per segment, as well as different parameters in the Hazen-Williams equation and the same rounded costs as in [M&G85].

In the linear programming (LP) model developed by Schaake and Lai ([S&L69]) "the operating pressure head at each node must be specified in advance of computing the optimum pipe diameters." They analyze it as follows:

> "Operating pressures are often constrained by required pressures at the extremities and are essentially unconstrained in the interior of the system since the total head-loss along a given path is fixed. To see if a given head-loss distribution along different paths is near optimal, the LP model can be used successively for different distributions. If small changes in the pressure pattern have little effect on cost, the pattern is near optimal."

This consideration explains the minimum head bounds used later on. For the demand nodes, they are 255 meter for all nodes except for nodes 16 and 17, where they have values of 260 and 272.8 respectively. These two values were part of the iterative input data of the solution published by Schaake and Lai in [S&L69]. As no information about a maximum head is available in [S&a01] or [S01], in WaTerInt the minimum head of the supply node $H_{1L} = 300$ is used in accordance to the other test networks.

Subramanian [S01] describes that "the initial flow bounds for the arcs are simply set equal to $\pm b_{i^*}$, where i^* is the source node index." These bounds are used analogously in WaTerInt. The fact that this water distribution system is nearly gravity fed is modeled with the head bounds. A Hazen-Williams coefficient of 100 is used for all links and all nodes are assumed to be at the same elevation.

In addition, a 4-inch pipe increment for the set of available pipe diameters is considered as suggested in [S&a01]. With WaTerInt, using floating point calculations, a feasible solution of 37,896,903.2702 is obtained after 1,173 seconds.

The optimal solution published in [S&a01] contains the diameters, segment lengths and new costs reprinted in Table 5.10. Regarding the correct sum of new costs according to the published optimal vector, the solution of WaTerInt improved about $3,473.4713$ US$.

Then, the original problem with 12"-increments for the pipe diameter is calculated according to the majority of published results. The vector of admissible pipe diameters shrinks to

$$(36,\ 48,\ 60,\ 72,\ 84,\ 96,\ 108,\ 120,\ 132,\ 144,\ 156,\ 168,\ 180,\ 192,\ 204)^{\top}\ .$$

With WaTerInt, a solution of $38,049,864.8659$ is obtained after a CPU time of 420 seconds. This improves the solution of $38,067,935$ US$ published in [S&a01] about $18,070$ US$.

Arc	Diameter	Length (ft)	New Costs (US $)
7	112	2408.899652	920,962.065953
7	116	7191.100348	2,871,547.079089
16	100	26399.999870	8,769,939.752862
16	104	0.000130	0.045337
17	96	0.000063	0.019895
17	100	31199.999937	10,364,474.283491
18	76	9745.580952	2,303,611.685506
18	80	14254.419048	3,590,655.926313
19	72	0.195798	43.280596
19	76	14399.804202	3,403,753.700483
21	68	10595.684537	2,181,889.521376
21	72	15804.315463	3,493,499.380606
Sum of detailed new costs:			37,900,376.741507
New costs published:			37,878,581

Table 5.10: Extracts of the data of the optimal solution for the New York City Network using 4" increments as published in [S01] and with higher accuracy in [S&a01]. A discrepancy of 21,795.74 between addition of costs per newly to be designed pipe segment and the optimal value is noted.

Comparing the floating point result obtained with `cplex` as part of Table D.5 with the one obtained with `linprog`, see Table D.6, it occurs that for one pipe several different diameters are part of the optimal solution of `linprog`. This can be regarded as further numerical artifact that may be improved when adding procedures guaranteeing the adjacency property to the floating point algorithm.

When comparing the corresponding Tables D.3 and D.4 for the New York 12" network, again the solution obtained with `cplex` has an upper bound being about 2, 415 US$ lower than the lower bound obtained with `linprog`.

5.2.4 Hanoi Network

The planned water distribution trunk network in Hanoi, Vietnam, is introduced by Fujiwara and Khang [F&K90]. This single source network with thirty-two nodes and thirty-four arcs is illustrated in Figure 5.4. The direction of the links is chosen according to [S01], which differs in some arcs from the one originally used by Fujiwara and Khang.

With Kruskal's algorithm in WaTerInt, the arcs 19, 23 and 34 are identified as non-tree arcs.

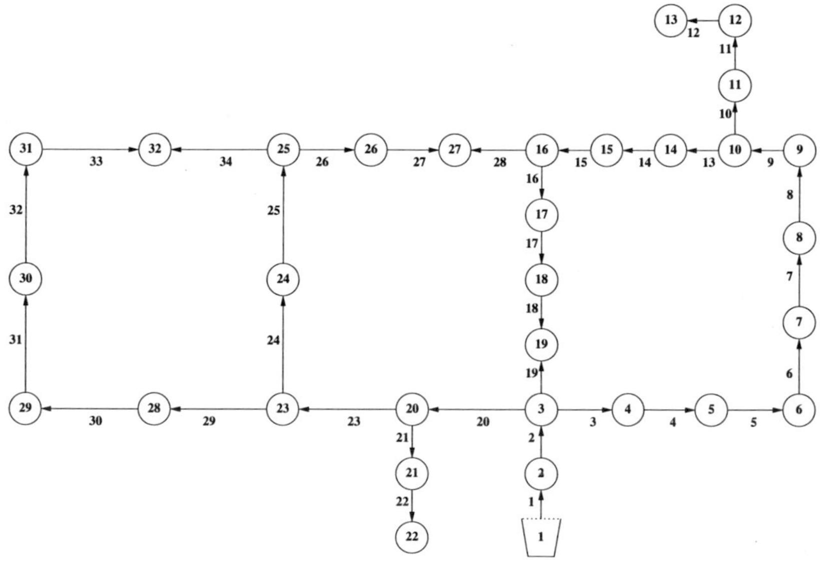

Figure 5.4: Layout of the Hanoi network

For comparability with recent published results as already presented in Section 2.2.1, a head-loss coefficient of $15,200$ and a discharge coefficient of 1.852 is used. Originally Fujiwara and Khang regarded the diameter in inch and used $162,5$ and 1.85 respectively. The first would correspond to a head-loss coefficient of $15,219$.

The costs for the available diameters were originally calculated with $c_k = 1.1\, d_k^{1.5}$. For comparability reasons, in the following the rounded values of Subramanian [S01] are used even if no consistent rounding method can be understood. They are reproduced in A.1.

The head bounds are taken from [S&a98], originally only minimum bounds for the head were considered. The flow bounds used are heuristically ascertained by Sherali et. al. ([S&a98]) and presented along with the detailed node and arc data in Table A.5. The nodes are all located at the same elevation, and as in the two-loop test network a Hazen-Williams coefficient of 130 is used for all links.

With WaTerInt, an optimal feasible solution of 6,055,542.3685 is obtained after a CPU time of 185 seconds with floating point calculations which corresponds to the solution published in [S&a01]. This solutions is presented in detail in Table D.7.

The maximum available diameter of $40''$ results, for the first pipe, in a velocity of approximately $6.83\,\frac{\mathrm{m}}{\mathrm{s}}$, and $6.53\,\frac{\mathrm{m}}{\mathrm{s}}$ for the second, see Table D.43. Following best engineering practices, one normally tries to obtain velocities between 1 and $2\,\frac{\mathrm{m}}{\mathrm{s}}$ [D99], as approximately obtained in the optimal solutions of the other networks. Thus, it probably would make sense to allow larger diameters for these pipes.

1991	Fujiwara and Khang	[F&K91]	6,320,000†
1994	Eiger et al.	[E&a94]	6,026,660‡
1998	Sherali et al.	[S&a98]	6,058,976
2000	Geem	[G00]	6,056,000♯̸
2001	Cunha and Sousa	[C&S01]	6,093,000♯̸
2001	Subramanian	[S01]	6,055,542
	present solution		6,055,542

Table 5.11: Computational results for the minimum costs of the Hanoi test network. (†) Indicates the use of different values for discard and head-loss coefficient, (‡) indicates a solution with violations in the constraints, (♯̸) the solution considers only discrete diameters in the pipe segments, and different parameters in the Hazen-Williams equation are used, additionally in [G00] the cost function is rounded to three decimal places.

5.3 Results for Different Hydraulic Parameters

5.3.1 Results for Different Head-Loss Formulae

Any comparison of the head-loss formulae depends on the choice of the Hazen-Williams coefficient and equivalent sand roughness. Figure 5.5 contains the results for different parameters for the two-loop and the New York network. Obviously, for these networks the costs seem to be monotone decreasing in C_{HW} and increasing in ε.

For the default value, the optimal costs of the two-loop network significantly decreases from $403,386$ US$ for the Hazen-Williams formula to $350,626$ US$ for Darcy-Weisbach, whereas the costs for Hanoi increase $6,055,543$ US$ from the first to $6,641,613$ US$ in the second. For the New York networks the Darcy-Weisbach problem is infeasible, whereas with the Hazen-Williams formula a solution for both versions, using 12" or 4" increments in the diameter is obtained.

Thus, it is reasonable to compare the values of the constants by a backward analysis. The optimal solutions due to Hazen-Williams as presented in Table D.1, D.3, D.5 and D.7 are used as basis for the following comparison of head-loss calculation. Pipe segments with length shorter than one meter, or for the New York network shorter than one foot, are not considered. For all pipe segments of these optimal solutions, the head-loss Φ_{HW} according to Hazen-Williams is calculated. Then the equivalent sand roughness which results in the determination of the same head-loss when using Darcy-Weisbach is identified as

$$\epsilon = 3.71\, d \left(10^{-\frac{|q|\sqrt{2\, c_{DW}\, l}}{\pi \sqrt{|\Phi_{HW}|\, g\, d^5}}} - \frac{2\, c_{CW}\, \sqrt{2\, c_{DW}\, l}}{\pi \sqrt{|\Phi_{HW}|\, g\, d^3}} \right) .$$

The results are summarized in Figure 5.6, and the mean value with and without outliers is provided, i.e. in the case of the two-loop test network the two lowest values are not considered for the determination of the mean value and for the New York network the highest two values are neglected.

For the real networks of New York City and Hanoi the values correspond to the expected values as discussed in Section 3.1. In the first example, the value of 0.02 foot corresponds to the default value used in WaTerInt. For the Hanoi network 0.23 mm is close to the value of 0.25 mm selected for cast iron.

For the two-loop network, pipes up to 24 inch are considered. The equivalent sand roughness of 0.19 mm does not correspond to the value for PVC pipes, this explains the differences obtained for the optimal solution.

Furthermore, comparing the optimal solutions obtained with Darcy-Weisbach to those obtained with adjusted Hazen-Williams shows that the maximum deviation is

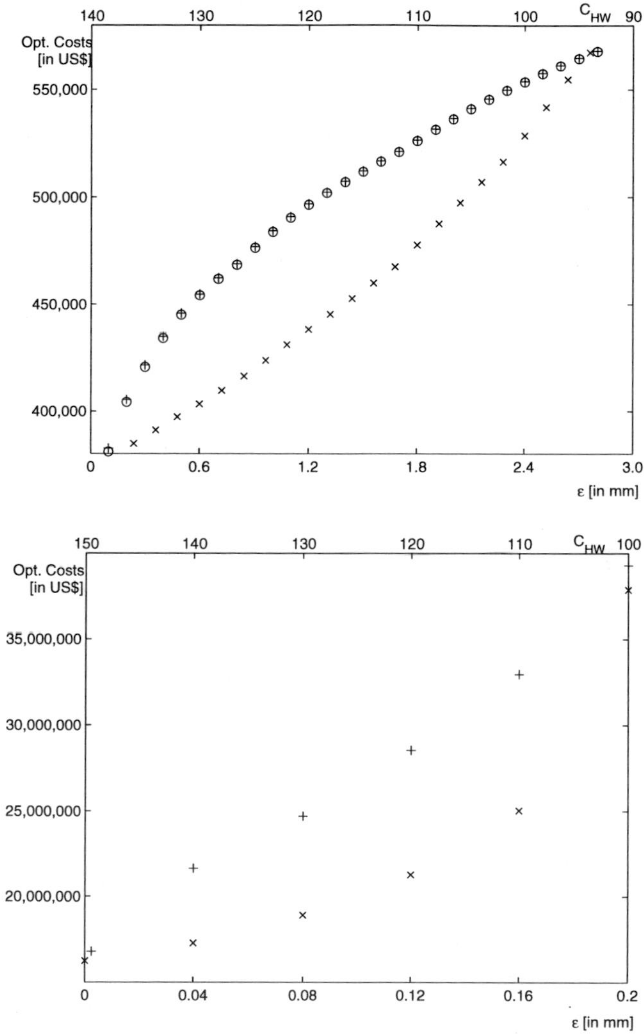

Figure 5.5: Comparison of head-loss formula for the two-loop and New-York (4") networks. The plots are based on the values in Tables D.46 and D.47. Head-loss calculations with Hazen-Williams are plotted with 'x', with Darcy-Weisbach with ' o ' and with adjusted Hazen-Williams as ' + '.

less than 1 % for the two-loop and Hanoi networks, both agree that the New York examples are infeasible.

Table D.47 indicates that for a too low value of C_{HW} or too high value of ε the problem becomes infeasible, probably due to the constraints of the fixed pipes. For Hanoi, as shown in Table D.48, for a low C_{HW} the source head is not high enough.

It should be noted in the comparisons of the optimal solution vectors part of Tables D.40 to D.44 that the optimum found is not necessarily unique. Nevertheless, when comparing the optimal solution vectors for the head-loss formulae for default values for the two-loop network, the deviation of the optimal flow speeds and head between Hazen-Williams and Darcy-Weisbach are relatively small, whereas the diameters selected differ more. For Darcy-Weisbach and adjusted Hazen-Williams lower

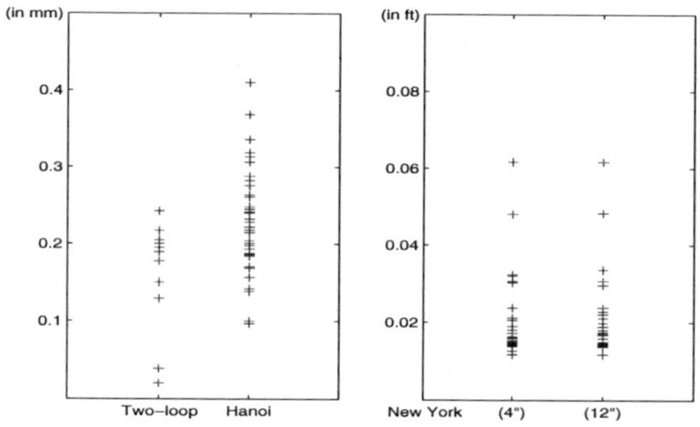

Test Network	Minimum Value	Maximum Value	Mean Value	without outlier	Value used in WaTerInt	rounded to mm
Two-loop	0.0203	0.2425	0.1597	0.1857	0.003 mm	0.003 mm
New York (4")	0.1395	0.7401	0.2433	0.2188	0.240 inch	6.096 mm
New York (12")	0.1400	0.7401	0.2427	0.2182	0.240 inch	6.096 mm
Hanoi	0.0966	0.4099	0.2280		0.250 mm	0.250 mm

Figure 5.6: Determination of the equivalent sand roughness for the test networks: For equal head-loss according to Hazen-Williams and Darcy-Weisbach for all pipe segments of the optimal solution of Table D.8, D.10, D.11 and D.12 the corresponding equivalent sand roughness is plotted, and a number of basic values is provided. Mean without outliers is the mean value not considering the two smallest values for the two-loop network or the two largest for the New York network, respectively.

deviations in the diameters occur. For Hanoi, differences in the flow speeds and head distributions can be observed as well.

In the comparison of two selected solutions for the New York (12") network with Hazen-Williams and adjusted Hazen-Williams formula in Tables D.41 and D.42, a reduced equivalent sand roughness of 0.01 ft has been used to ensure feasibility for both problems. The optimal solutions differs significantly by approximately 24% and thus, as expected, different diameters are selected for the newly to be designed pipes. Interestingly, the head distribution across the network does not have significant differences.

For the New York network mainly surface water is used. Therefore differences in the temperature have to be considered. For this reason, results for water of $4°$, $10°$ and $20°$ celsius are calculated. As summarized in Table D.49 the influence of the temperature is small compared to the results when using different material, i.e. changing the equivalent sand roughness, and thus may be neglected.

5.3.2 Results concerning Initial Head and Energy Costs

The effect of different initial head is part of Table D.50. For all three test networks, a minimal head exists below which the problem is infeasible. For the New York (12") network the problem also becomes infeasible for large initial head, resulting from the structure of the constraints, especially those for the fixed arcs. As seen in Figure 5.7, the results also show that the optimal cost is high at the onset of feasibility, reflecting a tendency to use large diameters in the optimal solution, and decreases as more initial head is available.

A comparison including different head-loss formula is part of Table D.51, again, according to the choice of default values for C_{HW} and ϵ, the results of Darcy-Weisbach is close to adjusted Hazen-Williams, whereas significant differences are obtained compared to Hazen-Williams. Interestingly, the relative deviation of Hazen-Williams to Darcy-Weisbach increases from 8% for an initial head of $240\,\mathrm{m}$ to 26% for $200\,\mathrm{m}$.

Optimal solutions for different energy costs calculated with the adjusted Hazen-Williams formula are part of Table D.52. For interpretation purposes these are split into energy and pipe costs in Table D.53 and D.54. All results for the discrete points of energy costs are calculated verified and thus reliable up to the used accuracy of 10^{-6}. Probably one can assume the following, even if the behavior between the points calculated cannot be guaranteed to follow simple interpolation.

For the two-loop network obviously the energy costs do not influence the optimal value, since this example is mainly gravity fed. Considering unrealistically small energy costs as in Figure D.6, insights can be gained on the energy cost – pipe cost tradeoffs. Stepwise increases from zero in the energy costs demonstrate how the optimal solution moves through three distinct regimes.

At low energy costs, the optimal solution is driven mainly by pipe cost. Small pipe diameters are chosen with corresponding low pipe costs, and additional initial head is used to ensure the demand is met. As unit energy costs increase, the optimal solution remains the same, until the additional energy costs lead to a configuration with larger diameters. From this point on, the optimal solution depends on a trade-off between costs for energy and costs for larger pipe diameters, until the point where energy costs are so large that they would dominate. In this case no additional head is further used.

For the New York network, energy costs are sufficiently low that additional head is used at the source node. As can be seen in Table D.54, the pipe costs do not vary in the range of energy costs considered. This appears to be due to the constraints of the fixed part of the network. For all these optimal solutions the head at the source node has a value of 306 ft , compared to 300 ft that are already provided by the natural elevation. As expected, energy costs for raising the head by 6 ft increase proportional to the unit costs.

Hanoi is in a regime where the optimal solution consists of a trade-off between pipe and energy costs. These examples confirm the engineering experience that except when just relying on gravity, the construction costs for the pipes are small compared to the energy costs.

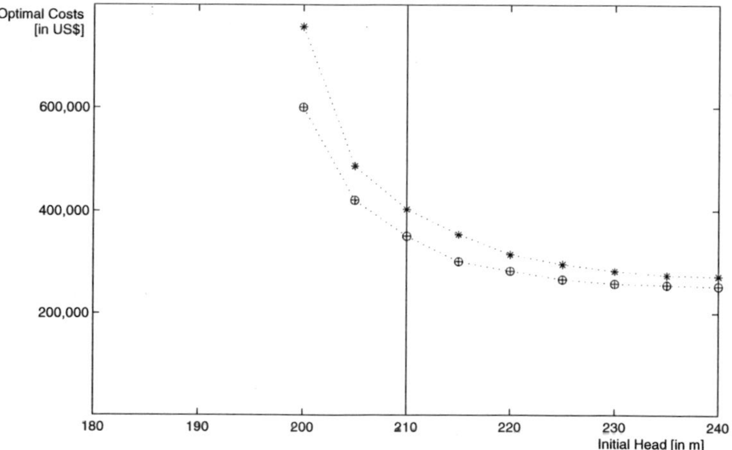

Figure 5.7: Optimal value depending on fixed initial head for the two-loop network according to Table D.51. The optimal value obtained for the Hazen-Williams problem is marked with '*', the one for Darcy-Weisbach with a circle and for adjusted Hazen-Williams as '+'.

Summary and Conclusion

With known interval analysis techniques, it is possible to verify the determination of the optimal solution of the nonlinear network optimization problem used for water distribution design. Essentially, the verification is based on two results from interval analysis. On one hand, outwardly rounded interval arithmetic provides the possibility to gain rigorous bounds for the zero of a nonlinear function together with a guarantee of existence and uniqueness. And on the other hand, optimal solutions determined with a linear floating point solver can a posteriori be made rigorous. This a posteriori verification of the lower bound of a linear optimization problem forms a central part of the verification of a global optimal solution for a nonlinear problem with a branch and bound algorithm.

One major advantage of interval analysis is the possibility to obtain error estimates along with the solution of a numerical problem. For this purpose, a number of procedures have to be adjusted and expanded. For obtaining a verified upper bound, feasibility of the solution vector has to be guaranteed. This part of the verification process depends essentially on the approximate solution and therefore on the linear solver used during the branch and bound algorithm.

At the same time, for the verification of feasibility, a detailed investigation of the structure of the problem becomes necessary, which remains a major challenge. In this thesis, the algorithms use a number of special properties of the constraints. Further research could focus on more general methods to verify feasibility, especially on obtaining inclusions for the solution of a nonlinear system on the boundary of an interval.

The underlying floating point algorithms have been improved using the structure of the constraints, mainly the head constraint propagation for expansion networks has been added which decreases the computational time by approximately one third to one half.

The floating point algorithm could be further adjusted analogous to the verified version, for example, a numerically determined solution could easily be adjusted to fulfill the adjacency property. Thus the application of interval techniques to an existing pure floating point algorithm can identify parts of the algorithm which may be numerically improved.

The computational results using non-verified floating point calculations are found to show numerical artifacts. For example, it is possible that a problem is identified as infeasible that in reality does have a solution, as shown by the two-loop expansion network on an HP Superdome server. Furthermore, there are examples where the lower bound for the optimal solution is larger than the upper bound, for the same calculations, for calculations on different computers, or for calculations using different linear solvers.

Using verified calculations, all these artifacts are avoided, and obtained results are always reliable. Nevertheless, in the current implementation the computational time for verified calculations takes on average fifteen times the time needed for the calculation without error analysis when relying just on pure floating point arithmetic. This factor could probably be further decreased when not depending on interpretation overhead of MATLAB, especially for object orientation.

In addition to the models discussed so far based on the Hazen-Williams formula, the algorithm has been expanded to be able to solve the more accurate Darcy-Weisbach optimization problem as well. Essentially, detailed investigation of the nonlinearity of the implicitly defined Colebrook-White equation have been necessary to retain certain monotonicity and convexity arguments and the adjacency property is proved for this new problem as well. However, the computational effort is around forty times higher than for the Hazen-Williams problem.

It is possible to improve the coefficients of the Hazen-Williams equation for more closely approximating the Darcy-Weisbach formula, as part of the adjusted Hazen-William problem. The computational results show that this problem provides improved accuracy while retaining approximately the same computational times as the originally used Hazen-Williams problem.

Finally, verified results have been used to compare hydraulic parameters for the test networks. The influence of the equivalent sand roughness cannot be neglected, but when regarding the whole life span of a pipe network it occurs that energy costs form a significant part of the overall costs, except when water is mainly delivered by gravity.

The underlying optimization problems are challenging insofar as the constraints are highly nonlinear. The largest calculated test problem contain 1,478 variables, with the developed package WaTerInt about 10 hours are needed on an HP Superdome server for determination of a rigorous optimal solution for the Hazen-Williams problem.

For one standard problem from literature, the New York City network, which contains 554 decision variables, 58 linear and 33 nonlinear equality constraints, for the Hazen-Williams problem less than 4 hours and for the Darcy-Weisbach problem approximately 54 hours on a notebook are needed to obtain a verified solution. The time needed when using adjusted Hazen-Williams is of the same order of magnitude as the one needed for Hazen-Williams.

According to Schaake and Lai the introductory citation of Hamming, "The purpose of computing is insight, not numbers", might be applied to mathematical programming. They identify one of the important merits of their model to be "an improved understanding of water distribution design which may result from these insights" [S&L69].

With WaTerInt, a software package is provided which, according to its modular structure, can easily be adjusted to further problems. On the one hand it can support engineers in the design process of water distribution networks, and on the other hand it can form a basis for further research on verification of optimization algorithms in connection with INTLAB. So it may be used in the scope of Hamming [H62]: "Computing is a tool that supplies numerical answers, but it is also an intellectual tool for examining the world."

Appendix A

Data of the Water Distribution Design Test Examples

The following tables contain the data for the test examples according to [S&a01], [A&S77] and [F&K90].

Two-Loop	
Diameter	Cost
1	2
2	5
3	8
4	11
6	16
8	23
10	32
12	50
14	60
16	90
18	130
20	170
22	300
24	550

New York	
Diameter	Cost
36	$1.1 \cdot 36^{1.24}$
40	$1.1 \cdot 40^{1.24}$
44	$1.1 \cdot 44^{1.24}$
$\vdots$	$\vdots$
192	$1.1 \cdot 192^{1.24}$
196	$1.1 \cdot 196^{1.24}$
200	$1.1 \cdot 200^{1.24}$
204	$1.1 \cdot 204^{1.24}$

Hanoi	
Diameter	Cost
12	45.73
16	70.40
20	98.39
24	129.33
30	180.74
40	278.28

Table A.1: Available pipe diameters in inch and cost data

Arc	Length	$q_{\min}$	$q_{\max}$
1	1000	1120	1120
2	1000	0	1020
3	1000	0	1020
4	1000	-650	900
5	1000	-120	900
6	1000	-450	570
7	1000	-100	920
8	1000	-370	650

Node	Demand	E_i	H_{iL}	H_{iU}
1	1120	210	210	210
2	-100	150	180	210
3	-100	160	190	210
4	-120	155	185	210
5	-270	150	180	210
6	-330	165	195	210
7	-200	160	190	210

Table A.2: Two-Loop test network data

Fixed Arcs	Diameter	Length
(1, 2)	18	1000.0000
(2, 3)	10	795.4084
(2, 3)	12	204.5916
(2, 4)	16	1000.0000
(4, 5)	1	1000.0000
(4, 6)	14	310.3541
(4, 6)	16	689.6459

Table A.3: Additional data for the Two-Loop expansion test network

Arc	Length	q_{min}	q_{max}	Fixed Diam.
1	11600	-2017.5	2017.5	180
2	19800	-2017.5	2017.5	180
3	7300	-2017.5	2017.5	180
4	8300	-2017.5	2017.5	180
5	8600	-2017.5	2017.5	180
6	19100	-2017.5	2017.5	180
7	9600	-2017.5	2017.5	
8	12500	-2017.5	2017.5	132
9	9600	-2017.5	2017.5	180
10	11200	-2017.5	2017.5	204
11	14500	-2017.5	2017.5	204
12	12200	-2017.5	2017.5	204
13	24100	-2017.5	2017.5	204
14	21100	-2017.5	2017.5	204
15	15500	-2017.5	2017.5	204
16	26400	-2017.5	2017.5	
17	31200	-2017.5	2017.5	
18	24000	-2017.5	2017.5	
19	14400	-2017.5	2017.5	
20	38400	-2017.5	2017.5	60
21	26400	-2017.5	2017.5	
22	4800	-2017.5	2017.5	132
23	4800	-2017.5	2017.5	132
24	13200	-2017.5	2017.5	72
25	13200	-2017.5	2017.5	72
26	13200	-2017.5	2017.5	72
27	13200	-2017.5	2017.5	72
28	7200	-2017.5	2017.5	60
29	7200	-2017.5	2017.5	60
30	15600	-2017.5	2017.5	72
31	15600	-2017.5	2017.5	72
32	12000	-2017.5	2017.5	60
33	12000	-2017.5	2017.5	60

Node	Demand	E_i	H_{iL}	H_{iU}
1	2017.5	0	300	300
2	-92.4	0	255	300
3	-92.4	0	255	300
4	-88.2	0	255	300
5	-88.2	0	255	300
6	-88.2	0	255	300
7	-88.2	0	255	300
8	-88.2	0	255	300
9	-170.0	0	255	300
10	-1.0	0	255	300
11	-170.0	0	255	300
12	-117.1	0	255	300
13	-117.1	0	255	300
14	-92.4	0	255	300
15	-92.4	0	255	300
16	-170.0	0	260	300
17	-57.5	0	272.8	300
18	-117.1	0	255	300
19	-117.1	0	255	300
20	-170.0	0	255	300
21	0.0	0	255	300
22	0.0	0	255	300
23	0.0	0	255	300
24	0.0	0	255	300
25	0.0	0	255	300
26	0.0	0	255	300

Table A.4: New York network data

Arc	Length	$q_{\min}$	$q_{\max}$
1	100	19940	19940
2	1350	19050	19050
3	900	4885	9455
4	1150	4755	9325
5	1450	4030	8600
6	450	3025	7595
7	850	1675	6245
8	850	1125	5695
9	800	600	5170
10	950	2000	2000
11	1200	1500	1500
12	3500	940	940
13	800	-1925	2645
14	500	-2540	2030
15	550	-2820	1750
16	2730	-3070	0
17	1750	-3935	-865
18	800	-5280	-2210
19	400	2270	5340
20	2200	6475	8255
21	1500	1415	1415
22	500	485	485
23	2650	3785	5565
24	1230	1265	4230
25	1300	445	3410
26	850	-3240	4400
27	300	-4140	3500
28	750	-3130	4510
29	1500	290	1920
30	2000	0	1630
31	1600	-360	1270
32	150	-720	910
33	860	-825	805
34	950	0	1630

Node	Demand	Elevation	H_{iL}	H_{iU}
1	19940	0	100	100
2	-890	0	30	100
3	-850	0	30	100
4	-130	0	30	100
5	-725	0	30	100
6	-1005	0	30	100
7	-1350	0	30	100
8	-550	0	30	100
9	-525	0	30	100
10	-525	0	30	100
11	-500	0	30	100
12	-560	0	30	100
13	-940	0	30	100
14	-615	0	30	100
15	-280	0	30	100
16	-310	0	30	100
17	-865	0	30	100
18	-1345	0	30	100
19	-60	0	30	100
20	-1275	0	30	100
21	-930	0	30	100
22	-485	0	30	100
23	-1045	0	30	100
24	-820	0	30	100
25	-170	0	30	100
26	-900	0	30	100
27	-370	0	30	100
28	-290	0	30	100
29	-360	0	30	100
30	-360	0	30	100
31	-105	0	30	100
32	-805	0	30	100

Table A.5: Hanoi network data

Appendix B

Extracts of WaTerInt - Sources

“Die Ansicht, daß Programme nur für den Computer geschrieben werden, ist leider weit verbreitet. Programme sollten (auch) für den menschlichen Leser zugänglich sein, ...”[1]

B.1 Extracts of the Main Routine “wds.m”

```
function [opt, acc] = wds ( town , head_loss_formula, k_value, viscosity ,...
                           source_head , head_cost , lp_solver ,...
                           verification , accuracy , no_relax ,...
                           branching , max_iter , verbose)
%
% WDS - Calculation of the Water Design Network Optimization Problem
%
% [opt, acc] = wds (town, head_loss_formula, k_value, viscosity,...
%                   source_head, head_cost, lp_solver, verification,
%                   accuracy, no_relax, branching, max_iter, verbose)
%
% town              = 'twoloop', 'twoloop_exp', 'newyork_12', 'newyork_4'
%                      or 'hanoi'
% head_loss_formula = 'HW' (Hazen-Williams), 'DW' (Darcy-Weisbach),
%                     'AHW' (adjusted Hazen-Williams)
% k_value           = Hazen-Williams coefficient or equivalent sand roughness
%                     in [cm] or [inch]
% viscosity         = kinematic viscosity (only used for Darcy-Weisbach and
%                     for adjusted Hazen-Williams) -- in [m^2 /s] or [ft^2/s]
% source_head       = fixed value for head at source node
%                     if empty, default value according to network data is used
% head_cost         = energy cost to raise head at source node in [US$/m]
% lp_solver         = 'linprog' or 'cplex'
```

[1] “Unfortunately there is a common misconception, that programs are written to be read by the computer only. Programs should be accessible to the human reader (as well), ... ”
Niklaus Wirth. *Gedanken zur Software-Explosion.* Informatik-Spektrum 17, 5 - 10, 1994.

```
% verification     = (0) floating point calculation -- (1) verified calc. --
%                    (2) two phase b&b -- (3) verified calc. in every step --
%                    (4) two phase b&b with adjustments
% accuracy         = floating point number out of [0,1], accuracy used for
%                    pruning subproblems; when calulations are verified
%                    accuracy for termination of branch and bound algorithm
% no_relax         = 4 or 6 (number of tangents for relaxation)
%                    or 0 (no relaxation)
% branching        = (1) default strategy -- (2) strategy according to delta's
%                    (3) default + head mstr -- (4) s2 + head mstr --
%                    (5) default + adjusted head mstr -- (6) s2 + adj.head mstr
% max_iter         = maximum number of iterations during branch and bound
%                    (default value: 70,000)
% verbose          = 0 (few output) -- 1 (full output) -- 2 (full output +
%                    additional information about relaxation status)
%
%
% opt              = best known upper bound for optimal value
% acc              = best known upper bound minus best known lower bound
%
% WDS is the main function for the optimization routines of WaTerInt,
% including those according to NOP defined in Sherali, Subramanian, Loganathan:
% "Effective Relaxations and Partitioning Schemes for Solving Water
% Distribution Network Design Problems to Global Optimality"
% Journal of Global Optimization 19: 1-26,2001.
%
% MODULES NEEDED: TWOLOOP,TWOLOOP_EXP, NEWYORK, HANOI, NOP, BB, VNOP,VBB, OUTPUT
% [...]
% Author:     A. C. Hailer
% WaTerInt:   version 2.1
% [...]

% Constants
HMAX = 1000;     % Maximum head above elevation when considering energy costs
PREC = '%.4f ';  % Precision string for floating point conversion

% Argument processing and setting of default parameters
if (nargin < 1)
  err_str = ['Usage: [opt, acc ] = '...
            'WDS (testexample, [head_loss_formula], [k_value], [viscosity] '...
            '[source_head], [head_costs], [lp_solver], [verification] ',...
            '[accuracy], [no_relax], [branching], [max_iter] [verbose])\n',...
            'e.g. wds(''twoloop'',[],[],[],[],[],''cplex'',0,[],[],[],[],1)'];
  error('WaTerInt:usage',err_str);
end
% [...]

global wds_parameter_str wds_error wds_filename
wds_error = @ wds_error_function;

% Save all parameters to the string 'wds_parameter_str'
% [...]
```

```
% Directories for calculated results
if ~exist('results','dir'), mkdir results; end
if ~exist('calculated','dir'), mkdir calculated; end

fp_calc = filename( town, inch, head_loss_formula, k_value, viscosity,...
                    source_head, head_cost, lp_solver, verification,...
                    accuracy, no_relax, branching, max_iter);
wds_filename = fp_calc;

% Initialization of the test problem data [...]
  [avail_diams, pipe_cost, arcs, arc_length, arcs_fixed, length_fixed,...
  unit, demand,elevation, flow_min,flow_max, head_min,head_max] = feval(town);
% [...]

if (head_cost ~= 0)
  head_min(1)  = elevation(1);
  head_max(1)  = max(head_min(1) + HMAX, head_max(1));
  head_cost    = head_cost * demand(1);
  % conversion to [US$/ft]
  if strcmp (town ,'newyork'), head_cost = head_cost * 0.3048; end
else
  if ~isempty(source_head)
    head_min(1) = source_head; head_max(1) = source_head;
  end
end

infeasible = 0; error_str = [];

% Initialization of global variables
global No_arcs No_nodes No_dia No_seg No_relax No_lambda No_narcs No_nseg

No_arcs   = size(arc_length,1);
No_nodes  = size(demand,1);
No_dia    = size(avail_diams,1);

No_seg    = No_arcs * No_dia;
No_relax  = no_relax ;

No_narcs  = No_arcs - size(arcs_fixed,1);
No_nseg   = No_narcs * No_dia;

if ~strcmp(head_loss_formula, 'DW')
  No_lambda = No_arcs;
else
  No_lambda = sum(length_fixed ~= 0);
end

% Initialization of the optimization problem
[ahw, dischard, ccw, costs, alpha, seg_fixed, node_edge, non_tree, post_tree,...
 adjust_lbeq, relax, select_idx, upper_bounding, adjust_head, adjust_flow,...
 u_fkt_l, Aeq, beq, lb, ub, lb_ub, ub_ub]...
   = nop (head_loss_formula, k_value, viscosity, avail_diams, pipe_cost,...
          arcs, arc_length, arcs_fixed, unit, demand, elevation,...
          flow_min, flow_max, head_min, head_max);
```

```
% Calculation of optimal solution - or if exists, just read it
if (~exist(fp_calc) || verbose)
  if (verification ~= 1) && (verification ~= 3)
    % Calculation of optimal floating-point solution ...
    if (verification ~= 0)
      bb_accuracy = 1e-6;
      if strcmp(town ,'twoloop_exp'), bb_branching = 2;else bb_branching = 5;end
    else
      bb_accuracy   = accuracy;       bb_branching  = branching;
    end
    fp_fl = filename( town, inch, head_loss_formula, k_value, viscosity,...
                      source_head, head_cost, lp_solver, 0, bb_accuracy,...
                      no_relax, bb_branching, max_iter);
    wds_filename = fp_fl;

    % Branch and Bound for optimal floating-point solution
    if (~exist(fp_fl) || verbose)
      tic; cpu = cputime;
      [opt_vec , global_lb , global_ub , infeasible , no_subs, no_subs_left]...
        = bb (head_loss_formula, k_value, lp_solver, bb_accuracy,...
              bb_branching, max_iter, dischard, ccw, avail_diams, costs,...
              head_cost, alpha, arcs, arc_length, arcs_fixed, length_fixed,...
              seg_fixed, node_edge, demand, elevation, flow_min, flow_max,...
              head_min, head_max, non_tree, post_tree, adjust_lbeq, relax,...
              select_idx, upper_bounding, adjust_flow, adjust_head,...
              Aeq, beq, lb, ub, lb_ub, ub_ub, verbose);
      cpu = cputime-cpu; time = toc;
      if (~strcmp (fp_fl, fp_calc) && ~verbose)
        fprintf('WDS info: %f s needed for first non-verified b&b.\n', cpu);
        save (fp_fl,'opt_vec', 'global_lb', 'global_ub', 'infeasible',...
              'no_subs','no_subs_left','time','cpu','s_town','error_str');
      end
    else
      % ... or if floating point solution exists, just read it
      if ~verbose, load(fp_fl); end
      if ~isempty(error_str), wds_error(error_str); end
      fprintf('WDS info: first part already calculated in %f seconds.\n',cpu);
    end
    wds_filename = fp_calc;
  end
  if (verification ~= 0)
    if infeasible
      wds_error('WDS - No solution found in unverified first part, aborted.');
    end
    % Calculation of verified optimal solution
    k_value         = intval( sprintf(PREC, k_value));
    arc_length      = intval( sprintf(PREC, arc_length));
    length_fixed    = intval( sprintf(PREC, length_fixed));
    demand          = intval( sprintf(PREC, demand));
    elevation       = intval( sprintf(PREC, elevation));
    flow_min        = inf_(intval( sprintf(PREC, flow_min)));
    flow_max        = sup(intval( sprintf(PREC, flow_max)));
    head_min        = inf_(intval( sprintf(PREC, head_min)));
    head_max        = sup(intval( sprintf(PREC, head_max)));
```

```
    [dischard, ccw, alpha, seg_fixed, adjust_lbeq, relax,...
     adjust_head, adjust_flow, test_adjacency, verify_lb, verify_ub,...
     verify_infeasibility, Aeq, beq, lb, ub]...
       = vnop (head_loss_formula, k_value, viscosity, dischard, avail_diams,...
               ahw, arcs, arc_length, arcs_fixed, seg_fixed, node_edge,...
               unit, demand, elevation);
    if (verification == 1) || (verification == 3)
      global_ub = []; opt_vec = []; cpu = 0; time = 0; no_subs = 0;
    end
    % Branch and Bound for verified optimal solution
    tic; cpu = cpu - cputime;
    [opt_vec , global_lb , global_ub , infeasible , no_subs_vbb,no_subs_left]...
       = vbb (opt_vec, head_loss_formula, k_value, verification,... % [...]
               Aeq, beq, lb, ub, lb_ub, ub_ub, verbose);
    cpu = cputime + cpu; time = time + toc;
    no_subs = no_subs + no_subs_vbb;
  end
  if ~verbose
    save (fp_calc, 'opt_vec', 'global_lb', 'global_ub', 'infeasible',...
                   'no_subs', 'no_subs_left','time','cpu','s_town','error_str');
  end
else % ... or if optimal solution exists, just read it
  if ~verbose, load(fp_calc); end
  if ~isempty(error_str), wds_error(error_str); end
end
% Structured output of computational results
output (infeasible, opt_vec, global_lb, global_ub, no_subs,... %[...]
       pipe_cost, alpha, arcs, arcs_fixed, length_fixed,seg_fixed, elevation);
if (verification == 0)
  acc = global_ub - global_lb;
else
  acc = sup(intval(global_ub) - intval(global_lb));
end
opt = global_ub;

% ---------------------------------------------------------------------------
% GENERAL FUNCTIONS FOR FILE HANDLING
% ---------------------------------------------------------------------------
function [name] = filename (town, inch, formula, k_value, viscosity,...
                           head, cost, lp_solver, verification,...
                           accuracy, no_relax, branching, max_iter)

name = strcat('./calculated/', town ,'_', sprintf('%d',inch),'_',... %[...]
      '.mat');
% ---------------------------------------------------------------------------
function wds_error_function (error_str)

global wds_parameter_str wds_filename

mdate = datestr(now,0); mdate(findstr(' ',mdate)) = '-';
out_file = fopen(strcat('./results/',mdate,'.txt'), 'w');
wds_file = fopen(strcat('./results/',mdate,'.wds'), 'w');
% Output of parameters used
for fp = [1, out_file, wds_file]
  fprintf(fp,wds_parameter_str); fprintf(fp,error_str); fprintf(fp,'\n\n\n\n');
end
fclose(out_file); fclose(wds_file);
save (wds_filename, 'error_str'); error('WaTerInt:err', error_str);
```

B.2 Extracts of "twoloop.m" as Example of Data Structure

```
function [avail_diams, pipe_cost, arcs, arc_length, arcs_fixed, length_fixed,...
         unit, demand, elevation, flow_min, flow_max, head_min, head_max]...
         = twoloop
%
% TWOLOOP Initialization of data for Two Loop Network.

% [...]

% ------------------------------------------------------------------------
% Commercially available pipes, costs given in $ per meter
unit = 'm';

avail_diams  = [1; 2; 3;  4;   6;  8; 10; 12; 14; 16;  18;   20;   22;  24];
pipe_cost    = [2; 5; 8; 11; 16; 23; 32; 50; 60; 90; 130; 170; 300; 550];

% Before calculations are made conversion inch => cm is necessary
avail_diams = avail_diams * 2.54;

% ------------------------------------------------------------------------
% Arc data for the network
No_arcs = 8; % No_nodes = 7;
arc_length = repmat(1000, No_arcs, 1); arcs = zeros(No_arcs,1,2);
arcs = [[ 1, 2]; ...
        [ 2, 3]; ...
        [ 2, 4]; ...
        [ 4, 5]; ...
        [ 4, 6]; ...
        [ 6, 7]; ...
        [ 3, 5]; ...
        [ 5, 7]];

arcs_fixed = []; length_fixed = repmat(0, No_arcs*size(avail_diams,1), 1);

% ------------------------------------------------------------------------
% Flow and head data of the network
flow_min = [1120;    0;    0; -650; -120; -450; -100; -370];
flow_max = [1120; 1020; 1020;  900;  900;  570;  920;  650];

head_min = [210; 180; 190; 185; 180; 195; 190];
head_max = [210; 210; 210; 210; 210; 210; 210];

% ------------------------------------------------------------------------
% Node data for the network
elevation = [ 210;  150;  160;  155;  150;  165;  160];
demand =    [1120; -100; -100; -120; -270; -330; -200];
```

B.3 Extracts of "nop.m" containing Central Functions

```
function [ahw, dischard, ccw, costs, alpha, seg_fixed, node_edge, non_tree,...
        post_tree, f_adjust_lbeq, f_relax, f_select_idx, f_upper_bounding,...
        f_adjust_head, f_adjust_flow, f_u_fkt_l,...
        Aeq, beq, lb, ub, lb_ub, ub_ub]...
          = nop (head_loss_formula, k_value, viscosity, avail_diams,...
                pipe_cost, arcs, arc_length, arcs_fixed, unit, demand,...
                elevation, flow_min, flow_max, head_min, head_max)
%
% NOP - Initialization of variables and functions needed for the branch and
%       bound algorithm
% [...]
global No_arcs No_nodes No_dia No_seg No_lambda No_narcs No_nseg wds_error
D_EXP = 4.87;

% Ensure ascending order of indices and build auxiliary index vector
if any(arcs_fixed ~= sort(arcs_fixed))
  feval(wds_error, 'NOP: vector arcs_fixed has to be sorted')
end
seg_fixed = []; % vector containing the indices of fixed segments
for i = arcs_fixed'
  seg_fixed = [seg_fixed; transpose((i-1)*No_dia+1:i*No_dia)];
end
% Node-edge matrix describing network structure
node_edge = repmat(0, No_nodes, No_arcs);
for i = 1:No_nodes
  node_edge (i,:) = (arcs(:,1)==i)'-(arcs(:,2)==i)';
end

costs = repmat (pipe_cost, No_narcs,1);

% Determination of additionally needed hydraulic constants
if strcmp (unit , 'm')
  dischard = 1.852;       headcoefficient = 15200;
  cdw      = 1e6/1296;    gravitation = 9.81;
  if ~isempty(viscosity), ccw(1) = 22.59 * pi * viscosity; end
elseif strcmp( unit , 'inch')
  dischard = 1.85;        headcoefficient = 851500;
  cdw      = 248832;      gravitation = 32.2;
  if ~isempty(viscosity), ccw(1) = (2.51 / 48) * pi * viscosity; end
else
  feval(wds_error, 'NOP: unit must be either "m" or "inch"')
end
ccw(2)   = 3.71;
if strcmp (head_loss_formula, 'HW')
  ccw = []; cdw = []; gravitation = [];
elseif strcmp (head_loss_formula, 'DW')
  headcoefficient = [];
  dischard = 2;
end
```

```
% ---------------------------------------------------------------------------
% Function handles needed during branch & bound

f_adjust_lbeq          = @ adjust_lbeq;
f_relax                = @ relax;
f_select_idx           = @ select_idx;
f_upper_bounding       = @ upper_bounding;

f_adjust_head          = @ adjust_head;
f_adjust_flow          = @ adjust_flow;
f_u_fkt_l              = @ u_fkt_l;

% ---------------------------------------------------------------------------
% Initialization of factor alpha or beta, respectively
ahw = [];
if strcmp (head_loss_formula, 'HW')
  alpha = transpose( headcoefficient * ( k_value .^(-dischard))...
                                    .* (avail_diams.^(-D_EXP)) );
elseif strcmp (head_loss_formula, 'AHW')
  [dischard, ahw]...
   = adjust_hazen_williams (k_value, headcoefficient, ccw, cdw, gravitation,...
                             avail_diams, flow_min, flow_max);

  alpha = transpose( headcoefficient * (ahw .^(-dischard))...
                                        .* (avail_diams.^(-D_EXP)) );

elseif strcmp (head_loss_formula, 'DW') % Darcy-Weisbach: beta
  alpha = transpose( 8 .* cdw ./ (gravitation .* avail_diams.^5 .* pi^2) );
end

% ---------------------------------------------------------------------------
% Initialization of fixed parts of lower bounding problem
% Aeq, beq :      Aeq x = beq
% lb, ub :        lower resp. upper bound for the lower bounding problem
% lb_ub, ub_ub:   lower resp. upper bound for the upper bounding problem

if ~strcmp (head_loss_formula, 'DW')
  aggregate_relation =...
    [kron(eye(No_arcs),ones(1,No_dia)), repmat(0,No_arcs,No_seg),...
     -diag(arc_length), repmat(0,No_arcs,No_arcs+No_nodes)];
else
  aggregate_relation = [];
end
% first No_arcs rows of A will be initialized with function "adjust_lb"
Aeq = [repmat(0,No_arcs,2*No_seg+No_lambda+No_arcs+No_nodes);...
       kron(eye(No_arcs),ones(1,No_dia)), kron(eye(No_arcs),ones(1,No_dia)),...
         repmat(0,No_arcs,No_lambda+No_arcs+No_nodes);...
       aggregate_relation;...
       repmat(0,No_nodes-1,2*No_seg+No_lambda),node_edge(2:end,:),...
         repmat(0,No_nodes-1,No_nodes)];

beq = [(elevation(arcs(:,1)) - elevation(arcs(:,2)));...
       repmat(0,No_arcs-No_narcs,1); arc_length;...
       repmat(0,size(aggregate_relation,1),1); demand(2:end)];
```

```
if (No_narcs ~= No_arcs)
  % Restriction to columns: x is fixed in A/P
  Aeq(:, [seg_fixed; No_seg+seg_fixed]) = [];
  beq([arcs_fixed]) = [];
  if ~strcmp (head_loss_formula, 'DW')
    % Restriction to P instead of A
    beq([No_arcs+arcs_fixed; 2*No_arcs+arcs_fixed]) = [];
    Aeq([No_arcs+arcs_fixed; 2*No_arcs+arcs_fixed],:) = [];
  else
    beq([No_arcs+arcs_fixed])  = [];
    Aeq([No_arcs+arcs_fixed],:) = [];
  end
end

% Elements 2*No_nseg+No_lambda+1 : 2*No_nseg+No_lambda+No_arcs*No_nodes
% are initialized during branch and bound
ub_aux = kron(arc_length,ones(No_dia,1)); ub_aux(seg_fixed) = [];

lb = [repmat(0,2*No_nseg+No_lambda+No_arcs+No_nodes,1)];
ub = [ub_aux; ub_aux; repmat(1,No_lambda,1); repmat(0,No_arcs+No_nodes,1)];

% Lower and upper bounds for the upper bounding problem
lb_ub = [repmat(0,No_nseg,1); head_min(1:No_nodes) - elevation(1:No_nodes)];
ub_ub = [ub_aux;              head_max(1:No_nodes) - elevation(1:No_nodes)];


% ---------------------------------------------------------------------------
% Non-Tree arcs for MSTR
non_tree = []; [weights, idx] = sort(flow_min - flow_max); weights = [];

if strcmp (head_loss_formula, 'DW') && (No_arcs ~= No_narcs)
  narcs = (1:No_arcs)'; narcs(arcs_fixed) = []; idx = [arcs_fixed; narcs];
end

% Nodes with the same number belong to the same subtree
blue_tree = 1:No_nodes;

for i = 1:No_arcs
  % Descending order of arc weights (flow_max - flow_min) given by idx(i)
  if (blue_tree (arcs(idx(i),1)) == blue_tree (arcs(idx(i),2)))
    % edges are in the same subtree
    non_tree = [non_tree, idx(i)];
  else
    % connect subtrees
    blue_tree (blue_tree == blue_tree (arcs(idx(i),2)))...
      = blue_tree (arcs(idx(i),1));
  end
end

% POST-ORDER TREE TRAVERSAL: ordered indices are written to vector post_tree
% first delete (mark as zero) all non-tree arcs
tree = arcs;
tree(non_tree , :) = zeros( size(tree( non_tree ,:)) );
tree = [tree, [1; repmat(0,No_arcs-1,1)]]; % add an auxiliary column
```

```
% Initialization of stack and solution vector
stack = [1]; post_tree = [];

while ~isempty(stack)
  succ = find(tree(:,1) == tree(stack(1),1) | tree(:,2) == tree(stack(1),1)...
           | tree(:,1) == tree(stack(1),2) | tree(:,2) == tree(stack(1),2) );

  % Delete already worked out arcs from list of successors and mark new succ's
  succ ( tree(succ,3)==1 ) = []; tree(succ,3) = 1;

  if isempty(succ)
    post_tree = [post_tree , stack(1)];
    stack (1) = [];
  else
    stack = [succ; stack];
  end
end



% ------------------------------------------------------------------------
% SUBFUNCTIONS
% ------------------------------------------------------------------------
function [Aeq_aux, beq_aux]...
        = adjust_lbeq (head_loss_formula, k_value, dischard, ccw,...
                       avail_diams, alpha, arcs, arcs_fixed, length_fixed,...
                       seg_fixed, node_edge, elevation, flow_min, flow_max)
%
% ADJUST_EQUALITIES - Determination of the adjusted equations for the
%                     linear lower bounding problem
%
% Returns first No_arcs rows of matrix Aeq and (No_narc+1:No_arcs) element of
% vector beq with Aeq x = beq

global No_arcs No_dia No_seg No_lambda No_narcs No_nseg

if ~strcmp (head_loss_formula, 'DW')
  vmin         = sign(flow_min) .* (abs(flow_min).^dischard);
  vmax         = sign(flow_max) .* (abs(flow_max).^dischard);

  vmin_aux     = kron(diag(vmin), alpha);
  vmax_aux     = kron(diag(vmax), alpha);

  if (No_arcs ~= No_narcs)
    vminmax_aux  = diag(kron(diag(vmin-vmax), alpha) * length_fixed);
    vminmax_aux  = vminmax_aux(arcs_fixed,:);
  end
else
  wmin         = repmat(0, 1, No_seg);   wmax = repmat(0, 1, No_seg);

  wmin         = (kron(abs(flow_min) .* flow_min, ones(No_dia,1))...
                    .* u_fkt_l (abs(flow_min), avail_diams, k_value, ccw))';
  wmax         = (kron(abs(flow_max) .* flow_max, ones(No_dia,1))...
                    .* u_fkt_l (abs(flow_max), avail_diams, k_value, ccw))';
```

```
  vmin_aux    = kron(eye(No_arcs), alpha)...
                .* repmat(reshape(wmin, No_dia, No_arcs)', 1, No_arcs);
  vmax_aux    = kron(eye(No_arcs), alpha)...
                .* repmat(reshape(wmax, No_dia, No_arcs)', 1, No_arcs);
  if (No_arcs ~= No_narcs)
    vminmax_aux = kron(eye(No_arcs), alpha)...
                  .* repmat(reshape((wmin-wmax) .* length_fixed',...
                                    No_dia, No_arcs)', 1, No_arcs);
    vminmax_aux = vminmax_aux(arcs_fixed, length_fixed ~= 0);
  end
end

Aeq_aux = [vmin_aux, vmax_aux, repmat(0,No_arcs,No_lambda+No_arcs),...
           - node_edge'];
beq_aux = [];

if (No_arcs ~= No_narcs)
  Aeq_aux(arcs_fixed,:) = []; Aeq_aux(:,[seg_fixed; No_seg+seg_fixed]) = [];

  Aeq_aux = [Aeq_aux; repmat(0,No_arcs-No_narcs, 2*No_nseg), vminmax_aux,...
             repmat(0,No_arcs-No_narcs, No_arcs), - (node_edge(:,arcs_fixed))'];

  beq_aux = [(elevation(arcs(:,1)) - elevation(arcs(:,2)))...
             - vmax_aux*length_fixed];
  beq_aux = beq_aux(arcs_fixed);
end


% -----------------------------------------------------------------------
function [Ar,br] = relax (idx, q_idx, head_loss_formula, flow_min, flow_max,...
                          dischard, diam, k_value, ccw, verbose)
%
% RELAX - Determination of relaxation to the nonlinear constraint
%         for the doubles flow_min and flow_max
%
% Returns matrix Ar and vector br with Ar x <= br

global No_arcs No_nodes No_dia No_relax No_lambda No_nseg wds_error

if ~strcmp (head_loss_formula, 'DW')
  BEPS = 1e-10; % accuracy around lambda star for searching zero
  q_fkt        = @ q_fkt_hw;          q_dt          = @ q_dt_hw;
  rel_eq_bar   = @ rel_eq_bar_hw;     rel_eq_tilde  = @ rel_eq_tilde_hw;

  vmin = sign(flow_min) * (abs(flow_min)^dischard);
  vmax = sign(flow_max) * (abs(flow_max)^dischard);
else
  BEPS = 1e-5; % accuracy around lambda star for searching zero
  q_fkt        = @ p_fkt;             q_dt          = @ p_dx;
  rel_eq_bar   = @ rel_eq_bar_dw;     rel_eq_tilde  = @ rel_eq_tilde_dw;

  vmin = abs(flow_min).* flow_min .* u_fkt (abs(flow_min), diam, k_value, ccw);
  vmax = abs(flow_max).* flow_max .* u_fkt (abs(flow_max), diam, k_value, ccw);
end
```

```
if (flow_min >= 0)
  if (verbose == 2), fprintf('relax: "case (ii)"\n'); end

  if (No_relax == 6)
    aux_vec = [0; 0.25; 0.5; 0.8; 0.9];
  elseif (No_relax == 4)
    aux_vec = [0; 0.5; 0.9];
  end
  aux_vec_one = repmat (1,No_relax,1); aux_vec_one(1) = -1;

  aux_vec_qdt = repmat (0, No_relax, 1);
  aux_vec_qdt(1) = flow_min - flow_max;

  aux_vec_qdt(2:No_relax)...
    = - feval(q_dt, aux_vec,vmin,vmax,dischard,diam,k_value,ccw);

  br = [-flow_max;...
        feval(q_fkt, aux_vec,vmin,vmax,dischard,diam,k_value,ccw)...
        - aux_vec .* feval(q_dt, aux_vec,vmin,vmax,dischard,diam,k_value,ccw)];

elseif (flow_max <= 0)
  if (verbose == 2), fprintf('relax: "case (iii)"\n'); end

  if (No_relax == 6)
    aux_vec = [0.1; 0.2; 0.5; 0.75; 1];
  elseif (No_relax == 4)
    aux_vec = [0.1; 0.5; 1];
  end
  aux_vec_one = repmat(-1,No_relax,1); aux_vec_one(1) = 1;

  aux_vec_qdt = repmat(0,No_relax,1);
  aux_vec_qdt(1) = flow_max - flow_min;

  aux_vec_qdt(2:No_relax)...
    = feval(q_dt, aux_vec, vmin, vmax, dischard, diam, k_value, ccw);

  br = [flow_max;...
        aux_vec .* feval(q_dt, aux_vec,vmin,vmax,dischard,diam,k_value,ccw)...
        - feval(q_fkt, aux_vec,vmin,vmax,dischard,diam,k_value,ccw)];

else
  star = 1 / (1 - (vmin/vmax));

  if ~strcmp (head_loss_formula, 'DW')
    inf_star = star - BEPS;
    sup_star = star + BEPS;
  else
    r = (ccw(1) .* diam ./ (1 - k_value / ccw(2) ./ diam) ).^2;
    inf_star = (vmax - r) / (vmax - vmin);
    sup_star = (vmax + r) / (vmax - vmin);

    inf_star = min(star - BEPS, inf_star);
    sup_star = max(star + BEPS, sup_star);
  end
```

```
if feval(rel_eq_bar ,0, flow_min, vmin,vmax, 1/dischard,diam,k_value,ccw) > 0
 bar_exists = 0;
else
 try
   [bar, fval, found]...
     = fzero(rel_eq_bar , [0, inf_star],[],...
                      flow_min, vmin, vmax, 1/dischard, diam,k_value,ccw);
     if (found < 0), error('NOP - relaxations: bar not found!'); end
     bar_exists = 1;
 catch
   fprintf('NOP-relax: omitted due to error while determining bar\n');
   Ar = [repmat(0, No_relax, 2*No_nseg + No_arcs + No_lambda + No_nodes)];
   br = repmat (0, No_relax, 1);
   return;
 end
end
if feval(rel_eq_tilde ,1, flow_max,vmin,vmax, 1/dischard,diam,k_value,ccw) < 0
 tilde_exists = 0;
else
 try
   [tilde, fval, found]...
     = fzero(rel_eq_tilde , [sup_star, 1],[], flow_max, vmin, vmax,...
                                      1/dischard, diam, k_value, ccw);
   if (found < 0), error('NOP - relaxations: tilde not found!'); end
   tilde_exists = 1;
 catch
   fprintf('NOP-relax: omitted due to error while determining tilde\n');
   Ar = [repmat(0, No_relax, 2*No_nseg + No_arcs + No_lambda + No_nodes)];
   br = repmat (0, No_relax, 1);
   return;
 end
end

if (~bar_exists) && (~tilde_exists)
 feval(wds_error,'NOP-error constructing relaxations, q not convex-concave?');

elseif (bar_exists) && (tilde_exists)
 if (verbose == 2), fprintf('relax: "case (i) a" \n'); end

 if (No_relax == 6)
   aux_vec = [0; bar/2; bar; tilde; (tilde+1)/2; 1];
 elseif (No_relax == 4)
   aux_vec = [0; bar; tilde; 1];
 end
 aux_vec_one = repmat(1,No_relax,1);
 aux_vec_one(No_relax/2+1:No_relax) = -1;

 aux_vec_qdt = repmat(0,No_relax,1);
 aux_vec_qdt = -1 .* aux_vec_one...
             .* feval(q_dt, aux_vec,vmin,vmax,dischard,diam,k_value,ccw);

 br = [aux_vec_one .*...
      ( feval(q_fkt, aux_vec,vmin,vmax,dischard,diam,k_value,ccw)...
      -aux_vec.*feval(q_dt, aux_vec,vmin,vmax,dischard,diam,k_value,ccw))];
```

```
 if strcmp (head_loss_formula, 'DW')...
  && ((feval(q_fkt, sup_star,vmin,vmax,dischard,diam,k_value,ccw)...
        > feval(q_fkt, bar,vmin,vmax,dischard,diam,k_value,ccw)...
            + (sup_star - bar)...
                * feval(q_dt, bar,vmin,vmax,dischard,diam,k_value,ccw))...
      || (feval(q_fkt, inf_star,vmin,vmax,dischard,diam,k_value,ccw)...
            < feval(q_fkt, tilde,vmin,vmax,dischard,diam,k_value,ccw)...
                + (inf_star - tilde)...
                  * feval(q_dt, tilde,vmin,vmax,dischard,diam,k_value,ccw)))

     fprintf('relax: case (i) a, relaxation omitted for flow [%f, %f]\n',...
                                                   flow_min, flow_max);
    aux_vec_one  = repmat (0, No_relax, 1);
    aux_vec_qdt  = repmat (0, No_relax, 1);
    br           = repmat (0, No_relax, 1);
 end


elseif ~tilde_exists
 if (verbose == 2), fprintf('relax: "case (i) b" \n'); end

 if (No_relax == 6)
  aux_vec = [0; bar/4; bar/2; bar*3/4; bar];
 elseif (No_relax == 4)
  aux_vec = [0; bar/2; bar];
 end
 aux_vec_one = repmat(1,No_relax,1);
 aux_vec_one(No_relax) = -1;

 aux_vec_qdt = repmat(0,No_relax,1);
 aux_vec_qdt(1:No_relax-1)...
  = - feval(q_dt, aux_vec, vmin, vmax, dischard, diam, k_value, ccw);

 aux_vec_qdt(No_relax) = flow_min - flow_max;

 br = [feval(q_fkt, aux_vec,vmin,vmax,dischard,diam,k_value,ccw)...
      - aux_vec...
          .*feval(q_dt, aux_vec,vmin,vmax,dischard,diam,k_value,ccw);...
      - flow_max];

 if strcmp (head_loss_formula, 'DW')...
  && ((feval(q_fkt, inf_star,vmin,vmax,dischard,diam,k_value,ccw)...
        < (inf_star * flow_min + (1-inf_star) * flow_max))...
      || (feval(q_fkt, sup_star,vmin,vmax,dischard,diam,k_value,ccw)...
            > feval(q_fkt, bar,vmin,vmax,dischard,diam,k_value,ccw)...
                + (sup_star - bar)...
                    * feval(q_dt, bar,vmin,vmax,dischard,diam,k_value,ccw)))

     fprintf('relax: case (i) b, relaxation omitted for flow [%f, %f]\n',...
                                                   flow_min, flow_max);
    aux_vec_one  = repmat (0, No_relax, 1);
    aux_vec_qdt  = repmat (0, No_relax, 1);
    br           = repmat (0, No_relax, 1);
 end
```

```
  elseif ~bar_exists
    if (verbose == 2), fprintf('relax: "case (i) c" \n'); end

    if (No_relax == 6)
      aux_vec = [tilde; (tilde+1)/4; (tilde+1)/2; (tilde+1)*3/4; 1];
    elseif (No_relax == 4)
      aux_vec = [tilde; (tilde+1)/2; 1];
    end
    aux_vec_one = repmat(-1,No_relax,1); aux_vec_one(1) = 1;

    aux_vec_qdt = repmat(0,No_relax,1);
    aux_vec_qdt(1) = flow_max - flow_min;
    aux_vec_qdt(2:No_relax)...
      = feval(q_dt, aux_vec,vmin,vmax,dischard,diam,k_value,ccw);

    br = [flow_max;...
          aux_vec.* feval(q_dt, aux_vec,vmin,vmax,dischard,diam,k_value,ccw)...
          - feval(q_fkt, aux_vec,vmin,vmax,dischard,diam,k_value,ccw)];

    if strcmp (head_loss_formula, 'DW')...
      && ((feval(q_fkt, sup_star,vmin,vmax,dischard,diam,k_value,ccw)...
            > (sup_star * flow_min + (1-sup_star) * flow_max))...
          || (feval(q_fkt, inf_star,vmin,vmax,dischard,diam,k_value,ccw)...
               < feval(q_fkt, tilde,vmin,vmax,dischard,diam,k_value,ccw)...
                 + (inf_star - tilde)...
                   * feval(q_dt, tilde,vmin,vmax,dischard,diam,k_value,ccw)))

       fprintf('relax: case (i) c, relaxation omitted for flow [%f, %f]\n',...
                                                        flow_min, flow_max);
      aux_vec_one  = repmat (0, No_relax, 1);
      aux_vec_qdt  = repmat (0, No_relax, 1);
      br           = repmat (0, No_relax, 1);
    end
  end
end

Ar = [repmat(0, No_relax, 2*No_nseg + idx - 1), aux_vec_qdt,...
      repmat(0, No_relax, No_lambda - idx + q_idx - 1), aux_vec_one,...
      repmat(0, No_relax, No_arcs - q_idx + No_nodes)];

% ---------------------------------------------------------------------------
function [idx, d_q, max_delta]...
        = select_idx (length, flow, head, glb, head_loss_formula, k_value,...
                      branching, dischard, ccw, avail_diams, alpha,...
                      arcs, elevation, non_tree, flow_min, flow_max)
%
% SELECT_IDX - Determination of branch and bound index
% Returns delta (max of discrepant arcs) and its index as well as the
% value of the flow (d_q) for bisecting the flow interval

global No_arcs No_dia No_seg
MAX_DIFF = 100;  % bound for the maximum flow difference used to select
                 % the branching variable
ZMAX = 1e-4;     % in case of DW flow lower than ZMAX is assumed to be zero
```

```
if ~strcmp (head_loss_formula, 'DW')
  head_loss = sign(flow) .* (abs(flow).^dischard)...
              .* (kron(eye(No_arcs), alpha) * length);
else
  flow( abs(flow) < ZMAX ) = 0;
  head_loss = kron(eye(No_arcs), alpha)...
              * ( (kron(abs(flow).* flow, ones(No_dia,1))...
                .* u_fkt_l (abs(flow), avail_diams, k_value, ccw)) .* length);
end

delta = abs(( elevation(arcs(:,1)) - elevation(arcs(:,2)) +...
             head(arcs(:,1)) - head(arcs(:,2))...
            ) - head_loss );
max_delta = max(delta);


if (branching == 1)
  if glb
    dummy = min([flow_max(non_tree) - flow(non_tree);...
                 flow(non_tree) - flow_min(non_tree)]);
    idx = [non_tree( (flow_max(non_tree) - flow(non_tree)) == dummy ),...
           non_tree( (flow(non_tree) - flow_min(non_tree)) == dummy )];
    idx = idx(1);

    if (flow_min(idx) >= flow(idx)) || (flow_max(idx) <= flow(idx))
      d_q = (flow_min(idx)+flow_max(idx)) / 2;
    else
      d_q = flow(idx);
    end
  else
    idx = non_tree( (flow_max(non_tree) - flow_min(non_tree))...
                    == max(flow_max(non_tree) - flow_min(non_tree)) );
    idx = idx(1);

    if (flow_min(idx)*flow_max(idx) >= 0)
      d_q = (flow_min(idx)+flow_max(idx)) / 2;
    else
      d_q = 0;
    end
  end
else % Second branching-strategy
  idx = non_tree( delta(non_tree) == max(delta( non_tree )) );
  dummy = max(flow_max(non_tree) - flow_min(non_tree));
  if dummy >= MAX_DIFF
    idx = non_tree( (flow_max(non_tree)-flow_min( non_tree ) == dummy) );
  end
  idx = idx(1);

  if (flow_min(idx)*flow_max(idx) >= 0)
    d_q = (flow_min(idx)+flow_max(idx)) / 2;
  else
    d_q = 0;
  end
end
```

```
% ---------------------------------------------------------------------------
function [Aeq, beq]...
        = upper_bounding (flow, head_loss_formula, k_value, dischard, ccw,...
                          avail_diams, alpha, arcs, arc_length,...
                          arcs_fixed, length_fixed, seg_fixed, node_edge,...
                          elevation)
%
% UPPER_BOUNDING - Determination of parameters for the linear upper
%                  bounding problem (NOP for given flow)
%
% Returns matrix Aeq and vector beq with Aeq x = beq

global No_arcs No_nodes No_dia No_seg No_narcs No_nseg
ZMAX = 1e-10;    % in case of DW flow lower than ZMAX is assumed to be zero

if ~strcmp (head_loss_formula, 'DW')
  v_flow = kron(diag (sign(flow).* (abs(flow).^dischard) ), alpha);
else
  w_aux = repmat(0, 1, No_seg);
  flow( abs(flow) < ZMAX ) = 0;
  w_aux = kron(abs(flow).* flow, ones(No_dia,1))...
          .* u_fkt_l (abs(flow), avail_diams, k_value, ccw);
  v_flow = kron(eye(No_arcs),alpha)...
           .* repmat(reshape(w_aux, No_dia, No_arcs)', 1, No_arcs);
end
Aeq = [v_flow, - node_edge';...
      kron(eye(No_arcs),ones(1,No_dia)), repmat(0,No_arcs,No_nodes)];
beq = [(elevation(arcs(:,1)) - elevation(arcs(:,2))); arc_length];

if (No_narcs ~= No_arcs)
  Aeq(:, seg_fixed) = [];
  Aeq ([arcs_fixed; No_arcs+arcs_fixed],:) = [];
  beq ([arcs_fixed; No_arcs+arcs_fixed]) = [];

  Aeq_aux = [repmat(0,No_arcs, No_nseg), - node_edge'];
  beq_aux = [(elevation(arcs(:,1)) - elevation(arcs(:,2)))...
            - v_flow * length_fixed];
  Aeq = [Aeq; Aeq_aux(arcs_fixed,:)]; beq = [beq; beq_aux(arcs_fixed)];
end

% ---------------------------------------------------------------------------
function [head_min, head_max]...
       = adjust_head (head_min, head_max, flow_min, flow_max,...
                      head_loss_formula, k_value, dischard, ccw,...
                      avail_diams, alpha, post_tree, non_tree, arcs,...
                      arc_length, arcs_fixed, length_fixed, seg_fixed)
%
% ADJUST_HEAD - Determination of adjusted head bounds according to
%               constraint propagation using actual flow bounds
%
% Returns adjusted head bounds

global No_arcs No_dia No_seg
HEPS = 1e-2;    % absolute minimum of flow to calculate head-loss
```

```
head_diff_min = head_min(arcs(:,1)) - head_max(arcs(:,2));
head_diff_max = head_max(arcs(:,1)) - head_min(arcs(:,2));

new_arcs = 1:No_arcs; new_arcs(arcs_fixed) = [];

if ~strcmp (head_loss_formula, 'DW')
  seg_min = repmat(0,No_seg,1); seg_max = repmat(0,No_seg,1);
  for i = new_arcs
    if (flow_min(i) > 0),       seg_min(i*No_dia)       = arc_length(i);
    elseif (flow_min(i) < 0),   seg_min((i-1)*No_dia+1) = arc_length(i);
    end
    if (flow_max(i) > 0),       seg_max((i-1)*No_dia+1) = arc_length(i);
    elseif (flow_max(i) < 0),   seg_max(i*No_dia)       = arc_length(i);
    end
  end
  seg_min(seg_fixed) = length_fixed(seg_fixed);
  seg_max(seg_fixed) = length_fixed(seg_fixed);

  phi_min = sign(flow_min) .* abs(flow_min).^(dischard)...
            .* (kron(eye(No_arcs), alpha) * seg_min);
  phi_max = sign(flow_max) .* abs(flow_max).^(dischard)...
            .* (kron(eye(No_arcs), alpha) * seg_max);

else % Darcy-Weisbach
  dw_min  = repmat(0,No_arcs,1); dw_max = repmat(0,No_arcs,1);
  for i = new_arcs
    if (flow_min(i) > 0),  dw_min(i) = No_dia;
    else                   dw_min(i) = 1;
    end
    if (flow_max(i) > 0),  dw_max(i) = 1;
    else                   dw_max(i) = No_dia;
    end
  end
  for i = arcs_fixed'
    d_idx = find(length_fixed((i-1)*No_dia+1:i*No_dia) ~= 0);
    d_idx = mod(d_idx,No_dia); d_idx(d_idx==0) = No_dia;

    if (flow_min(i) > 0),  dw_min(i) = d_idx(end);
    else                   dw_min(i) = d_idx(1);
    end
    if (flow_max(i) > 0),  dw_max(i) = d_idx(1);
    else                   dw_max(i) = d_idx(end);
    end
  end
  diam_min = avail_diams(dw_min); diam_max = avail_diams(dw_max);

  flow_max ((flow_max < HEPS) & (flow_max > 0)) = HEPS;
  flow_min ((flow_min > -HEPS) & (flow_min < 0)) = -HEPS;

  phi_min = abs(flow_min) .* flow_min .* arc_length...
            .* u_fkt (abs(flow_min), diam_min, k_value, ccw).* alpha(dw_min)';
  phi_max = abs(flow_max) .* flow_max .* arc_length...
            .* u_fkt (abs(flow_max), diam_max, k_value, ccw).* alpha(dw_max)';
end
```

```
head_diff_min = max(head_diff_min, phi_min);
head_diff_max = min(head_diff_max, phi_max);

for i = [non_tree , post_tree]
  head_min(arcs(i,1)) = max ([head_min(arcs(i,1));...
                              head_min(arcs(i,2)) + head_diff_min(i)]);
  head_min(arcs(i,2)) = max ([head_min(arcs(i,2));...
                              head_min(arcs(i,1)) - head_diff_max(i)]);
  head_max(arcs(i,1)) = min ([head_max(arcs(i,1));...
                              head_max(arcs(i,2)) + head_diff_max(i)]);
  head_max(arcs(i,2)) = min ([head_max(arcs(i,2));...
                              head_max(arcs(i,1)) - head_diff_min(i)]);
end


% ---------------------------------------------------------------------------
function [flow_min, flow_max]...
        = adjust_flow (flow_min, flow_max, head_min, head_max,...
                       head_mstr, head_loss_formula, k_value, dischard,...
                       ccw, avail_diams, alpha, arcs, arc_length,...
                       arcs_fixed, length_fixed, seg_fixed)
%
% ADJUST_FLOW - Determination of adjusted flow bounds according to
%               constraint propagation using actual head bounds
%
% Returns adjusted flow bounds

global No_arcs No_dia No_seg
MEPS = 1e-6; % head loss lower than this value is regarded as zero
FEPS = 1e-2; % if possible, flow bounds below this value are set to zero

head_diff_min = head_min(arcs(:,1)) - head_max(arcs(:,2));
head_diff_max = head_max(arcs(:,1)) - head_min(arcs(:,2));

new_arcs = 1:No_arcs; new_arcs(arcs_fixed) = [];

if ~strcmp (head_loss_formula, 'DW')
  seg_min = repmat(0,No_seg,1); seg_max = repmat(0,No_seg,1);
  for i = new_arcs
   if (head_diff_min(i) > 0),     seg_min((i-1)*No_dia+1) = arc_length(i);
   elseif (head_diff_min(i) < 0), seg_min(i*No_dia)       = arc_length(i);
   end
   if (head_diff_max(i) > 0),     seg_max(i*No_dia)       = arc_length(i);
   elseif (head_diff_max(i) < 0), seg_max((i-1)*No_dia+1) = arc_length(i);
   end
  end
  seg_min(seg_fixed) = length_fixed(seg_fixed);
  seg_max(seg_fixed) = length_fixed(seg_fixed);

  flow_min_h = sign(head_diff_min) .* abs(head_diff_min).^(1/dischard)...
               .* (kron(eye(No_arcs), alpha) * seg_min).^(-1/dischard);
  flow_max_h = sign(head_diff_max) .* abs(head_diff_max).^(1/dischard)...
               .* (kron(eye(No_arcs), alpha) * seg_max).^(-1/dischard);
else
```

```
  dw_min = repmat(0, No_arcs,1); dw_max = repmat(0, No_arcs,1);

  for i = new_arcs
    if (head_diff_min(i) > 0),  dw_min(i) = 1;
    else                        dw_min(i) = No_dia;
    end
    if (head_diff_max(i) > 0),  dw_max(i) = No_dia;
    else                        dw_max(i) = 1;
    end
  end
  for i = arcs_fixed'
    d_idx = find(length_fixed((i-1)*No_dia+1:i*No_dia) ~= 0);
    d_idx = mod(d_idx,No_dia);  d_idx(d_idx==0) = No_dia;
    if (head_diff_min(i) > 0),  dw_min(i) = d_idx(1);
    else                        dw_min(i) = d_idx(end);
    end
    if (head_diff_max(i) > 0),  dw_max(i) = d_idx(end);
    else                        dw_max(i) = d_idx(1);
    end
  end
  diam_min = avail_diams(dw_min); diam_max = avail_diams(dw_max);

  min_idx       = (abs(head_diff_min) > MEPS);
  max_idx       = (abs(head_diff_max) > MEPS);

  head_diff_min = head_diff_min ./ alpha(dw_min)' ./arc_length;
  head_diff_max = head_diff_max ./ alpha(dw_max)' ./arc_length;
  head_diff_min = head_diff_min(min_idx);
  head_diff_max = head_diff_max(max_idx);

  flow_min_h = repmat(0, No_arcs,1);  flow_max_h = repmat(0, No_arcs,1);
  diam_min   = diam_min(min_idx);     diam_max   = diam_max(max_idx);

  flow_min_h(min_idx) = - 2 * sign(head_diff_min) .* sqrt(abs(head_diff_min))...
                      .* log10(ccw(1)*diam_min./ sqrt(abs(head_diff_min))...
                             + k_value / ccw(2) ./ diam_min );
  flow_max_h(max_idx) = - 2 * sign(head_diff_max) .* sqrt(abs(head_diff_max))...
                      .* log10(ccw(1)*diam_max./ sqrt(abs(head_diff_max))...
                             + k_value / ccw(2) ./ diam_max );

  % to avoide to deal in subsequent calls with tiny flow bounds
  flow_min_h ((flow_min_h < FEPS) & (flow_min_h > 0)) = 0;
  flow_max_h ((flow_max_h > -FEPS) & (flow_max_h < 0)) = 0;

end

if (head_mstr ~= 2)
  flow_min = max(flow_min, flow_min_h); flow_max = min(flow_max, flow_max_h);
else
  % workaround for floating point errors within two-loop expansion network
  idx           = (flow_min ~= flow_max);
  flow_min(idx) = max(flow_min(idx), flow_min_h(idx));
  flow_max(idx) = min(flow_max(idx), flow_max_h(idx));
end
```

```
% ---------------------------------------------------------------------------
function [dischard, ahw]...
         = adjust_hazen_williams (k_value, headcoefficient, ccw, cdw,...
                              gravitation, avail_diams, flow_min, flow_max);
%
% ADJUST_HAZEN_WILLIAMS - Adjust Hazen-Williams coefficients to conform
%                         with Darcy-Weisbach results
%
% Returns adjusted dischard and vector of Hazen-Williams coefficients (ahw)
% for all available diameters
global No_dia
NMAX    = 100;  % Number of data points within flow bounds
QMIN    = 1;    % Minimal flow to be regarded when determining AHW-coefficients
D_EXP   = 4.87;

ahw        = repmat(0,No_dia,1);

flow_max   = max(abs([flow_min;flow_max]));
flow       = linspace(QMIN , flow_max, NMAX)';

diams      = kron(avail_diams, ones(NMAX ,1));
flow       = repmat(flow, No_dia,1);

friction   = u_fkt (flow, diams, k_value, ccw);

% Determination of dischard, assume: u(q,d) = x * q^y * d^z
dischard = [];
for k = 1:No_dia
  for i = (k-1)*NMAX+1 : k*NMAX
    dischard = [dischard; 2 + log(friction(i) ./ friction(i+1:k*NMAX))...
                          ./ log(flow(i) ./ flow(i+1:k*NMAX))];
  end
end
dischard = median(dischard);

% Determination of AHW-coefficients
hazen_williams = (headcoefficient * gravitation * pi^2 ./ (8 * cdw)...
                 .* diams.^(5-D_EXP)...
                 .* flow.^(dischard-2) ./ friction).^(1/dischard);
for k = 1:No_dia
  ahw(k) = median(hazen_williams((k-1)*NMAX+1 : k*NMAX));
end



% ---------------------------------------------------------------------------
% LOCAL SUBFUNCTIONS FOR RELAXATIONS (HAZEN-WILLIAMS)
% ---------------------------------------------------------------------------
function [fktvalue] = q_fkt_hw (x, vmin, vmax, dischard, varargin)
%
% Q_FKT_HW - Flow: q(x)

y = x .* vmin + (1-x) .* vmax;
fktvalue = sign (y) .* (abs(y).^(1/dischard));
```

```
% ---------------------------------------------------------------------------
function [fktvalue] = q_dt_hw (x, vmin, vmax, dischard, varargin)
%
% Q_DT_HW - Flow derivative: q'(x)

fktvalue = ((vmin - vmax) ./ dischard)...
           .* (abs(x.*vmin + (1-x).*vmax)).^(1/dischard - 1);


% ---------------------------------------------------------------------------
function [fktvalue] = rel_eq_bar_hw (x,flow_min,vmin,vmax,inv_dischard,varargin)
%
% REL_EQ_BAR_HW - Flow equation (overline)

y = x * vmin + (1-x) * vmax;
fktvalue = flow_min - sign (y) * abs(y)^(inv_dischard)...
          - (1-x)*((vmin - vmax) * inv_dischard) * (abs(y))^(inv_dischard-1);


% ---------------------------------------------------------------------------
function [fktvalue] = rel_eq_tilde_hw (x, flow_max, vmin, vmax,...
                                       inv_dischard, varargin)
% REL_EQ_TILDE_HW - Flow equation (tilde)

y = x * vmin + (1-x) * vmax;
fktvalue = flow_max - sign (y) * abs(y)^(inv_dischard)...
          + x * ((vmin - vmax) * inv_dischard) * (abs(y))^(inv_dischard - 1);



% ---------------------------------------------------------------------------
% LOCAL SUBFUNCTIONS FOR DARCY-WEISBACH
% ---------------------------------------------------------------------------
function [fkt_value] = psi_fkt (x, flow, diam, k_value, ccw)
%
% PSI_FKT - Friction Factor (Colebrook-White), f is zero of psi_fkt
% called only by u_fkt and u_fkt_1, hence flow ~= 0 and x > 0 can be assumed

fkt_value = 1 ./ (sqrt(x)) + ...
            2 * log10( ccw(1) .* diam ./ (abs(flow) .* sqrt(x))...
                       + k_value ./(ccw(2) .* diam) );


% ---------------------------------------------------------------------------
function [friction] = u_fkt (flow, diam, k_value, ccw)
% U_FKT
global wds_error

FMIN = 1e-4; FMAX = 1e4; % starting values for bisection
UEPS = 1e-10;            % points lower than UEPS are assumed to be zero

friction   = repmat(0, size(flow));
f_max      = max(size(friction));
d_max      = max(size(diam));
```

```
if (d_max == 1)
  diam = repmat(diam, f_max);
elseif (f_max ~= d_max)
feval(wds_error,'NOP - u_fkt: dimensions of flow and diameter do not coincide');
end

for i = 1 : f_max
  if (flow(i) < 0)
    feval(wds_error, 'NOP - u_fkt: not defined for negative flow');

  elseif (flow(i) > UEPS)
    [friction(i), fval, exitflag]...
      = fzero (@ psi_fkt ,[FMIN FMAX],[], flow(i), diam(i), k_value, ccw);

    if (exitflag ~= 1)
      feval(wds_error, 'NOP - u_fkt: no zero of colebrook-white');
    end
  else
    friction(i) = 0; % not defined but used only in connection of w_fkt
  end
end


% -----------------------------------------------------------------------------
function [friction] = u_fkt_l (flow, diam, k_value, ccw)
%
% U_FKT_L - The long version of u_fkt, where every flow-diameter combination
%              is determined
global wds_error

FMIN = 1e-4; FMAX = 1e4; % starting values for bisection
UEPS = 1e-10;              % points lower than UEPS are assumed to be zero

f_max      = max(size(flow));
d_max      = max(size(diam));
friction   = repmat(0, f_max * d_max, 1);

for i = 1:f_max
  for k = 1:d_max
    if (flow(i) < 0)
      feval(wds_error, 'NOP - u_fkt_l: not defined for negative flow');

    elseif (flow(i) > UEPS)
      [friction((i-1)*d_max + k), fval, exitflag]...
        = fzero (@ psi_fkt ,[FMIN FMAX],[], flow(i), diam(k), k_value, ccw);

      if (exitflag ~= 1)
        feval(wds_error, 'NOP - u_fkt_l: no zero of colebrook-white');
      end
    else
      friction((i-1)*d_max + k) = 0; % not defined but used only for w_fkt
    end
  end
end
```

```
% ---------------------------------------------------------------------------
function [flow] = p_fkt (x, wmin, wmax, dummy, diam, k_value, ccw)
%
% P_FKT - Function for relaxation of fixed arcs for Darcy-Weisbach: p(lambda)

y = x .* wmin + (1-x) .* wmax;
r = (ccw(1) * diam / (1 - k_value / ccw(2) / diam) )^2;

flow = repmat(0, size(y));
idx = ((y <= -r) | (r <= y));

flow(idx) = -2 * sign(y(idx)) .* sqrt(abs(y(idx)))...
          .* log10(ccw(1) * diam./ sqrt(abs(y(idx))) + k_value /ccw(2) /diam);


% ---------------------------------------------------------------------------
function [flow] = p_dx (x, wmin, wmax, dummy, diam, k_value, ccw)
%
% P_DX - Function for relaxation of fixed arcs for Darcy-Weisbach: p'(lambda)
global wds_error

y = x .* wmin + (1-x) .* wmax;
r = (ccw(1) * diam / (1 - k_value / ccw(2) / diam) )^2;

if all(abs(y) > r)
  z = ccw(1) .* diam ./ sqrt(abs(y)) + k_value / (ccw(2) * diam);
  flow = (log10(z) ./ sqrt(abs(y)) - ccw(1) * diam ./abs(y) ./ z ./ log(10))...
       .* (wmax - wmin);
else
  feval(wds_error, 'NOP - p_dx: not defined for intervals close to 0');
end


% ---------------------------------------------------------------------------
function [fktvalue] = rel_eq_bar_dw (x, flow_min, wmin, wmax, dummy,...
                                     diam, k_value, ccw)
% REL_EQ_BAR_DW - Flow equation (overline)

fktvalue = flow_min - p_fkt (x, wmin, wmax, [], diam, k_value, ccw)...
           - (1-x) .* p_dx (x, wmin, wmax, [], diam, k_value, ccw);


% ---------------------------------------------------------------------------
function [fktvalue] = rel_eq_tilde_dw (x, flow_max, wmin, wmax, dummy,...
                                       diam, k_value, ccw)
% REL_EQ_TILDE_DW - Flow equation (tilde)

fktvalue = flow_max - p_fkt (x, wmin, wmax, [], diam, k_value, ccw)...
           + x * p_dx (x, wmin, wmax, [], diam, k_value, ccw);
```

B.4 Extracts of "bb.m" containing the Branch and Bound Algorithm

```
function [opt_vec, global_lb, global_ub, infeasible, no_subs, no_subs_left]...
        = bb (head_loss_formula, k_value, lp_solver, accuracy, branching,...
              max_iter, dischard, ccw, avail_diams, costs, head_cost,...
              alpha, arcs, arc_length, arcs_fixed, length_fixed,...
              seg_fixed, node_edge, demand, elevation, flow_min, flow_max,...
              head_min, head_max, non_tree, post_tree, adjust_lbeq, relax,...
              select_idx, upper_bounding, adjust_flow, adjust_head,...
              Aeq, beq, lb, ub, lb_ub, ub_ub, verbose);
%
% BB - Branch and bound algorithm of water design network optimization
% [...]
global No_arcs No_nodes No_dia No_seg No_relax No_lambda No_narcs No_nseg...
       wds_error
% Constants
MAX_SUBS   = 1000;  % preallocation size of list of subproblems
DELTA_ACC  = 1e-6;  % maximal deviation of non-linear constraints for
                    % accepting a solution as feasible
BS_P       = 0.9;   % selection percentage used for branching strategy 1 and 2

% Initialization of B&B variables
infeasible  = 1;             no_subs     = 1;
global_lb   = inf;           global_ub   = inf;
opt_vec     = repmat(0, No_seg+No_arcs+No_nodes, 1);
nseg = 1:No_seg; nseg(seg_fixed) = []; seg_fixed_idx = find(length_fixed ~= 0);

status = 1; exitflag = 1;
if strcmp (lp_solver , 'linprog')
  warning('off','MATLAB:divideByZero'); linprog_options = optimset('linprog');
  linprog_options = optimset(linprog_options, 'Display','off');
end
lb_costs = [costs; costs; zeros(No_lambda+No_arcs,1);...
            head_cost; zeros(No_nodes-1,1)];
ub_costs = [costs; head_cost; zeros(No_nodes-1,1)];

if ((branching == 1) || (branching == 2)), head_mstr = 0;
elseif (branching == 3),                   head_mstr = 1; branching = 1;
elseif (branching == 4),                   head_mstr = 1; branching = 2;
elseif (branching == 5),                   head_mstr = 2; branching = 1;
elseif (branching == 6),                   head_mstr = 2; branching = 2;
end
% Definition and preallocation of data structure for subproblems
subproblem = repmat( struct('d_idx', 0,...
                            'd_q', 0,...
                            'qmin', repmat(0, No_arcs, 1),...
                            'qmax', repmat(0, No_arcs, 1),...
                            'hmin', repmat(0, No_nodes, 1),...
                            'hmax', repmat(0, No_nodes, 1),...
                            'lb', -inf),...
                    MAX_SUBS, 1);
```

```
subproblem(1).qmin = flow_min; subproblem(1).qmax = flow_max;
subproblem(1).hmin = head_min; subproblem(1).hmax = head_max;
sub2 = 1; sub1=[]; % Data for first loop

for bb_idx = 1: max_iter  % BRANCH & BOUND LOOP
  for s_idx = [sub2, sub1]
    if verbose
      fprintf('-----------------------------------------------------\n');
      fprintf('Global upper bound: \t \t \t %8.4f\n', global_ub);
      fprintf('Subproblem's old lower bound:\t\t %8.4f\n',subproblem(s_idx).lb);
    end
    % Adjustment of flow bounds according to MSTR
    for i = post_tree
      subproblem(s_idx).qmax(i) = min([subproblem(s_idx).qmax(i);...
              - demand(arcs(i,2))...
              - sum(subproblem(s_idx).qmin ( (arcs(:,2) == arcs(i,2))...
                                    & (arcs(:,1) ~= arcs(i,1)) ))...
              + sum(subproblem(s_idx).qmax ( arcs(:,1) == arcs(i,2) ));...
              + demand(arcs(i,1))...
              + sum(subproblem(s_idx).qmax ( arcs(:,2) == arcs(i,1) ))...
              - sum(subproblem(s_idx).qmin ( (arcs(:,1) == arcs(i,1))...
                                    & (arcs(:,2) ~= arcs(i,2)) ))]);
      subproblem(s_idx).qmin(i) = max([subproblem(s_idx).qmin(i);...
              - demand(arcs(i,2))...
              - sum(subproblem(s_idx).qmax ( (arcs(:,2) == arcs(i,2))...
                                    & (arcs(:,1) ~= arcs(i,1)) ))...
              + sum(subproblem(s_idx).qmin ( arcs(:,1) == arcs(i,2) ));...
              + demand(arcs(i,1))...
              + sum(subproblem(s_idx).qmin ( arcs(:,2) == arcs(i,1) ))...
              - sum(subproblem(s_idx).qmax ( (arcs(:,1) == arcs(i,1))...
                                    & (arcs(:,2) ~= arcs(i,2)) ))]);
    end
    if any(subproblem(s_idx).qmin > subproblem(s_idx).qmax)
      if verbose
        fprintf('-----------------------------------------------------\n');
        fprintf('Subproblem pruned by flow MSTR \n');
      end
      subproblem(s_idx) = []; continue;
    end
    % Adjustment of flow bounds according to head constraint propagation
    if head_mstr
      [subproblem(s_idx).hmin, subproblem(s_idx).hmax]...
        = feval( adjust_head , subproblem(s_idx).hmin,subproblem(s_idx).hmax,...
                      subproblem(s_idx).qmin,subproblem(s_idx).qmax,...
                      head_loss_formula, k_value, dischard, ccw,...
                      avail_diams, alpha, post_tree, non_tree, arcs,...
                      arc_length, arcs_fixed, length_fixed, seg_fixed);
      if any(subproblem(s_idx).hmin > subproblem(s_idx).hmax)
        if verbose
          fprintf('-----------------------------------------------------\n');
          fprintf('Subproblem pruned by negative length of head interval\n');
        end
        subproblem(s_idx) = []; continue;
      end
```

```
  [subproblem(s_idx).qmin, subproblem(s_idx).qmax]...
    = feval( adjust_flow , subproblem(s_idx).qmin, subproblem(s_idx).qmax,...
                           subproblem(s_idx).hmin, subproblem(s_idx).hmax,...
                           head_mstr, head_loss_formula, k_value,...
                           dischard, ccw, avail_diams, alpha, arcs,...
                           arc_length, arcs_fixed, length_fixed, seg_fixed);
   if any(subproblem(s_idx).qmin > subproblem(s_idx).qmax)
     if verbose
       fprintf('-----------------------------------------------------\n');
       fprintf('Subproblem pruned by negative length of flow interval\n');
     end
     subproblem(s_idx) = []; continue;
   end
 end % if head_mstr

 % Computation of adjusted equalities and relaxations for subproblems
 [Aeq(1:No_arcs,:), beq(No_narcs+1:No_arcs)]...
   = feval( adjust_lbeq , head_loss_formula, k_value, dischard, ccw,...
                          avail_diams, alpha, arcs, arcs_fixed,...
                          length_fixed, seg_fixed, node_edge, elevation,...
                          subproblem(s_idx).qmin, subproblem(s_idx).qmax);
 if (No_relax ~= 0)
   A = repmat(0, No_relax*No_lambda, 2*No_nseg+No_lambda+No_arcs+No_nodes);
   b = repmat(0, No_relax*No_lambda, 1);
   if ~strcmp (head_loss_formula, 'DW')
     for i = 1:No_arcs
       if (subproblem(s_idx).qmin(i) < subproblem(s_idx).qmax(i))
         [A((i-1)*No_relax+1:i*No_relax,:),b((i-1)*No_relax+1:i*No_relax)]...
           = feval( relax , i, i, head_loss_formula,...
                    subproblem(s_idx).qmin(i), subproblem(s_idx).qmax(i),...
                    dischard, [],[],[], verbose);
       end
     end
   else
     for i = 1:No_lambda
       j  = ceil(seg_fixed_idx(i) / No_dia);  % index for arcs_fixed
       k  = mod(seg_fixed_idx(i), No_dia);    % index for diameter
       if (k==0), k = No_dia; end

       if (subproblem(s_idx).qmin(j) < subproblem(s_idx).qmax(j))
         [A((i-1)*No_relax+1:i*No_relax,:),b((i-1)*No_relax+1:i*No_relax)]...
           = feval(relax , i, j, head_loss_formula,...
                   subproblem(s_idx).qmin(j), subproblem(s_idx).qmax(j),...
                   dischard, avail_diams(k), k_value, ccw, verbose);
       end
     end
   end
 else A = []; b = []; % No_relax = 0
 end
 % Adjustment of lower and upper bounds
 lb(2*No_nseg+No_lambda+1 : 2*No_nseg+No_lambda+No_arcs+No_nodes)...
   = [subproblem(s_idx).qmin; subproblem(s_idx).hmin-elevation];
 ub(2*No_nseg+No_lambda+1 : 2*No_nseg+No_lambda+No_arcs+No_nodes)...
   = [subproblem(s_idx).qmax; subproblem(s_idx).hmax-elevation];
```

```
lb_old = subproblem(s_idx).lb;% needed for selection of index to be branched
if strcmp (lp_solver , 'cplex')
  [subproblem(s_idx).lb, x_sub, lambda, status]...
    = lp_cplex_mex (lb_costs, [A; Aeq], [b; beq],...
                     lb, ub, 1, transpose(1:(size(A,1))));
else
  [x_sub, subproblem(s_idx).lb, exitflag, infos]...
    = linprog(lb_costs, A, b, Aeq, beq, lb, ub,[], linprog_options );
end
% Prune by infeasibility: admissible region of subproblem is empty
if (exitflag <= 0) || (status ~= 1)
  if verbose, fprintf('B&B info: pruned by infeasibility of LB!\n'); end
  subproblem(s_idx) = []; continue;
end
if verbose
  fprintf('Subproblem's new lower bound:\t\t %8.4f\n',subproblem(s_idx).lb);
  fprintf('---------------------------------------------------\n');
end

% Find index to be branched
if (subproblem(s_idx).lb * BS_P > lb_old), glb = 1; else glb = 0; end
x_sub = [x_sub(1:No_nseg)+x_sub(No_nseg+1:2*No_nseg);...
         x_sub(2*No_nseg+No_lambda+1:end)];

x_aux = x_sub(1:No_nseg);
x_sub = [length_fixed; x_sub(No_nseg+1:end)]; x_sub(nseg) = x_aux;

[subproblem(s_idx).d_idx, subproblem(s_idx).d_q, delta]...
   = feval(select_idx , x_sub(1:No_seg), x_sub(No_seg+1: No_seg+No_arcs),...
                       x_sub(No_seg+No_arcs+1: No_seg+No_arcs+No_nodes),...
                       glb, head_loss_formula, k_value, branching,...
                       dischard, ccw, avail_diams, alpha, arcs,...
                       elevation, non_tree,...
                       subproblem(s_idx).qmin, subproblem(s_idx).qmax);
if (delta < DELTA_ACC)
  % Accept solution as feasible
  if verbose, fprintf('B&B info: "DELTA < %f !"\n', DELTA_ACC ); end
  if subproblem(s_idx).lb < global_ub
    global_ub = subproblem(s_idx).lb;
    opt_vec = x_sub;
    infeasible = 0;
  end
  global_lb = min(global_lb , subproblem(s_idx).lb);
  subproblem(s_idx) = []; continue;
end
Aeq_ub = repmat(0, No_narcs+No_arcs, No_nseg+No_nodes);
beq_ub = repmat(0, No_narcs+No_arcs, 1);
% Computation of upper bound
flow_sub = x_sub(No_seg+1:No_seg+No_arcs);
[Aeq_ub, beq_ub]...
  = feval( upper_bounding , flow_sub, head_loss_formula, k_value,...
                           dischard, ccw, avail_diams, alpha, arcs,...
                           arc_length, arcs_fixed, length_fixed,...
                           seg_fixed, node_edge, elevation);
```

```
    if strcmp (lp_solver , 'cplex')
      [fval_ub, x_sub, lambda, status]...
        = lp_cplex_mex (ub_costs, Aeq_ub, beq_ub, lb_ub, ub_ub);
    else
      [x_sub, fval_ub, exitflag, infos]...
        = linprog(ub_costs, [],[], Aeq_ub, beq_ub,...
                                            lb_ub, ub_ub, [], linprog_options);
    end
    if (exitflag <= 0) || (status ~= 1)
      if verbose
        fprintf('B&B info: no solution of upper bounding problem!\n');
      end
      continue;
    end
    if (fval_ub < global_ub)
      opt_vec         = [length_fixed; flow_sub; x_sub(No_nseg+1:end)];
      opt_vec(nseg)   = x_sub(1:No_nseg);
      infeasible      = 0;
      global_ub       = fval_ub;
    end
  end % SUBPROBLEM'S-FOR-LOOP


  % Clean list according to obsolete subproblems
  del_idx         = ([subproblem.lb] >= global_ub * (1 - accuracy));
  global_lb       = min ([global_lb , [subproblem(del_idx).lb]]);
  no_subs_left    = sum(del_idx);
  subproblem(del_idx) = [];

  % Break if work for all subproblems is done
  if isempty(subproblem) || (sum([subproblem.d_idx]~=0) == 0)
    if verbose, fprintf('B&B info: "No subproblems left!"\n'); end
    return;
  end
  % Node Selection Step - sub_idx = subproblem to be worked out
  sub_idx = ( [subproblem.lb] ~= -inf );
  sub_idx = find( [subproblem.lb] == min( [subproblem(sub_idx).lb] ));
  sub_idx = sub_idx(end);

  % Partitioning Step - branch subproblem
  no_subs             = no_subs+2;
  sub1                = (sum([subproblem.d_idx]~=0)) + 1;
  sub2                = sub1 + 1;
  subproblem(sub1)    = subproblem(sub_idx);
  subproblem(sub2)    = subproblem(sub_idx);
  d_idx               = subproblem(sub_idx).d_idx;

  subproblem(sub1).qmax(d_idx) = subproblem(sub_idx).d_q;
  subproblem(sub2).qmin(d_idx) = subproblem(sub_idx).d_q;

  subproblem(sub_idx) = []; sub1 = sub1-1; sub2 = sub2-1;
  if verbose, fprintf('Actual number of subproblems:\t %d\n', sub2); end
end % B&B-FOR-LOOP


error_str = sprintf(['BB: maximum number of iterations reached ! \n',...
                    'max_iter = %d\n'], max_iter);
feval(wds_error, error_str);
```

B.5 Extracts of "vnop.m" containing Verification of Central Functions

```
function [dischard, ccw, alpha, seg_fixed, f_adjust_lbeq, f_relax,...
        f_adjust_head, f_adjust_flow, f_test_adjacency,...
        f_verify_lb, f_verify_ub, f_verify_infeasibility, Aeq, beq, lb, ub]...
          = vnop (head_loss_formula, k_value, viscosity, dischard,...
                  avail_diams, ahw, arcs, arc_length, arcs_fixed,...
                  seg_fixed, node_edge, unit, demand, elevation)
%
% VNOP - Initialization of variables and functions needed for the verified
%        version of the branch and bound algorithm
% [...]
global No_arcs No_nodes No_dia No_seg No_lambda No_narcs No_nseg

% Determination of interval variables and constants
ZERO     = intval(0);
PI       = acos(intval(-1));
D_EXP    = intval('4.87'); % Hazen-Williams exponent for diameters

% Determination of additionally needed hydraulic constants
if strcmp (head_loss_formula, 'HW')
  if strcmp (unit , 'm')
    dischard        = intval('1.852');
    headcoefficient = intval(15200);
  elseif strcmp (unit , 'inch')
    dischard        = intval('1.85');
    headcoefficient = intval(851500);
  end
  ccw = [];
else % AHW or DW
  if strcmp (unit , 'm')
    cdw                = 1e6/intval(1296);
    gravitation        = intval('9.81');
    ccw(1)             = intval('22.59') * PI * viscosity;
  elseif strcmp (unit , 'inch')
    cdw                = intval(248832);
    gravitation        = intval('32.2');
    ccw(1)             = (intval('2.51') / 48) * PI * viscosity;
  end
  ccw(2)               = intval('3.71');
  if strcmp (head_loss_formula, 'AHW')
    dischard = intval(dischard);
    if strcmp (unit , 'm')
      headcoefficient = intval(15200);
    elseif strcmp (unit , 'inch')
      headcoefficient = intval(851500);
    end
  else % DW
    dischard           = intval(2);
  end
end
```

```
% ---------------------------------------------------------------------------
% Initialization of factor alpha or beta, respectively
if strcmp (head_loss_formula, 'HW')
  alpha = transpose( headcoefficient * (k_value.^(-dischard))...
                                      .* (avail_diams.^(-D_EXP)) );
elseif strcmp (head_loss_formula, 'AHW')
  ahw   = intval(ahw);
  alpha = transpose( headcoefficient * (ahw.^(-dischard))...
                                      .* (avail_diams.^(-D_EXP)) );
else % Darcy-Weisbach: beta
  alpha = transpose( 8.* cdw./ (gravitation * avail_diams.^intval(5).* PI^2) );
end


% ---------------------------------------------------------------------------
% Initialization of fixed parts of lower bounding problem
% Aeq, beq :      Aeq x = beq
% lb, ub :        lower resp. upper bound for the lower bounding problem

if ~strcmp (head_loss_formula, 'DW')
  aggregate_relation =...
    [intval([kron(eye(No_arcs), ones(1,No_dia))]),...
       repmat(ZERO, No_arcs,No_seg), -diag(arc_length),...
       repmat(ZERO, No_arcs,No_arcs+No_nodes)];
else
  aggregate_relation = [];
end

% first No_arcs rows of A will be initialized with function "adjust_lb"
Aeq = [repmat(ZERO, No_arcs,2*No_seg+No_lambda+No_arcs+No_nodes);...
       intval([kron(eye(No_arcs), ones(1,No_dia))]),...
       intval([kron(eye(No_arcs), ones(1,No_dia))]),...
         repmat(ZERO,No_arcs,No_lambda+No_arcs+No_nodes);...
       aggregate_relation;...
       repmat(ZERO,No_nodes-1,2*No_seg+No_lambda),...
         intval(node_edge(2:end,:)), repmat(ZERO,No_nodes-1,No_nodes)];

beq = [(elevation(arcs(:,1)) - elevation(arcs(:,2)));...
       repmat(ZERO,No_arcs-No_narcs,1); arc_length;...
       repmat(ZERO,size(aggregate_relation,1),1); demand(2:end)];

if (No_narcs ~= No_arcs)
  % Restriction to columns: x is fixed in A/P
  Aeq(:, [seg_fixed; No_seg+seg_fixed]) = [];
  beq([arcs_fixed]) = [];
  if ~strcmp (head_loss_formula, 'DW')
    % Restriction to P instead of A
    Aeq([No_arcs+arcs_fixed; 2*No_arcs+arcs_fixed],:) = [];
    beq([No_arcs+arcs_fixed; 2*No_arcs+arcs_fixed]) = [];
  else
    Aeq([No_arcs+arcs_fixed],:) = [];
    beq([No_arcs+arcs_fixed]) = [];
  end
end
```

```
% Lower and upper bounds
ub_aux = kron(arc_length,ones(No_dia,1)); ub_aux(seg_fixed) = [];

% elements 2*No_nseg+No_lambda+1 : 2*No_nseg+No_lambda+No_arcs*No_nodes
% will be initialized during branch and bound
lb = [repmat(0,2*No_nseg+No_lambda+No_arcs+No_nodes,1)];

ub = [repmat(sup(ub_aux),2,1); repmat(1,No_lambda,1);...
      repmat(0,No_arcs+No_nodes,1)];

% ---------------------------------------------------------------------------
% Function handles needed during branch & bound

f_adjust_lbeq            = @ adjust_lbeq;
f_relax                  = @ relax;

f_adjust_head            = @ adjust_head;
f_adjust_flow            = @ adjust_flow;

f_test_adjacency         = @ test_adjacency;
f_verify_lb              = @ verify_lb;
f_verify_ub              = @ verify_ub;
f_verify_infeasibility = @ verify_infeasibility;


% ---------------------------------------------------------------------------
% SUBFUNCTIONS
% ---------------------------------------------------------------------------
function [Aeq_aux, beq_aux]...
        = adjust_lbeq (head_loss_formula, k_value, dischard, ccw,...
                       avail_diams, alpha, arcs, arcs_fixed, length_fixed,...
                       seg_fixed, node_edge, elevation, flow_min, flow_max)
%
% ADJUST_EQUALITIES - Determination of the adjusted equations for the
%                     linear lower bounding problem
%
% Returns first No_arcs rows of matrix Aeq and (No_narc+1:No_arcs) element of
% vector beq with Aeq x = beq

global No_arcs No_dia No_seg No_lambda No_narcs No_nseg
ZERO      = intval(0);

if ~strcmp (head_loss_formula, 'DW')
  vmin          = inf_( sgn_pow (flow_min, dischard));
  vmax          = sup( sgn_pow (flow_max, dischard));
  vmin_aux      = kron(diag(vmin), alpha);
  vmax_aux      = kron(diag(vmax), alpha);
  if (No_arcs ~= No_narcs)
    vminmax_aux   = diag(kron(diag(vmin-vmax),alpha) * length_fixed);
    vminmax_aux   = vminmax_aux(arcs_fixed,:);
  end
else
  wmin          = (w_fkt_l (intval(flow_min), avail_diams, k_value, ccw))';
  wmax          = (w_fkt_l (intval(flow_max), avail_diams, k_value, ccw))';
```

```
   vmin_aux      = kron(eye(No_arcs), alpha)...
                     .* repmat(reshape(wmin, No_dia, No_arcs)', 1, No_arcs);
   vmax_aux      = kron(eye(No_arcs), alpha)...
                     .* repmat(reshape(wmax, No_dia, No_arcs)', 1, No_arcs);
   if (No_arcs ~= No_narcs)
     aux           = (wmin-wmax) .* length_fixed';
     vminmax_aux   = kron(eye(No_arcs), alpha)...
                       .* repmat(reshape(aux, No_dia, No_arcs)', 1, No_arcs);
     vminmax_aux   = vminmax_aux(arcs_fixed, length_fixed ~= 0);
   end
end
Aeq_aux = [vmin_aux, vmax_aux, repmat(ZERO,No_arcs,No_lambda+No_arcs),...
           - node_edge'];
beq_aux = [];

if (No_arcs ~= No_narcs)
  Aeq_aux(arcs_fixed,:) = []; Aeq_aux(:,[seg_fixed; No_seg+seg_fixed]) = [];

  Aeq_aux = [Aeq_aux; repmat(ZERO,No_arcs-No_narcs, 2*No_nseg), vminmax_aux,...
             repmat(ZERO,No_arcs-No_narcs, No_arcs),...
             (-node_edge(:,arcs_fixed))'];

  beq_aux = [(elevation(arcs(:,1)) - elevation(arcs(:,2)))...
             - vmax_aux*length_fixed];
  beq_aux = beq_aux(arcs_fixed);
end

% ---------------------------------------------------------------------------
function [Ar,br] = relax (idx, q_idx, head_loss_formula, flow_min, flow_max,...
                          dischard, diam, k_value, ccw, verbose)
%
% RELAX - Determination of relaxation to the nonlinear constraint
%         for the doubles flow_min and flow_max,
%         dischard is assumed to be of type intval
%
% Returns matrix Ar and vector br with Ar x <= br

global No_arcs No_nodes No_dia No_relax No_lambda No_nseg wds_error

REPS = 1e-2; % accuracy used for bisection

if ~strcmp (head_loss_formula, 'DW')
  q_fkt          = @ q_fkt_hw;          q_dt          = @ q_dt_hw;
  rel_eq_bar     = @ rel_eq_bar_hw;     rel_eq_tilde  = @ rel_eq_tilde_hw;

  vmin = sgn_pow (flow_min, dischard);
  vmax = sgn_pow (flow_max, dischard);
else
  q_fkt          = @ p_fkt;             q_dt          = @ p_dx;
  rel_eq_bar     = @ rel_eq_bar_dw;     rel_eq_tilde  = @ rel_eq_tilde_dw;

  vmin = w_fkt (intval(flow_min), diam, k_value, ccw);
  vmax = w_fkt (intval(flow_max), diam, k_value, ccw);
end
```

```
if (flow_min >= 0)
  if (verbose == 2), fprintf('relax: "case (ii)"\n'); end
  if (No_relax == 6)
    aux_vec = [0; 0.25; 0.5; 0.8; 0.9];
  elseif (No_relax == 4)
    aux_vec = [0; 0.5; 0.9];
  end
  aux_vec_one = repmat (1,No_relax,1); aux_vec_one(1) = -1;

  aux_vec_qdt = repmat (0, No_relax, 1);
  aux_vec_qdt(1) = inf_(intval(flow_min) - intval(flow_max));
  aux_vec_qdt(2:No_relax)...
    = inf_(- feval(q_dt, aux_vec,vmin,vmax,dischard,diam,k_value,ccw));

  br = [-flow_max;...
        sup( feval(q_fkt, aux_vec,vmin,vmax,dischard,diam,k_value,ccw)...
        - aux_vec.* feval(q_dt, aux_vec,vmin,vmax,dischard,diam,k_value,ccw))];

elseif (flow_max <= 0)
  if (verbose == 2), fprintf('relax: "case (iii)"\n'); end
  if (No_relax == 6)
    aux_vec = [0.1; 0.2; 0.5; 0.75; 1];
  elseif (No_relax == 4)
    aux_vec = [0.1; 0.5; 1];
  end
  aux_vec_one = repmat(-1,No_relax,1); aux_vec_one(1) = 1;

  aux_vec_qdt = repmat(0,No_relax,1);
  aux_vec_qdt(1) = inf_(intval(flow_max) - intval(flow_min));
  aux_vec_qdt(2:No_relax)...
    = inf_( feval(q_dt, aux_vec,vmin,vmax,dischard,diam,k_value,ccw));

  br = [flow_max;...
        sup(aux_vec.*feval(q_dt, aux_vec,vmin,vmax,dischard,diam,k_value,ccw)...
              - feval(q_fkt, aux_vec,vmin,vmax,dischard,diam,k_value,ccw))];

else
  if ~strcmp (head_loss_formula, 'DW')
    star = 1 / (1 - (vmin/vmax)); inf_star = inf_(star); sup_star = sup(star);
  else
    r = (ccw(1) .* diam ./ (1 - k_value / ccw(2) ./ diam) ).^2;
    inf_star_int = (vmax - r) / (vmax - vmin); inf_star = inf_(inf_star_int);
    sup_star_int = (vmax + r) / (vmax - vmin); sup_star = sup(sup_star_int);
  end
  inv_dischard = 1 / dischard;

  bar = feval( rel_eq_bar ,0, flow_min,vmin,vmax,inv_dischard,diam,k_value,ccw);
  if (bar > 0)
    bar_exists = 0;
  else
    bar = bisect_vlb ( rel_eq_bar ,0, inf_star, REPS , flow_min, vmin, vmax,...
                                                inv_dischard, diam, k_value, ccw);
    bar_exists = 1;
  end
```

```
tilde = feval ( rel_eq_tilde ,1, flow_max, vmin, vmax, inv_dischard, diam,...
                                                        k_value, ccw);
if (tilde < 0)
  tilde_exists = 0;
else
  tilde = bisect_vub ( rel_eq_tilde , sup_star, 1, REPS , flow_max, vmin,...
                                   vmax, inv_dischard, diam, k_value, ccw);
  tilde_exists = 1;
end


if (~bar_exists) && (~tilde_exists)
feval(wds_error,'VNOP-error constructing relaxations, q not convex-concave?');

elseif (bar_exists) && (tilde_exists)
  if (verbose == 2), fprintf('relax: "case (i) a" \n'); end

  if (No_relax == 6)
    aux_vec = [0; bar/2; bar; tilde; (tilde+1)/2; 1];
  elseif (No_relax == 4)
    aux_vec = [0; bar; tilde; 1];
  end

  aux_vec_one = repmat(1,No_relax,1);
  aux_vec_one(No_relax/2+1:No_relax) = -1;

  aux_vec_qdt = repmat(0,No_relax,1);
  aux_vec_qdt = -1 .* aux_vec_one...
                .* feval(q_dt, aux_vec,vmin,vmax,dischard,diam,k_value,ccw);
  aux_vec_qdt = inf_(aux_vec_qdt);

  br = [sup(aux_vec_one...
       .* (feval(q_fkt, aux_vec,vmin,vmax,dischard,diam,k_value,ccw)...
       -aux_vec.*feval(q_dt, aux_vec,vmin,vmax,dischard,diam,k_value,ccw)))];

  if strcmp (head_loss_formula, 'DW')
    bar_int = intval(bar);      tilde_int = intval(tilde);

    if (~(feval(q_fkt, sup_star_int,vmin,vmax,[],diam,k_value,ccw)...
         <= feval(q_fkt, bar_int,vmin,vmax,[],diam,k_value,ccw)...
            + (sup_star_int - bar_int)...
              * feval(q_dt, bar_int,vmin,vmax,[],diam,k_value,ccw))...
        || ~(feval(q_fkt, inf_star_int,vmin,vmax,[],diam,k_value,ccw)...
           >= feval(q_fkt, tilde_int,vmin,vmax,[],diam,k_value,ccw)...
              + (inf_star_int - tilde_int)...
                * feval(q_dt, tilde_int,vmin,vmax,[],diam,k_value,ccw)))

      fprintf('relax: case (i) a, relaxation omitted for flow [%f, %f]\n',...
                                                    flow_min, flow_max);
      aux_vec_one  = repmat (0, No_relax, 1);
      aux_vec_qdt  = repmat (0, No_relax, 1);
      br           = repmat (0, No_relax, 1);
    end
  end
```

```
elseif (~tilde_exists)
 if (verbose == 2), fprintf('relax: "case (i) b" \n'); end
 if (No_relax == 6)
  aux_vec = [0; bar/4; bar/2; bar*3/4; bar];
 elseif (No_relax == 4)
  aux_vec = [0; bar/2; bar];
 end
 aux_vec_one = repmat(1,No_relax,1);
 aux_vec_one(No_relax) = -1;

 aux_vec_qdt = repmat(0,No_relax,1);
 aux_vec_qdt(1:No_relax-1)...
  = inf_(- feval(q_dt, aux_vec,vmin,vmax,dischard,diam,k_value,ccw));
 aux_vec_qdt(No_relax) = inf_(intval(flow_min) - intval(flow_max));

 br = [sup( feval(q_fkt, aux_vec,vmin,vmax,dischard,diam,k_value,ccw)...
           - aux_vec...
             .* feval(q_dt, aux_vec,vmin,vmax,dischard,diam,k_value,ccw));...
       - flow_max];
 if strcmp (head_loss_formula, 'DW')
  bar_int = intval(bar);
  if (~(feval(q_fkt, inf_star_int,vmin,vmax,[],diam,k_value,ccw)...
        >= (inf_star_int * flow_min + (1-inf_star_int) * flow_max))...
      || ~(feval(q_fkt, sup_star_int,vmin,vmax,[],diam,k_value,ccw)...
           <= feval(q_fkt, bar_int,vmin,vmax,[],diam,k_value,ccw)...
              + (sup_star_int - bar_int)...
                 * feval(q_dt, bar_int,vmin,vmax,[],diam,k_value,ccw)))

    fprintf('relax: case (i) b, relaxation omitted for flow [%f, %f]\n',...
                                              flow_min, flow_max);
    aux_vec_one  = repmat (0, No_relax, 1);
    aux_vec_qdt  = repmat (0, No_relax, 1);
    br           = repmat (0, No_relax, 1);
  end
 end

elseif (~bar_exists)
 if (verbose == 2), fprintf('relax: "case (i) c" \n'); end
 if (No_relax == 6)
  aux_vec = [tilde; (tilde+1)/4; (tilde+1)/2; (tilde+1)*3/4; 1];
 elseif (No_relax == 4)
  aux_vec = [tilde; (tilde+1)/2; 1];
 end
 aux_vec_one = repmat(-1,No_relax,1); aux_vec_one(1) = 1;

 aux_vec_qdt = repmat(0,No_relax,1);
 aux_vec_qdt(1) = inf_(intval(flow_max) - intval(flow_min));
 aux_vec_qdt(2:No_relax)...
  = inf_( feval(q_dt, aux_vec,vmin,vmax,dischard,diam,k_value,ccw));

 br = [flow_max;...
       sup( aux_vec...
             .* feval(q_dt, aux_vec,vmin,vmax,dischard,diam,k_value,ccw)...
            - feval(q_fkt, aux_vec,vmin,vmax,dischard,diam,k_value,ccw))];
```

```
    if strcmp (head_loss_formula, 'DW')
      tilde_int = intval(tilde);
      if (~(feval(q_fkt, sup_star_int,vmin,vmax,[],diam,k_value,ccw)...
            <= (sup_star_int * flow_min + (1-sup_star_int) * flow_max))...
         || ~(feval(q_fkt, inf_star_int,vmin,vmax,[],diam,k_value,ccw)...
              >= feval(q_fkt, tilde_int,vmin,vmax,[],diam,k_value,ccw)...
                 + (inf_star_int - tilde_int)...
                   * feval(q_dt, tilde_int,vmin,vmax,[],diam,k_value,ccw)))

        fprintf('relax: case (i) c, relaxation omitted for flow [%f, %f]\n',...
                                                       flow_min, flow_max);
        aux_vec_one  = repmat (0, No_relax, 1);
        aux_vec_qdt  = repmat (0, No_relax, 1);
        br           = repmat (0, No_relax, 1);
      end
    end
  end
end
Ar = [repmat(0, No_relax, 2*No_nseg + idx - 1), aux_vec_qdt,...
      repmat(0, No_relax, No_lambda - idx + q_idx - 1), aux_vec_one,...
      repmat(0, No_relax, No_arcs - q_idx + No_nodes)];


% ---------------------------------------------------------------------------
function [head_min, head_max]...
        = adjust_head (head_min, head_max, flow_min, flow_max,...
                       head_loss_formula, k_value, dischard, ccw,...
                       avail_diams, alpha, post_tree, non_tree, arcs,...
                       arc_length, arcs_fixed, length_fixed, seg_fixed)
%
% ADJUST_HEAD - Determination of adjusted head bounds according to
%               constraint propagation using actual flow bounds
%
% Returns adjusted head bounds

global No_arcs No_dia No_seg
HEPS = 1e-2;    % absolute minimum of flow to calculate head-loss
ZERO = intval(0);

head_diff_min = inf_(intval(head_min(arcs(:,1))) - head_max(arcs(:,2)));
head_diff_max = sup(intval(head_max(arcs(:,1))) - head_min(arcs(:,2)));

new_arcs = 1:No_arcs;            new_arcs(arcs_fixed) = [];

if ~strcmp (head_loss_formula, 'DW')
  seg_min = repmat(ZERO,No_seg,1); seg_max = repmat(ZERO,No_seg,1);
  for i = new_arcs
    if (flow_min(i) > 0),       seg_min(i*No_dia)        = arc_length(i);
    elseif (flow_min(i) < 0),   seg_min((i-1)*No_dia+1)  = arc_length(i);
    end
    if (flow_max(i) > 0),       seg_max((i-1)*No_dia+1)  = arc_length(i);
    elseif (flow_max(i) < 0),   seg_max(i*No_dia)        = arc_length(i);
    end
  end
```

```
  seg_min(seg_fixed) = length_fixed(seg_fixed);
  seg_max(seg_fixed) = length_fixed(seg_fixed);

  phi_min = inf_( sgn_pow (intval(flow_min), dischard)...
                 .* (kron(eye(No_arcs), alpha) * seg_min));
  phi_max = sup( sgn_pow (intval(flow_max), dischard)...
                 .* (kron(eye(No_arcs), alpha) * seg_max));

else % Darcy-Weisbach
  dw_min = repmat(0,No_arcs,1); dw_max = repmat(0,No_arcs,1);
  for i = new_arcs
    if (flow_min(i) > 0),    dw_min(i) = No_dia;
    else                     dw_min(i) = 1;
    end
    if (flow_max(i) > 0),    dw_max(i) = 1;
    else                     dw_max(i) = No_dia;
    end
  end
  for i = arcs_fixed'
    d_idx = find(length_fixed((i-1)*No_dia+1:i*No_dia) ~= 0);
    d_idx = mod(d_idx,No_dia); d_idx(d_idx==0) = No_dia;
    if (flow_min(i) > 0),    dw_min(i) = d_idx(end);
    else                     dw_min(i) = d_idx(1);
    end
    if (flow_max(i) > 0),    dw_max(i) = d_idx(1);
    else                     dw_max(i) = d_idx(end);
    end
  end
  diam_min = avail_diams(dw_min);   diam_max = avail_diams(dw_max);
  flow_min = intval(flow_min);      flow_max = intval(flow_max);

  flow_max ((flow_max < HEPS) & (flow_max > 0)) = intval(HEPS);
  flow_min ((flow_min > -HEPS) & (flow_min < 0)) = intval(-HEPS);

  phi_min = inf_(arc_length .* alpha(dw_min)'...
                 .* w_fkt (flow_min, diam_min, k_value, ccw));
  phi_max = sup(arc_length .* alpha(dw_max)'...
                 .* w_fkt (flow_max, diam_max, k_value, ccw));
end


head_diff_min = intval( max(head_diff_min, phi_min) );
head_diff_max = intval( min(head_diff_max, phi_max) );

for i = [non_tree , post_tree]
  head_min(arcs(i,1)) = max ([head_min(arcs(i,1));...
                             inf_(head_min(arcs(i,2)) + head_diff_min(i))]);
  head_min(arcs(i,2)) = max ([head_min(arcs(i,2));...
                             inf_(head_min(arcs(i,1)) - head_diff_max(i))]);
  head_max(arcs(i,1)) = min ([head_max(arcs(i,1));...
                             sup(head_max(arcs(i,2)) + head_diff_max(i))]);
  head_max(arcs(i,2)) = min ([head_max(arcs(i,2));...
                             sup(head_max(arcs(i,1)) - head_diff_min(i))]);
end
```

```
% ----------------------------------------------------------------------
function [flow_min, flow_max]...
         = adjust_flow (flow_min, flow_max, head_min, head_max,...
                        head_loss_formula, k_value, dischard, ccw,...
                        avail_diams, alpha, arcs, arc_length, arcs_fixed,...
                        length_fixed, seg_fixed)
%
% ADJUST_FLOW - Determination of adjusted flow bounds according to
%               constraint propagation using actual head bounds
%
% Returns adjusted flow bounds
global No_arcs No_dia No_seg
FEPS = 1e-2;     % if possible, flow bounds below this value are set to zero
ZERO = intval(0);

head_diff_min  = inf_(intval(head_min(arcs(:,1))) - head_max(arcs(:,2)));
head_diff_max  = sup(intval(head_max(arcs(:,1))) - head_min(arcs(:,2)));
new_arcs       = 1:No_arcs; new_arcs(arcs_fixed) = [];
if ~strcmp (head_loss_formula, 'DW')
  seg_min = repmat(ZERO,No_seg,1); seg_max = repmat(ZERO,No_seg,1);
  for i = new_arcs
   if (head_diff_min(i) > 0),     seg_min((i-1)*No_dia+1) = arc_length(i);
   elseif (head_diff_min(i) < 0), seg_min(i*No_dia)       = arc_length(i);
   end
   if (head_diff_max(i) > 0),     seg_max(i*No_dia)       = arc_length(i);
   elseif (head_diff_max(i) < 0), seg_max((i-1)*No_dia+1) = arc_length(i);
   end
  end
  seg_min(seg_fixed) = length_fixed(seg_fixed);
  seg_max(seg_fixed) = length_fixed(seg_fixed);
  flow_min_h = inf_( sgn_pow (head_diff_min, 1/dischard)...
                  .* (kron(eye(No_arcs), alpha) * seg_min).^(-1/dischard));
  flow_max_h = sup( sgn_pow (head_diff_max, 1/dischard)...
                  .* (kron(eye(No_arcs), alpha) * seg_max).^(-1/dischard));
else % Darcy-Weisbach
  dw_min = repmat(0,No_arcs,1); dw_max = repmat(0,No_arcs,1);
  for i = new_arcs
    if (head_diff_min(i) > 0),  dw_min(i) = 1;
    else                        dw_min(i) = No_dia;
    end
    if (head_diff_max(i) > 0),  dw_max(i) = No_dia;
    else                        dw_max(i) = 1;
    end
  end
  for i = arcs_fixed'
    d_idx = find(length_fixed((i-1)*No_dia+1:i*No_dia) ~= 0);
    d_idx = mod(d_idx,No_dia);  d_idx(d_idx==0) = No_dia;
    if (head_diff_min(i) > 0),  dw_min(i) = d_idx(1);
    else                        dw_min(i) = d_idx(end);
    end
    if (head_diff_max(i) > 0),  dw_max(i) = d_idx(end);
    else                        dw_max(i) = d_idx(1);
    end
  end
```

```
  diam_min = avail_diams(dw_min); diam_max = avail_diams(dw_max);

  h_min        = inf_( head_diff_min ./ alpha(dw_min)' ./ arc_length );
  h_max        = sup( head_diff_max ./ alpha(dw_max)' ./ arc_length );
  flow_min_h   = inf_(inv_w_fkt (h_min, diam_min, k_value, ccw));
  flow_max_h   = sup(inv_w_fkt (h_max, diam_max, k_value, ccw));

  % to avoide to deal in subsequent calls with tiny flow bounds
  flow_min_h ((flow_min_h < FEPS) & (flow_min_h > 0)) = 0;
  flow_max_h ((flow_max_h > -FEPS) & (flow_max_h < 0)) = 0;
end
flow_min = max(inf_(flow_min), flow_min_h);
flow_max = min(sup(flow_max), flow_max_h);


% ------------------------------------------------------------------------
function [fulfilled, err_str]...
        = test_adjacency (head_loss_formula, avail_diams, pipe_cost, alpha)
%
% TEST_ADJACENCY - Check, if adjacency property is fulfilled
%
% Returns (1) if adjacency property is fulfilled,
%         (0) if not, and in this case an error string is provided

D_EXP         = intval('4.87');
avail_diams   = intval(avail_diams);
pipe_cost     = intval(pipe_cost);

fulfilled     = 1; err_str = [];

if strcmp (head_loss_formula, 'HW')
  div_diff = (pipe_cost(2:end) - pipe_cost(1:end-1))...
           ./(avail_diams(2:end).^(-D_EXP) - avail_diams(1:end-1).^(-D_EXP));

elseif strcmp (head_loss_formula, 'DW')
  div_diff = (pipe_cost(2:end) - pipe_cost(1:end-1))...
           ./ (avail_diams(2:end).^(-3) - avail_diams(1:end-1).^(-3));

elseif strcmp (head_loss_formula, 'AHW')

  if any( alpha(1:end-1) <= alpha(2:end) )
    fulfilled = 0;
    err_str = 'Adjusted Hazen Coefficients not valied, no adjacency property!';
  end
  div_diff = (pipe_cost(2:end) - pipe_cost(1:end-1))...
           ./ (alpha(2:end) - alpha(1:end-1))';
end

if any( pipe_cost(1:end-1) >= pipe_cost(2:end) )...
    | any( avail_diams(1:end-1) >= avail_diams(2:end) )...
    | any (div_diff(1:end-1) <= div_diff(2:end) )
  fulfilled = 0;
  err_str = 'Cost function not valid, no adjacency property!';
end
```

```
% ----------------------------------------------------------------------------
function [verified_lb] = verify_lb (c, A, a, B, b, lb, ub, y, z)
%
% VERIFY_LB - Verifies the lower bound for a linear optimization problem
%             in the form
%                               c'x -> min
%                               s.t.
%                               A x <= a
%                               B x = b
%                               lb <= x <= ub ,
%
%             where y, z are the Lagrange-paramters determined with
%             an approximate lp-solver
%
% Returns verified lower bound

% Guarantee that y <= 0
y( (y > 0) ) = 0;

% Calulation of d and lower bound according to C. Jansson
d      = c - A' * intval(y) - B' * intval(z);

d_pos  =  intersect (d, infsup(0,inf));   d_pos(isnan(d_pos)) = intval(0);
d_neg  = -intersect (d, infsup(-inf,0));  d_neg(isnan(d_neg)) = intval(0);


verified_lb = inf_(a' * intval(y) + b' * intval(z) + lb' * d_pos - ub' * d_neg);


% ----------------------------------------------------------------------------
function [infeasible] = verify_infeasibility (lp_solver, A, a, B, b, lb, ub)
%
% VERIFY_INFEASIBILITY - Verifies the infeasibility for a linear
%                        optimization problem (LP) in the form:
%
%                               c'x -> min
%                               s.t.
%                               A x <= a
%                               B x = b
%                               lb <= x <= ub .
%
% B being an interval matrix and b being an interval vector is possible.
%
% Returns (1) if the linear problem (LP) is infeasible,
%         (0) if no proof can be provided

OPT_MAX = 1e6; % restrict region for linear solver

p = size(a,1);
q = size(b,1);
n = size(lb,1);

lb_aux = [repmat(0, p,1); repmat(- OPT_MAX , q,1); repmat(0, 2*n,1)];
ub_aux = repmat( OPT_MAX , p+q+2*n,1);
```

```
% solve auxiliary problem
if strcmp (lp_solver , 'cplex')
  [opt, x, lambda, status] =...
    lp_cplex_mex ( mid([a; b; -lb; ub]), mid([A', B', -eye(n), eye(n)]),...
                  zeros(n,1), lb_aux, ub_aux, 1);
else
  linprog_options = optimset('linprog');
  linprog_options = optimset(linprog_options, 'Display','off');
  [x, opt, exitflag, infos, lambda] =...
    linprog( mid([a; b; -lb; ub]), [], [], mid([A', B', -eye(n), eye(n)]),...
            zeros(n,1), lb_aux, ub_aux, [], linprog_options );
end

y = intval(x(1:p)); z = intval(x(p+1:p+q));

y (~(y >= 0)) = intval(0); % ensure y to be within bounds

d = A'*y + B'*z;
u =  intersect (d, infsup(0,inf));      u(isnan(u))  = intval(0);
v = -intersect (d, infsup(-inf,0));     v(isnan(v))  = intval(0);

x = [y; z; u; v];

if (isempty( find(~(x >= lb_aux)) )) && (([a; b; -lb; ub]' * x) < 0)
  infeasible = 1;
else
  infeasible = 0;
end


% ----------------------------------------------------------------------------
function [verified_ub, x_vec, verified, error_str]...
        = verify_ub (x, head_loss_formula, k_value, lp_solver, dischard,...
                     ccw, avail_diams, ub_costs, alpha, arcs, arc_length,...
                     arcs_fixed, length_fixed, seg_fixed, node_edge,...
                     elevation, demand, flow_min, flow_max,...
                     head_min, head_max, verbose)
%
% VERIFY_UB - Verifies the upper bound for NOP
%
% Returns verified upper bound of NOP (verified_ub)
% and the inclusion for the solution vector used to determine
% this bound (x_vec)
% verified = (1) if verification succeeded
%            (0) if not, in this case an error string is provided

global No_arcs No_nodes No_dia No_seg No_narcs No_nseg
ETA = 1e-10;    % Factor for adjusting auxiliary bounds for finding verified
                % solution within bounds
MAX_UB_ITER = 10;

verified_ub = []; x_vec = []; verified = 0; error_str = [];

x = mid(x);
```

```
% According to the adjacency property except one or two adjacent x_ijk all
% others are set to zero
idx2 = [];

for i = 1:No_arcs
  if ~(any(i == arcs_fixed))
    seg = (i-1)*No_dia+1 : i*No_dia;
    k = find( x(seg) == max(x(seg))); k = k(1);
    if (k == 1)
      k = (i-1)*No_dia+2;
    elseif (k == No_dia)
      k = i*No_dia;
    elseif (x((i-1)*No_dia+k+1) > x((i-1)*No_dia+k-1))
      k = (i-1)*No_dia+k+1;
    else
      k = (i-1)*No_dia+k;
    end
    idx2 = [idx2; k];
  end
end
% Auxiliary variables and adjusted cost vector for x2
idx2s     = mod(idx2, No_dia);      idx2s (idx2s==0) = No_dia;
alpha_k1  = alpha(idx2s-1);         alpha_k2         = alpha(idx2s);
diam_k1   = avail_diams(idx2s-1);   diam_k2          = avail_diams(idx2s);

c_vec = [ub_costs(idx2s) - ub_costs(idx2s-1); ub_costs(No_nseg+1:end)];

new_arcs = transpose(1:No_arcs); new_arcs(arcs_fixed) = [];

% Determination of parts to calculate an inclusion of
length   = x (idx2);
flow     = x (No_seg+1:No_seg+No_arcs);
head     = x (No_seg+No_arcs+1:No_seg+No_arcs+No_nodes);


if No_narcs == No_arcs
  % Verification of F_1
  flow = verify_udlss ( node_edge(2:end,:), demand(2:end), flow);

  lb   = [repmat(0,No_narcs,1);sup(head_min-elevation)]; lb_org = lb;
  ub   = inf_([arc_length; head_max-elevation]); ub_org = ub;
  x    = infsup(min(max([length;head],lb),ub), max(min([length;head],ub),lb));

  % Verification of F_2
  if ~strcmp (head_loss_formula, 'DW')
    v_flow = sgn_pow (flow, dischard);
    rho = v_flow .* (alpha_k2 - alpha_k1)';
  else
    v_flow = w_fkt (flow, diam_k1, k_value, ccw);
    rho = alpha_k2' .* w_fkt(flow, diam_k2, k_value, ccw) - alpha_k1' .* v_flow;
  end
  b = elevation(arcs(:,1))-elevation(arcs(:,2))-v_flow.* alpha_k1'.* arc_length;
  F2 = [diag(rho), - node_edge'];
  i = 1; found = 0;
```

```
  while (i <= MAX_UB_ITER ) && ~found
    i = i+1;
    x = verify_udlss (F2 , b, x, (No_narcs+1: No_narcs+No_nodes)');

    if ~any(isnan(x)) & isempty( find((~(x >= lb_org)) | (~(x <= ub_org))) )
      length = x(1:No_narcs); head = x(No_narcs+1:end);
      if verbose
        fprintf('verify_ub: Verification of upper bound succeeded.\n');
      end
      found = 1;
    else
      lb_idx = (~(x >= lb)); ub_idx = (~(x <= ub));
      lb(lb_idx)   = lb(lb_idx) * (1+ETA) + ETA ;
      ub(ub_idx)   = ub(ub_idx) * (1-ETA) - ETA ;
      if strcmp (lp_solver , 'cplex')
        [dummy, x, lambda, status] =...
           lp_cplex_mex ( c_vec, mid(F2), mid(b), lb, ub);
      else
        linprog_options = optimset('linprog');
        linprog_options = optimset(linprog_options, 'Display','off');
        [x, dummy, exitflag, infos, lambda] =...
           linprog( c_vec, [],[], mid(F2), mid(b), lb, ub, [],linprog_options );
      end
    end
  end % loop for verifying [length; head]
  if ~found
    error_str = 'VNOP - verify_ub: no inclusion of solution vector!';
    return;
  end

else % No_narcs ~= No_arcs
  lb = [repmat(0,No_narcs,1); flow_min; sup(head_min-elevation)];
  ub = [inf_(arc_length(new_arcs)); flow_max; inf_(head_max-elevation)];

  i = 1; found = 0; non_idx = No_arcs+1;
  while (i <= MAX_UB_ITER ) && ~found
    i = i+1;
    % Ensure that x is within bounds
    x = [length; flow; head];
    x = infsup(min(max(x,lb),ub), max(min(x,ub),lb)); x = intval(mid(x));

    x(No_narcs+1:end)...
      = verify_udnlss (x(No_narcs+1:end), non_idx, head_loss_formula,...
                       k_value, dischard, ccw, avail_diams, alpha, arcs,...
                       arcs_fixed, length_fixed, node_edge, demand,elevation);

    if ~any(isnan(x(No_narcs+1:end)))
      lb_idx = find(~(x(No_narcs+No_arcs+1:end) >= lb(No_narcs+No_arcs+1:end)));
      ub_idx = find(~(x(No_narcs+No_arcs+1:end) <= ub(No_narcs+No_arcs+1:end)));

      if isempty(lb_idx) && isempty(ub_idx)
        flow = x(No_narcs+1:No_narcs+No_arcs); head = x(No_narcs+No_arcs+1:end);
        found = 1;
      else
```

```
        if ~isempty(lb_idx)
          head(lb_idx) = lb(No_narcs+No_arcs+lb_idx);
        elseif ~isempty(ub_idx)
          head(ub_idx) = ub(No_narcs+No_arcs+ub_idx);
        end
        for j = (No_arcs + [lb_idx; ub_idx])'
          if ~any(j == non_idx), non_idx = [non_idx; j]; end
        end
      end
    else
      error_str = 'VNOP - verify_ub: no inclusion of F3 could be determined!';
      return;
    end
  end % loop for verifying [flow; head]
  if ~found
    error_str = 'VNOP - verify_ub: no inclusion of F3 within bounds!';
    return;
  end
  head_aux = elevation(arcs(:,2)) - elevation(arcs(:,1))...
           + head(arcs(:,2)) - head(arcs(:,1));
  for i = 1:2
    if ~strcmp (head_loss_formula, 'DW')
      v_flow   = sgn_pow (flow(new_arcs), dischard);
      F4       = v_flow .* (alpha_k2 - alpha_k1)';
    else
      v_flow   = w_fkt (flow(new_arcs), diam_k1, k_value, ccw);
      F4       = alpha_k2' .* w_fkt (flow(new_arcs), diam_k2, k_value, ccw)...
                 - alpha_k1' .* v_flow;
    end
    b4 = - v_flow .* alpha_k1' .* arc_length(new_arcs) - head_aux(new_arcs);
    length = b4 ./ F4;
    k = find(~(length <= arc_length(new_arcs))); l = find(~(length >= 0));
    if isempty(k) && isempty(l)
      if verbose
        fprintf('verify_ub: Verification of upper bound succeeded.\n');
      end
      break;
    else
      k ( idx2s(k) == No_dia ) = [];        l ( idx2s(l) == 2 ) = [];
      idx2s( k )   = idx2s( k ) + 1;        idx2( k )   = idx2( k ) + 1;
      idx2s( l )   = idx2s( l ) - 1;        idx2( l )   = idx2( l ) - 1;
      alpha_k1     = alpha(idx2s-1);        alpha_k2    = alpha(idx2s);
      diam_k1      = avail_diams(idx2s-1); diam_k2     = avail_diams(idx2s);
      c_vec = [ub_costs(idx2s) - ub_costs(idx2s-1); ub_costs(No_nseg+1:end)];
    end
  end
end % No_narcs == resp. ~= No_arcs

% Doublecheck solution bo be within bounds
if ~(all(head_min-elevation <= head) && all(head <= head_max-elevation) &&...
     all(repmat(0,No_narcs,1) <= length) && all(length <= arc_length(new_arcs)))
  error_str = 'VNOP - verify_ub: no inclusion of solution vector!';
  return;
end
```

```
% Determination of solution vector
length_org          = intval(length_fixed);
length_org(idx2)    = length;
length_org(idx2-1)  = arc_length(new_arcs) - length;

% Verified inclusion for the solution vector and upper bound
verified     = 1;
x_vec        = [length_org; flow; head]; length_org(seg_fixed) = [];
verified_ub  = sup(ub_costs' * [length_org;head]);


% ---------------------------------------------------------------------------
% LOCAL SUBFUNCTIONS FOR RELAXATIONS
% ---------------------------------------------------------------------------
function [x0] = bisect_vlb (f, xl, xu, beps, varargin)
%
% BISECT_VLB - Bisection of interval function f,
%              where f is strictly monotone increasing
global wds_error

f_xl = feval(f, xl, varargin{:});

if ~(f_xl < 0)
  feval(wds_error, 'VNOP - bisect_vlb: f(xl) < 0 does not hold true!');
end
h = xu - xl;

for n = 1 : ceil( log2( h/beps ));
  h = h/2; x = xl+h;
  f_x = feval(f, x, varargin{:});
  if (f_x < 0), xl = x; end
end
x0 = xl;

% ---------------------------------------------------------------------------
function [x0] = bisect_vub (f, xl, xu, beps, varargin)
%
% BISECT_VUB - Bisection of interval function f,
%              where f is strictly monotone increasing
global wds_error

f_xu = feval(f, xu, varargin{:});

if ~(f_xu > 0)
  feval(wds_error, 'VNOP - bisect_vub: f(xu) > 0 does not hold true!');
end
h = xu - xl;

for n = 1 : ceil( log2( h/beps ));
  h = h/2; x = xu-h;
  f_x = feval(f, x, varargin{:});
  if (f_x > 0), xu = x; end
end
x0 = xu;
```

```
% ------------------------------------------------------------------------
% LOCAL SUBFUNCTIONS FOR VERIFICATION OF UPPER BOUND
% ------------------------------------------------------------------------
function [y] = F3 (z, head_loss_formula, k_value, dischard, ccw, avail_diams,...
                   alpha, arcs, arcs_fixed, length_fixed, node_edge,...
                   demand, elevation)
%
% F3 - Constraint Function, always returns an interval

global No_arcs No_narcs No_nodes No_dia wds_error

if size(z,2) ~= 1, feval(wds_error, 'VNOP - F3: z is not a column-vector.'); end

y          = repmat(intval(0), No_arcs-No_narcs+No_nodes-1, 1);

flow       = z(1:No_arcs);
head       = z(No_arcs+1:No_arcs+No_nodes);
head_aux   = elevation(arcs(:,2)) - elevation(arcs(:,1)) - node_edge' * head;

% No_arcs-No_narcs elements of y corresponding to A\P
if (No_narcs ~= No_arcs)
  if ~strcmp (head_loss_formula, 'DW')
    v_flow = sgn_pow (flow, dischard);
    alpha_aux = (alpha * reshape(length_fixed,No_dia,No_arcs))';

    F_p2 = v_flow(arcs_fixed) .* alpha_aux(arcs_fixed);
  else
    v_flow = reshape(w_fkt_1 (flow(arcs_fixed), avail_diams, k_value, ccw),...
                     No_dia, No_arcs-No_narcs);
    length_aux = reshape(length_fixed,No_dia,No_arcs);

    F_p2 = (alpha * (v_flow .* length_aux(:,arcs_fixed)))';
  end
  y (1 : No_arcs-No_narcs) = F_p2 + head_aux(arcs_fixed);
end
% Last No_nodes-1 elements of y
y(No_arcs-No_narcs+1:end) = node_edge(2:end,:) * flow - demand(2:end);


% ------------------------------------------------------------------------
function [J] = F3_dx (z, head_loss_formula, k_value, dischard, ccw,...
                      avail_diams, alpha, arcs_fixed,length_fixed, node_edge)
%
% F3_DX - Jacobi Marix of F3 at [q,H], always returns an interval matrix

global No_arcs No_narcs No_nodes No_dia wds_error

if size(z,2) ~= 1, feval(wds_error, 'VNOP-F3_dx: z is not a column-vector.');end

J         = repmat(intval(0), No_arcs-No_narcs+No_nodes-1,...
                              No_narcs-No_narcs+No_arcs+No_nodes);
z         = intval(z);
flow      = z(1:No_arcs);
new_arcs  = transpose(1:No_arcs); new_arcs(arcs_fixed) = [];
```

```
% No_arcs-No_narcs elements of y corresponding to A\P
if No_narcs ~= No_arcs
  if ~strcmp (head_loss_formula, 'DW')
    v_flow_dx  = dischard .* (abs(flow(arcs_fixed)).^(dischard-1));
    alpha_aux = (alpha * reshape(length_fixed,No_dia,No_arcs))';

    J(1:No_arcs-No_narcs, arcs_fixed)...
      = diag (v_flow_dx .* alpha_aux(arcs_fixed));
  else
    v_flow_dx = reshape(w_fkt_dq (flow(arcs_fixed), avail_diams,k_value,ccw),...
                        No_dia, No_arcs-No_narcs);
    length_aux = reshape(length_fixed, No_dia, No_arcs);

    J(1:No_arcs-No_narcs, arcs_fixed)...
      = diag (alpha * (v_flow_dx .* length_aux(:,arcs_fixed)));
  end
  J(1:No_arcs-No_narcs, No_arcs+1:end) = - node_edge(:,arcs_fixed)';
end
% Last No_nodes-1 rows of J
J(No_arcs-No_narcs+1:end,:) = [node_edge(2:end,:), zeros(No_nodes-1,No_nodes)];


% -----------------------------------------------------------------------
function [x] = verify_udlss (A, b, approx, non_idx)
%
% VERIFY_UDLSS - Verified inclusion for a linear underdetermined
%                system A x = b close to double vector approx
%                non_idx column-vector containing indices for fixed values
%
% Returns inclusion vector if verification succeeded, else vector with NaN
global wds_error

if ~exist('non_idx'), non_idx=[]; end
try
  A * approx - b;
catch
  feval(wds_error,'VNOP - verify_udlss: matrix-vector dimensions must agree\n');
end

% lu returns lower triangular matrix L, upper triangular matrix U,
% and permutation matrix P such that P*X = L*U.
% The order implied by P is used for selecting the variables
% idx contains all indices used for the mxm-submatix
if isa (A, 'intval')
  [L,U,P] = lu(mid(A'));
else
  [L,U,P] = lu(A');
end
idx_lu = P * transpose(1:size(approx,1));
for (i = non_idx'), idx_lu (idx_lu == i) = []; end
idx = [idx_lu; non_idx(1:end)]; idx = idx(1:size(b,1));

x = intval(approx); x(idx) = intval(0);
x(idx) = verifylss (A(:,idx), b-A*x);
```

```
% -----------------------------------------------------------------------------
function [X] = verify_udnlss (x, non_idx, head_loss_formula, k_value,...
                              dischard, ccw, avail_diams, alpha, arcs,...
                              arcs_fixed, length_fixed, node_edge,...
                              demand, elevation)
%
% VERIFY_UDNLSS - Verified inclusion for nonlinear function F(x) = 0
%                 close to double vector approx
%                 non_idx column-vector containing indices for fixed values
%
% Returns inclusion vector if verification succeeded, else vector with NaN

VEPS         = 1e-12;      % factor for inflation of interval
MAX_UB_ITER  = 10;

if ~exist('non_idx'), non_idx=[]; end

% Adjustment and inflation of approximate solution
x        = intval(x);

idx      = 1:max(size(x)); idx(non_idx) = [];
x(idx)   = x(idx) + VEPS * infsup(-1,1);

F_dx = F3_dx (x, head_loss_formula, k_value, dischard, ccw, avail_diams,...
                 alpha, arcs_fixed, length_fixed, node_edge);

% Determination of indices for square function to calculate an inclusion
[L,U,P] = lu(mid( F_dx')); idx_lu = P*transpose(1:size(F_dx,2));
for i = non_idx'
  idx_lu (idx_lu == i) = [];
end
idx = [idx_lu; non_idx(1:end)]; idx = idx(1:size( F_dx ,1 ));

xs   = x(idx);

% The following lines are taken from the procedure "verifynlss" of
% INTLAB Version 4.1.2, S. M. Rump 2004, and adjusted to the syntax of F:

% Interval iteration
R = inv(mid( F_dx (:,idx)));

if rank(mid(F_dx(:,idx))) ~= min(size(mid(F_dx(:,idx))))
  fprintf('VNOP-verify_udnlss: gradient matrix seems to be singular.');
  X = intval(repmat(NaN,size(x))); % inclusion failed
  return
end

Z = - R * F3 (x, head_loss_formula, k_value, dischard, ccw, avail_diams,...
                 alpha, arcs, arcs_fixed, length_fixed, node_edge,...
                 demand, elevation);

X = Z;
E = 0.1*rad(X)*hull(-1,1) + midrad(0,realmin);
ready = 0; k = 0;
```

```
while ( ~ready ) && (k < MAX_UB_ITER) && ( ~any(isnan(X)) )
  k = k+1;
  Y = hull( X + E , 0 );     % epsilon inflation
  Yold = Y;
  x(idx) = xs+Y;

  F_dx = F3_dx (x, head_loss_formula, k_value, dischard, ccw, avail_diams,...
                 alpha, arcs_fixed, length_fixed, node_edge);

  C = eye(size(idx,1)) - R * F_dx (:,idx);
  i=0;

  while ( ~ready ) && (i < 2) % improved interval iteration
    i = i+1;
    X = Z + C * Y;
    ready = all(all(in0(X,Y)));
    Y = intersect(X,Yold);
  end
end

X = intval(x);

if ready
  X(idx) = xs+Y; % verified inclusion
else
  X(idx) = intval(repmat(NaN,size(idx,1),1)); % inclusion failed
end



% ---------------------------------------------------------------------------
function [X] = verify_psi (xs, flow, diam, k_value, ccw)
%
% VERIFY_PSI - Verified inclusion for nonlinear function psi(f) = 0
%               close to double vector approx
%
% Returns inclusion interval if verification succeeded, else NaN

PEPS        = 1e-10; % minimum value to accept gradient as non-zero
MAX_UB_ITER = 10;
LOG10       = log (intval(10));

y_dx = - 0.5 * xs^(-1.5)...
      - (1/ (log(10) * (xs + k_value * abs(mid(flow))...
                    / (ccw(2) * mid(diam)^2 * mid(ccw(1))) * xs^(1.5))));

if (abs(y_dx) < PEPS)
  fprintf('VNOP-verify_psi: gradient seems to be zero.');
  X = intval(NaN); % no inclusion obtained
  return
end

% The following lines are taken from the procedure "verifynlss" of
% INTLAB Version 4.1.2, S. M. Rump 2004, and adjusted to the syntax of psi:
```

```
% Interval iteration
R = 1 / y_dx;
xs_aux = sqrt(intval(xs));

Z = - R * ( 1 / xs_aux + 2 * log10( ccw(1).*diam ./ abs(flow)./ xs_aux...
                                    + k_value ./ (ccw(2) .* diam)));
X = Z;

E = 0.1*rad(X)*hull(-1,1) + midrad(0,realmin);
ready = 0; k = 0;

while ( ~ready ) && (k < MAX_UB_ITER) && ( ~any(isnan(X)) )
  k = k+1;
  Y = hull( X + E , 0 );          % epsilon inflation
  Yold = Y;
  x = xs+Y;

  x_aux = x^(1.5);
  y_dx = - 0.5 * 1/x_aux...
           - (1/ (LOG10 * (x + k_value * abs(flow)...
                               / (ccw(2) * diam * diam * ccw(1)) * x_aux)));
  C = 1 - R * y_dx;
  i=0;
  while ( ~ready ) && ( i < 2 ) % improved interval iteration
    i = i+1;
    X = Z + C * Y;
    ready = in0(X,Y);
    Y = intersect(X,Yold);
  end
end

if ready
  X = xs+Y;          % verified inclusion
else
  X = intval(NaN); % inclusion failed
end


% ---------------------------------------------------------------------------
% LOCAL SUBFUNCTIONS FOR HAZEN-WILLIAMS
% ---------------------------------------------------------------------------
function [y] = sgn_pow (q, dischard)
%
% SGN_POW - Function for nonlinear part of Hazen-Williams equation

if isa(q, 'intval')
  q_inf = inf_(q); q_sup = sup(q);

  y = infsup( inf_(sign(q_inf) .* (abs(q_inf).^dischard)),...
              sup(sign(q_sup) .* (abs(q_sup).^dischard)) );

else
  y = sign(q) .* (abs(q).^dischard);
end
```

```
% -----------------------------------------------------------------------------
function [fktvalue] = q_fkt_hw (x, vmin, vmax, dischard, varargin)
%
% Q_FKT_HW - Flow: q(lambda), vmin, vmax are of type intval

fktvalue = sgn_pow ( x.*vmin + (intval(1) - x).*vmax, (1/dischard) );


% -----------------------------------------------------------------------------
function [fktvalue] = q_dt_hw (x, vmin, vmax, dischard, varargin)
%
% Q_DT_HW - Flow derivative: q'(lambda)

% vmin, vmax and dischard are of type intval
fktvalue = ((vmin - vmax) ./ dischard)...
          .* abs(x.*vmin + (intval(1) - x).*vmax).^(1/dischard - 1);



% -----------------------------------------------------------------------------
function [fktvalue] = rel_eq_bar_hw (x, flow_min, vmin, vmax, inv_dischard,...
                                                                 varargin)
% REL_EQ_BAR_HW - Flow equation (overline)

y = x.*vmin + (intval(1) - x).*vmax;

fktvalue = flow_min - sgn_pow (y, (inv_dischard))...
           - (intval(1) - x) * ((vmin - vmax) * inv_dischard)...
                            * (abs(y))^(inv_dischard - 1);


% -----------------------------------------------------------------------------
function [fktvalue] = rel_eq_tilde_hw (x, flow_max, vmin, vmax, inv_dischard,...
                                                                 varargin)
% REL_EQ_TILDE_HW - Flow equation (widetilde)

y = x.*vmin + (intval(1) - x).*vmax;

fktvalue = flow_max - sgn_pow (y, inv_dischard)...
           + x * ((vmin-vmax) * inv_dischard) * (abs(y))^(inv_dischard - 1);




% -----------------------------------------------------------------------------
% LOCAL SUBFUNCTIONS FOR DARCY-WEISBACH
% -----------------------------------------------------------------------------
function [fktvalue] = psi_fkt (x, flow, diam, k_value, ccw)
%
% PSI_FKT - Friction factor (Colebrook-White), f is zero of psi_fkt
% called only by u_fkt and u_fkt_1, hence flow ~= 0 and x > 0 can be assumed

fktvalue = 1 /(sqrt(x)) + 2 * log10( ccw(1).*diam ./ (abs(flow) .* sqrt(x))...
                                     + k_value ./ (ccw(2) .* diam) );
```

```
% ---------------------------------------------------------------------------
function [friction] = u_fkt (flow, diam, k_value, ccw)
% U_FKT
global wds_error

FMIN = 1e-4;       FMAX = 1e4;
ZERO = intval(0);

friction    = repmat(ZERO, size(flow));
f_max       = max(size(friction));
d_max       = max(size(diam));

if (d_max == 1)
  diam = repmat(diam, size(flow));
elseif (f_max ~= d_max)
  feval(wds_error, 'VNOP - u_fkt: dimension of flow and diam do not coincide');
end

for i = 1: f_max
  if (flow(i) > 0)
    % floating point approximation
    [f_approx, fval, exitflag]...
      = fzero (@ psi_fkt ,[FMIN FMAX],[], mid(flow(i)), diam(i),...
                                              mid(k_value), mid(ccw));
    if exitflag ~= 1
      feval(wds_error, 'VNOP - u_fkt: no zero of colebrook-white');
    end
    % verification of approximate solution
    [ friction(i) ] = verify_psi (f_approx, flow(i), diam(i), k_value, ccw);

    if isnan(friction(i))
      feval(wds_error, 'VNOP - u_fkt: no inclusion for psi-fkt found');
    end

  elseif (flow(i) == 0)
    friction(i) = ZERO; % not defined but used only in connection of w_fkt
  else
    feval(wds_error,...
      'VNOP - u_fkt: not defined for intervals containing 0 or negative flow');
  end
end


% ---------------------------------------------------------------------------
function [friction] = u_fkt_l (flow, diam, k_value, ccw)
% U_FKT_L - The long version of u_fkt, where every flow-diameter combination
%           is determined
global wds_error

FMIN = 1e-4;       FMAX = 1e4;
ZERO = intval(0);

d_max = max(size(diam));
friction = repmat(ZERO, size(flow,1)*size(diam,1), 1);
```

```
for i = 1:max(size(flow))
  for k = 1:d_max
    if (flow(i) > 0)
      % floating point approximation
      [f_approx, fval, exitflag]...
        = fzero (@ psi_fkt ,[FMIN FMAX],[], mid(flow(i)), diam(k),...
                                                  mid(k_value), mid(ccw));
      if exitflag ~= 1
        feval(wds_error, 'VNOP - u_fkt_l: no zero of colebrook-white');
      end

      % verification of approximate solution
      [friction((i-1)*d_max + k)] = verify_psi (f_approx, flow(i),...
                                                  diam(k), k_value, ccw);
      if isnan(friction(i))
        feval(wds_error, 'VNOP - u_fkt_l: no inclusion for psi-fkt found');
      end

    elseif (flow(i) == 0)
      friction((i-1)*d_max + k) = ZERO; % not defined but used only for w_fkt
    else
     feval(wds_error,...
     'VNOP - u_fkt_l: not defined for intervals containing 0 or negative flow');
    end
  end
end


% ---------------------------------------------------------------------------
function [y] = w_fkt (flow, diam, k_value, ccw)
% W_FKT - only needed for interval flow

ZERO  = intval(0);
n_idx = (flow ~= 0);

if ~any(in(0, flow(n_idx)))

  y = repmat(ZERO, size(flow));

  y(n_idx) = abs(flow(n_idx)) .* flow(n_idx)...
             .* u_fkt (abs(flow(n_idx)), diam(n_idx), k_value, ccw);
else
  flow_inf = inf_(flow);                flow_sup = sup(flow);
  y_inf    = repmat(ZERO,size(flow_inf)); y_sup  = y_inf;
  l_idx    = (flow_inf ~= 0);           u_idx    = (flow_sup ~= 0);
  flow_inf = intval(flow_inf(l_idx));   flow_sup = intval(flow_sup(u_idx));

  y_inf(l_idx)...
    = abs(flow_inf).* flow_inf.* u_fkt (abs(flow_inf), diam(l_idx),k_value,ccw);
  y_sup(u_idx)...
    = abs(flow_sup).* flow_sup.* u_fkt (abs(flow_sup), diam(u_idx),k_value,ccw);

  y = infsup(inf_(y_inf), sup(y_sup));
end
```

```
% ------------------------------------------------------------------------------
function [y] = w_fkt_l (flow, diam, k_value, ccw)
% W_FKT_L - only needed for interval flow

flow_l     = kron( abs(flow) .* flow, ones(max(size(diam)), 1));
y          = flow_l .* u_fkt_l (abs(flow), diam, k_value, ccw);


% ------------------------------------------------------------------------------
function [phi] = w_fkt_dq (flow, diam, k_value, ccw)
% W_FKT_DQ - partial deviation dw/dq, only needed in the long version and
%            only defined for flow ~= 0
global wds_error

if any(in(0, flow))
  feval(wds_error, 'VNOP - w_fkt_dq: not defined for intervals containing 0');
end
flow_l   = kron( flow,   ones(max(size(diam)), 1));
diam_l   = repmat( diam, max(size(flow)), 1);
u        = u_fkt_l (abs(flow), diam, k_value, ccw);

phi = flow_l .* flow_l .* u...
      ./ (ccw(1).* diam_l./ log(10).* 10.^(-1./ (2*sqrt(u))) + abs(flow_l)./ 2);


% ------------------------------------------------------------------------------
function [flow] = inv_w_fkt (phi, diam, k_value, ccw)
% INV_W_FKT - w^(-1) (q) - range of definition of w^(-1) excludes the intervals
%             [-r,0) and (0,r], continuation set to zero for these values

r = (ccw(1) .* diam ./ (1 - k_value / ccw(2) ./ diam) ).^2; r_sup = sup(r);

q_aux = repmat (intval(0), size(phi)); idx = (abs(phi) > inf_(r));
if (max(size(diam)) > 1), diam = diam(idx); end

% first omitting sign(phi)
q_aux(idx)...
  = (-2) .* sqrt(abs(phi(idx)))...
    .* log10( ccw(1) * diam ./ sqrt(abs(phi(idx))) + k_value ./ ccw(2) ./ diam);

phi_inf = inf_(phi);  phi_sup = sup(phi);

idx_inf = ( ((-r_sup < phi_inf)&(phi_inf < r_sup)) & (phi_inf ~= 0));
idx_sup = ( ((-r_sup < phi_sup)&(phi_sup < r_sup)) & (phi_sup ~= 0));

flow = infsup( inf_(sign(phi_inf) .* q_aux), sup(sign(phi_sup) .* q_aux));

% usually these index vectors are expected to be empty
if ~isempty(idx_inf)
  flow(idx_inf) = infsup (min(inf_(flow(idx_inf)),0), sup(flow(idx_inf)));
end
if ~isempty(idx_sup)
  flow(idx_sup) = infsup (inf_(flow(idx_sup)), max(0,sup(flow(idx_sup))));
end
```

```
% ---------------------------------------------------------------------------
function [flow] = p_fkt (x, wmin, wmax, dummy, diam, k_value, ccw)
% P_FKT - needed for relaxation purposes, and based on continuation of w^(-1),
%          i.e. w^(-1) set to zero out of range of defintion

y      = x .* wmin + (intval(1) - x) .* wmax;

flow   = inv_w_fkt (y, diam, k_value, ccw);


% ---------------------------------------------------------------------------
function [flow] = p_dx (x, wmin, wmax, dummy, diam, k_value, ccw)
% P_DX - Partial derivative of p, i.e. dp /dlambda (lambda)
global wds_error

y = x .* wmin + (intval(1) - x) .* wmax;

r = sup((ccw(1) * diam / (1 - k_value / ccw(2) / diam) )^2);

if all(abs(y) > r)
  z      = ccw(1) * diam ./ sqrt(abs(y)) + k_value ./ (ccw(2) .* diam);
  flow   = (log10(z) ./ sqrt(abs(y)) - ccw(1) * diam./ abs(y)./ z./ log(10))...
             .* (wmax - wmin);
else
  feval(wds_error, 'VNOP - p_dx: not defined for intervals close to 0');
end


% ---------------------------------------------------------------------------
function [fktvalue] = rel_eq_bar_dw (x, flow_min, wmin, wmax, dummy,...
                                     diam, k_value, ccw)
% REL_EQ_BAR_DW - Flow equation (overline)

fktvalue = flow_min - p_fkt (x, wmin, wmax, [], diam, k_value, ccw)...
           - (intval(1) - x) .* p_dx (x, wmin, wmax, [], diam, k_value, ccw);


% ---------------------------------------------------------------------------
function [fktvalue] = rel_eq_tilde_dw (x, flow_max, wmin, wmax, dummy,...
                                       diam, k_value, ccw)
% REL_EQ_TILDE_DW - Flow equation (tilde)

fktvalue = flow_max - p_fkt (x, wmin, wmax, [], diam, k_value, ccw)...
           + x * p_dx (x, wmin, wmax, [], diam, k_value, ccw);
```

B.6 Extracts of "vbb.m" containing Verification of the Branch and Bound Algorithm

```
function [opt_vec, global_lb, global_ub, infeasible, no_subs, no_subs_left]...
          = vbb (opt_vec, head_loss_formula, k_value, verification,...
                 lp_solver, accuracy, branching, max_iter, dischard, ccw,...
                 avail_diams, costs, head_cost, alpha, arcs, arc_length,...
                 arcs_fixed, length_fixed, seg_fixed, node_edge, demand,...
                 elevation, flow_min, flow_max, head_min, head_max,...
                 non_tree, post_tree, adjust_lbeq, relax, select_idx,...
                 upper_bounding, adjust_flow, adjust_head, test_adjacency,...
                 verify_lb, verify_ub, verify_infeasibility,...
                 Aeq, beq, lb, ub, lb_ub, ub_ub, verbose);
%
% VBB - Verified branch and bound algorithm of the water design network
%       optimization problem. [...]
% verification: (1) - b&b includes upper bound search, but verifies at the end
%               (2) - b&b is based on fixed a priori verified upper bound
%               (3) - same as (1), but every upper bound is verified directly
%               (4) - uses (2) first and changes later to (3)
% [...]
global No_arcs No_nodes No_dia No_seg No_relax No_lambda No_narcs No_nseg...
       wds_error
% Constants
MAX_SUBS = 1000;  % preallocation size of list of subproblems
BS_P     = 0.9;   % selection percentage used for branching strategy 1 and 2
V3_P     = 0.5;   % percentage of max_iter where verification (4) changes
                  % from (2) to (3)
WEPS     = 0.1;   % miminum absolute flow to start verification of upper bound
if (verification == 1)
  BBUB = 10^-2;   % factor for a priori pruning of subproblems
else
  BBUB = 0;
end

% Initialization of B&B variables
no_subs       = 1;
no_subs_left  = 0;
infeasible    = 1;
global_lb     = -inf;

if (branching == 1) || (branching == 2),  head_mstr = 0;
elseif (branching == 3),                  head_mstr = 1; branching = 1;
elseif (branching == 4),                  head_mstr = 1; branching = 2;
end
if (verification == 4)
  change_verification = 1; change_idx = floor(V3_P * max_iter);
  verification = 2;
else
  change_verification = 0;
end
nseg = 1:No_seg; nseg(seg_fixed) = []; seg_fixed_idx = find(length_fixed ~= 0);
```

```
lb_costs = [costs; costs; zeros(No_lambda+No_arcs,1);...
            head_cost; zeros(No_nodes-1,1)];
ub_costs = [costs; head_cost; zeros(No_nodes-1,1)];

% Test, if adjacency property is fulfilled
[fulfilled, err_str] = feval(test_adjacency , head_loss_formula, avail_diams,...
                                                 costs(1:No_dia), alpha);
if (~fulfilled), feval(wds_error, err_str); end

% A priori verification of upper bound
if (verification == 2)
  if (strcmp (head_loss_formula, 'DW')...
       & (any(abs(opt_vec(No_seg+1:No_seg+No_arcs)) < WEPS)) )
    verified = 0;
    error_str = 'VBB - verify_ub not called as actual flow contains 0';
  else
    [global_ub , opt_vec , verified, error_str]...
      = feval(verify_ub , opt_vec, head_loss_formula, k_value, lp_solver,...
                        dischard, ccw, avail_diams, ub_costs, alpha, arcs,...
                        arc_length, arcs_fixed, length_fixed, seg_fixed,...
                        node_edge, elevation, demand, flow_min, flow_max,...
                        head_min, head_max, verbose);
  end
  if (~verified)
    feval(wds_error,...
       ['VBB - Floating point solution could not be verified. - ', error_str]);
  end
  infeasible = 0;
else
  global_ub = inf;
  opt_vec  = [];
end
% ... initialization of B&B variables
status = 1; exitflag = 1;
if strcmp (lp_solver , 'linprog')
  linprog_options = optimset('linprog');
  linprog_options = optimset(linprog_options, 'Display','off');
  warning('off','MATLAB:divideByZero');
end
% Definition and preallocation of data structure for subproblems
subproblem = repmat( struct('d_idx', 0,...
                            'd_q',    0,...
                            'qmin',   repmat(0, No_arcs, 1),...
                            'qmax',   repmat(0, No_arcs, 1),...
                            'hmin',   repmat(0, No_nodes, 1),...
                            'hmax',   repmat(0, No_nodes, 1),...
                            'lb',     -inf,...
                            'status', 0),...
                      MAX_SUBS, 1);
subproblem(1).qmin = flow_min;  subproblem(1).qmax = flow_max;
subproblem(1).hmin = head_min;  subproblem(1).hmax = head_max;

% Data for first loop
sub_idx = 1; sub2 = 1; sub1=[];
```

```
for bb_idx = 1: max_iter    % Branch and bound loop
 if ((change_verification) && (bb_idx == change_idx)), verification = 3; end

 for s_idx = [sub2, sub1]
  if verbose
   fprintf('--------------------------------------------------------\n');
   fprintf('Global upper bound: \t \t \t %8.4f\n', global_ub);
   fprintf('Subproblem's old lower bound:\t\t %8.4f\n',subproblem(s_idx).lb);
  end
  % Adjustment of flow bounds according to MSTR
  mstr_prune_subproblem = 0;
  q_mstr = infsup( subproblem(s_idx).qmin, subproblem(s_idx).qmax);
  for i = post_tree
   flowbounds_1 = q_mstr(i);
   flowbounds_2 = - demand(arcs(i,2))...
    - sum(q_mstr( arcs(:,2) == arcs(i,2) & arcs(:,1) ~= arcs(i,1)))...
    + sum(q_mstr( arcs(:,1) == arcs(i,2) ));
   flowbounds_3 = demand(arcs(i,1))...
    + sum(q_mstr ( arcs(:,2) == arcs(i,1) ))...
    - sum(q_mstr ( arcs(:,1) == arcs(i,1) & arcs(:,2) ~= arcs(i,2) ));

   fb_min =max([inf_(flowbounds_1), inf_(flowbounds_2), inf_(flowbounds_3)]);
   fb_max =min([sup(flowbounds_1), sup(flowbounds_2), sup(flowbounds_3)]);
   if (fb_min <= fb_max)
    q_mstr(i) = infsup (fb_min, fb_max);
   else
    mstr_prune_subproblem = 1;
   end
  end % i = post_tree
  if mstr_prune_subproblem
   if verbose
    fprintf('--------------------------------------------------------\n');
    fprintf('Subproblem pruned by flow MSTR \n');
   end
   subproblem(s_idx) = []; continue;
  end
  subproblem(s_idx).qmin = inf_(q_mstr); subproblem(s_idx).qmax = sup(q_mstr);
  % Adjustment of flow bounds according to head constraint propagation
  if head_mstr
   [hmin, hmax]...
    = feval( adjust_head , subproblem(s_idx).hmin,subproblem(s_idx).hmax,...
                          subproblem(s_idx).qmin,subproblem(s_idx).qmax,...
                          head_loss_formula, k_value, dischard, ccw,...
                          avail_diams, alpha, post_tree, non_tree, arcs,...
                          arc_length, arcs_fixed, length_fixed, seg_fixed);
   if all(hmin <= hmax)
    subproblem(s_idx).hmin = hmin;  subproblem(s_idx).hmax = hmax;
   else
    if verbose
     fprintf('--------------------------------------------------------\n');
     fprintf('Subproblem pruned by adjust_head\n');
    end
    subproblem(s_idx) = []; continue;
   end
```

```
    [qmin, qmax]...
      = feval( adjust_flow , subproblem(s_idx).qmin,subproblem(s_idx).qmax,...
                             subproblem(s_idx).hmin,subproblem(s_idx).hmax,...
                             head_loss_formula, k_value, dischard, ccw,...
                             avail_diams, alpha, arcs, arc_length,...
                             arcs_fixed, length_fixed, seg_fixed);
    if all(qmin <= qmax)
      subproblem(s_idx).qmin = qmin; subproblem(s_idx).qmax = qmax;
    else
      if verbose
        fprintf('-----------------------------------------------------\n');
        fprintf('Subproblem pruned by adjust_flow\n');
      end
      subproblem(s_idx) = []; continue;
    end
  end
  % Computation of adjusted equalities and relaxations for subproblems
  [Aeq(1:No_arcs,:), beq(No_narcs+1:No_arcs)]...
    = feval(adjust_lbeq , head_loss_formula, k_value, dischard, ccw,...
                          avail_diams, alpha, arcs, arcs_fixed,...
                          length_fixed, seg_fixed, node_edge, elevation,...
                          subproblem(s_idx).qmin, subproblem(s_idx).qmax);
  if (No_relax ~= 0)
    A = repmat(0, No_relax*No_lambda,2*No_nseg+No_lambda+No_arcs+No_nodes);
    b = repmat(0, No_relax*No_lambda,1);
    if ~strcmp (head_loss_formula, 'DW')
      for i = 1:No_arcs
        if (subproblem(s_idx).qmin(i) < subproblem(s_idx).qmax(i))

          [A((i-1)*No_relax+1:i*No_relax,:),b((i-1)*No_relax+1:i*No_relax)]...
            = feval( relax , i, i, head_loss_formula,...
                             subproblem(s_idx).qmin(i),...
                             subproblem(s_idx).qmax(i), dischard,...
                             [],[],[], verbose);
        end
      end
    else
      for i = 1:No_lambda
        j  = ceil(seg_fixed_idx(i) / No_dia);  % index for arcs_fixed
        k  = mod(seg_fixed_idx(i), No_dia);    % index for diameter
        if (k==0), k = No_dia; end

        if (subproblem(s_idx).qmin(j) < subproblem(s_idx).qmax(j))
          [A((i-1)*No_relax+1:i*No_relax,:),b((i-1)*No_relax+1:i*No_relax)]...
            = feval( relax , i, j, head_loss_formula,...
                             subproblem(s_idx).qmin(j),...
                             subproblem(s_idx).qmax(j),...
                             dischard, avail_diams(k),...
                             k_value, ccw, verbose);
        end
      end
    end
  else A = []; b = []; % No_relax = 0
  end
```

```
% Adjustment of lower and upper bounds
lb(2*No_nseg+No_lambda+1 : 2*No_nseg+No_lambda+No_arcs+No_nodes)...
  = [subproblem(s_idx).qmin; inf_(subproblem(s_idx).hmin - elevation)];
ub(2*No_nseg+No_lambda+1 : 2*No_nseg+No_lambda+No_arcs+No_nodes)...
  = [subproblem(s_idx).qmax; sup(subproblem(s_idx).hmax - elevation)];

lb_old = subproblem(s_idx).lb;% needed for selection of index to be branched
if strcmp (lp_solver , 'cplex')
  [fl_lb, x_sub, lambda, status]...
    = lp_cplex_mex (lb_costs, mid([A; Aeq]), mid([b; beq]), lb, ub,...
                    0, transpose(1:(size(A,1))));
else
  [x_sub, fl_lb, exitflag, infos, lambda]...
    = linprog(lb_costs, mid(A), mid(b), mid(Aeq), mid(beq), lb, ub,...
                                              [], linprog_options);
end
if verbose, fprintf('Subproblem's new lower bound:\t\t %8.4f\n', fl_lb); end
% Verification of lower bound
if strcmp (lp_solver , 'cplex')
  y = lambda(1: No_lambda*No_relax);
  z = lambda(No_lambda*No_relax+1: end);
else
  y = -lambda.ineqlin;
  z = -lambda.eqlin;
end
if isempty(A)
  A = zeros(1,size(Aeq,2)); b = 0; y = 0;
end
subproblem(s_idx).lb...
  = feval( verify_lb ,lb_costs, A, b, Aeq, beq, lb, ub, y, z);
if verbose
  fprintf('    ... verified lower bound:\t\t %8.4f\n',subproblem(s_idx).lb);
  fprintf('---------------------------------------------------\n');
end
% Ensure, that lower bound of branched problem does not decrease
if (subproblem(s_idx).lb < lb_old), subproblem(s_idx).lb = lb_old; end

% Prune by infeasibility: admissible region of subproblem is empty
if (exitflag <= 0) || (status ~= 1)
  if (subproblem(s_idx).lb > global_ub + BBUB )
    subproblem(s_idx) = []; continue;
  else
    subproblem_infeasible...
      = feval( verify_infeasibility , lp_solver, A, b, Aeq, beq, lb, ub);
    if subproblem_infeasible
      if verbose
        fprintf('VB&B info: subproblem pruned by infeasibility!\n');
      end
      subproblem(s_idx) = []; continue;
    else
      subproblem(s_idx).status = 1;
    end
  end
end
```

```
% Find index to be branched
if (subproblem(s_idx).lb * BS_P > lb_old), glb = 1; else glb = 0; end

x_sub = [x_sub(1:No_nseg) + x_sub(No_nseg+1:2*No_nseg);...
         x_sub(2*No_nseg+No_lambda+1:end)];
x_aux = x_sub(1:No_nseg);
x_sub = [mid(length_fixed); x_sub(No_nseg+1:end)]; x_sub(nseg) = x_aux;

[subproblem(s_idx).d_idx, subproblem(s_idx).d_q]...
   = feval( select_idx , x_sub(1:No_seg),...
                     x_sub(No_seg+1: No_seg+No_arcs),...
                     x_sub(No_seg+No_arcs+1: No_seg+No_arcs+No_nodes),...
                     glb, head_loss_formula, mid(k_value), branching,...
                     mid(dischard), mid(ccw), avail_diams,...
                     mid(alpha), arcs, mid(elevation), non_tree,...
                     subproblem(s_idx).qmin, subproblem(s_idx).qmax);

% Computation of upper bound, apart from branching-strategy (2)
if (verification ~= 2)
  Aeq_ub = repmat(0, No_narcs+No_arcs, No_nseg+No_nodes);
  beq_ub = repmat(0, No_narcs+No_arcs, 1);

  flow_sub = x_sub(No_seg+1: No_seg+No_arcs);

  [Aeq_ub, beq_ub]...
    = feval(upper_bounding , flow_sub, head_loss_formula, mid(k_value),...
                             mid(dischard), mid(ccw), avail_diams,...
                             mid(alpha), arcs, mid(arc_length),...
                             arcs_fixed, mid(length_fixed), seg_fixed,...
                             node_edge, mid(elevation));

  if strcmp (lp_solver , 'cplex')
    [fval_ub, x_sub, lambda, status]...
     = lp_cplex_mex (ub_costs, Aeq_ub, beq_ub, lb_ub, ub_ub);
  else
    [x_sub, fval_ub, exitflag, infos]...
     = linprog(ub_costs, [], [], Aeq_ub, beq_ub,...
                              lb_ub, ub_ub, [], linprog_options );
  end
  if (exitflag <= 0) || (status ~= 1)
    if verbose,
      fprintf('VB&B info: no solution of upper bounding problem!\n');
    end
    continue;
  end

  if (verification == 3)
    x = [mid(length_fixed); flow_sub; x_sub(No_nseg+1:end)];
    x(nseg) = x_sub(1:No_nseg);

    if (strcmp (head_loss_formula, 'DW') & (any(abs(flow_sub) < WEPS)))
      verified = 0;
      error_str = 'VBB - verify_ub not called as actual flow contains 0';
    else
```

```
        [fval_ub, x_sub, verified, error_str]...
          = feval( verify_ub , x, head_loss_formula, k_value, lp_solver,...
                             dischard, ccw, avail_diams, ub_costs, alpha,...
                             arcs, arc_length, arcs_fixed,...
                             length_fixed, seg_fixed, node_edge,...
                             elevation, demand, flow_min, flow_max,...
                             head_min, head_max, verbose);
      end
      if verified
        if (fval_ub < global_ub)
          infeasible = 0;        global_ub = fval_ub;  opt_vec = x_sub;
        end
      else % ~verified
        fprintf([error_str,'\n']); continue;
      end
    else % verification == 1
      if (fval_ub < global_ub)
        infeasible     = 0;      global_ub = fval_ub;
        opt_vec        = [mid(length_fixed); flow_sub; x_sub(No_nseg+1:end)];
        opt_vec(nseg)  = x_sub(1:No_nseg);
      end
    end
  end % (verification == 1) || (verification == 3)
end % SUBPROBLEM'S-FOR-LOOP


% Clean list according to obsolete subproblems
subproblem([subproblem.lb] > global_ub + BBUB ) = [];
sub_idx = ( [subproblem.lb] ~= -inf );

if (min([subproblem(sub_idx).lb]) > (global_ub * ( 1 - accuracy) ))
  global_lb = min([subproblem(sub_idx).lb]);
  no_subs_left = sum(sub_idx);
  subproblem = [];
end

% Break if work for all subproblems is done
if isempty(subproblem) || (sum([subproblem.d_idx]~=0) == 0)
  if verbose, fprintf('VB&B info: "No subproblems left!"\n'); end
  if (verification ~= 1) || (infeasible), return; end

  % Verification of upper bound
  if (strcmp (head_loss_formula, 'DW')...
      & (any(abs(opt_vec(No_seg+1:No_seg+No_arcs)) < WEPS)))
    verified = 0;
    error_str = 'VBB - verify_ub not called as actual flow contains 0';
  else
    [verified_ub , opt_vec , verified, error_str]...
      = feval( verify_ub , opt_vec , head_loss_formula, k_value, lp_solver,...
                           dischard, ccw, avail_diams, ub_costs, alpha,...
                           arcs, arc_length, arcs_fixed, length_fixed,...
                           seg_fixed, node_edge, elevation, demand,...
                           flow_min, flow_max, head_min, head_max, verbose);
  end
```

```
    if (~verified), feval(wds_error, error_str); end
    if verbose, fprintf('verified upper bound : %f \n',verified_ub); end

    % Check of a posteriori verification
    if (global_ub + BBUB) < verified_ub
      err_str =...
        ['--------------------------------------------------\n',...
         '--------------------------------------------------\n',...
         sprintf('Verified upper bound:  \t \t %8.4f\n', verified_ub),...
         sprintf('Upper bound used for B&B: \t \t %8.4f\n', global_ub + BBUB),...
         sprintf('--------------------------------------------------\n'),...
         'Accuracy for upper bound not large enough for verified result!'];
      feval(wds_error , err_str);
    end
    global_ub = verified_ub;
    return;
  end % if work for all subproblems is done



  % Node selection step - sub_idx = subproblem to be worked out
  sub_idx = find( ([subproblem.lb] ~= -inf) & ([subproblem.status] ~= 1));

  if (verification == 2) || isempty(sub_idx) ||...
     min([subproblem(sub_idx).lb]) > (global_ub * ( 1 - accuracy ))
    sub_idx = find( [subproblem.lb] ~= -inf );
  end
  aux_idx = find( [subproblem(sub_idx).lb] == min( [subproblem(sub_idx).lb] ));
  aux_idx = aux_idx(end);
  sub_idx = sub_idx(aux_idx);


  % Partitioning step - branching of subproblem
  no_subs = no_subs+2;
  sub1 = sum([subproblem.d_idx]~=0) + 1; sub2 = sub1 + 1;

  subproblem(sub1) = subproblem(sub_idx);
  subproblem(sub2) = subproblem(sub_idx);
  d_idx = subproblem(sub_idx).d_idx;

  subproblem(sub1).qmax(d_idx) = subproblem(sub_idx).d_q;
  subproblem(sub2).qmin(d_idx) = subproblem(sub_idx).d_q;

  subproblem(sub_idx) = []; sub1 = sub1 - 1; sub2 = sub2 - 1;
  if verbose, fprintf('Actual number of subproblems:\t %d\n', sub2); end

end % B&B-FOR-LOOP


error_str = sprintf(['VBB: maximum number of iterations reached ! \n',...
                     'max_iter = %d\n'], max_iter);
feval(wds_error , error_str);
```

B.7 Extracts of "output.m" containing Formatting of Calculated Results

```
function output (infeasible, opt_vec, global_lb, global_ub, no_sub,...
                 no_sub_left, time, cpu, u_fkt_l, town, s_town,...
                 head_loss_formula, k_value, viscosity, source_head,...
                 head_cost, verification, lp_solver, accuracy, branching,...
                 max_iter, dischard, ahw, ccw, avail_diams, pipe_cost,...
                 alpha, arcs, arcs_fixed, length_fixed, seg_fixed, elevation)
%
% OUTPUT - Structured output of the optimal solution of the water design
%          network optimization problem

% [...]
global No_arcs No_nodes No_dia No_seg No_relax No_narcs wds_parameter_str
LEPS = 1e-4; % floating point: length <= LEPS is regarded as zero

% Argument processing [...]

  % Write all actual variables to file
  mdate = datestr(now,0); mdate(findstr(' ',mdate)) = '-';
  save (strcat('./results/',mdate,'.mat'));
  out_file = fopen(strcat('./results/',mdate,'.txt'), 'w');
  wds_file = fopen(strcat('./results/',mdate,'.wds'), 'w');

% [...]
% Adjustments for parameters of verified calculation
if (verification ~= 0)
  k_value = mid(k_value); dischard = mid(dischard); ahw = mid(ahw);
end
% Output of parameters used
for fp = [1, out_file, wds_file]
  fprintf(fp, wds_parameter_str);
  if strcmp (head_loss_formula, 'AHW')
    fprintf(fp, 'Adjusted Hazen-Williams coefficients:\n');
    fprintf(fp, '   dischard    = %f \n', dischard);
    fprintf(fp, '   diameter AHW \n');
    fprintf(fp, '   %6.0f %8.2f \n', [avail_diams, ahw]');
  end
  fprintf(fp,'-----------------------------------------------------------\n');
  fprintf(fp,'-----------------------------------------------------------\n');
  fprintf(fp,'Number of subproblems calculated: %12d \n', no_sub);
  fprintf(fp,'Number of subproblems remaining: %12d \n\n', no_sub_left);
  fprintf(fp,'-----------------------------------------------------------\n');
  fprintf(fp,'Time needed for computation: %f seconds\n', time);
  fprintf(fp,'CPU Time:                    %f seconds\n\n', cpu);
  fprintf(fp,'-----------------------------------------------------------\n');
  if infeasible, fprintf(fp, '\n\n\n Problem is infeasible!\n\n'); end
end
if infeasible
  fclose(out_file); fclose(wds_file);
  return;
end
```

```
% Partition of opt_vec
length  = opt_vec (1 : No_seg);
flow    = opt_vec (No_seg+1 : No_seg+No_arcs);
head    = opt_vec (No_seg+No_arcs+1 : No_seg+No_arcs+No_nodes);

% New determination of maximal deviation of nonlinear constraints
if (verification == 0)
  if ~strcmp (head_loss_formula, 'DW')
    head_loss = sign(flow) .* (abs(flow).^dischard)...
                .* (kron(eye(No_arcs),alpha) * length);
  else
    w_aux = repmat(0, No_seg,1);
    w_aux = kron(abs(flow).*flow, ones(No_dia,1)) .*...
            feval( u_fkt_1, abs(flow), avail_diams, k_value, ccw);
    head_loss = kron(eye(No_arcs),alpha) * (w_aux .* length);
  end
  delta = abs(( elevation(arcs(:,1)) - elevation(arcs(:,2)) +...
             head(arcs(:,1)) - head(arcs(:,2))...
            ) - head_loss );
end
% Conversion of diameters to original unit (inch)
if strcmp (town ,'twoloop') || strcmp (town ,'twoloop_exp')...
   || strcmp (town ,'hanoi')
     avail_diams = avail_diams ./ 2.54;
elseif ~strcmp (town , 'newyork')
  fprintf('\nOUTPUT warning: unknown test case, use of standard units!\n');
  avail_diams = avail_diams ./ 2.54;
end

% Layout for Interpretation of Water Distribution System
% UNITS: diameter in [inch], flow in [m^3/s], fluid velocity in [m/s]
if strcmp (town ,'newyork')
 wds_flow = flow*(0.3048^3); wds_length = length*0.3048; wds_head = head*0.3048;
else
 wds_flow = flow / 3600; wds_length = length; wds_head = head;
end
wds_length(seg_fixed) = 0;

% Preparation for wds file
if (verification == 0)
  % IF lenght <= LEPS THEN pipe length "0" is assumed
  idx_n = find( wds_length >= LEPS );
  idx_f = find( length_fixed >= LEPS );
else
  idx_n = find( wds_length ~= 0);    wds_length  = mid(wds_length);
  idx_f = find( length_fixed ~= 0);  wds_flow    = mid(wds_flow);
  wds_head = mid(head);
end

% Newly to be designed arcs
d_idx = mod(idx_n,No_dia); d_idx(d_idx==0) = No_dia;
q_idx = ceil(idx_n./No_dia);
wds_layout_n = [arcs(q_idx,:), avail_diams(d_idx), wds_length(idx_n),...
                  4 * wds_flow(q_idx)./ (pi*(avail_diams(d_idx)*0.0254).^2)];
```

```
% Fixed arcs
d_idx = mod(idx_f,No_dia); d_idx(d_idx==0) = No_dia;
q_idx = ceil(idx_f./No_dia);

wds_layout_f = [arcs(q_idx,:), avail_diams(d_idx), mid(length_fixed(idx_f)),...
                4 * wds_flow(q_idx)./ (pi*(avail_diams(d_idx)*0.0254).^2)];

length(seg_fixed) = []; new_arcs = arcs; new_arcs(arcs_fixed,:) = [];

% Preparation for txt file: extract all indices where length is significant
if (verification == 0)
  % IF lenght <= LEPS THEN pipe length "0" is assumed
  l_idx    = find( length >= LEPS );
  diam_idx = mod(l_idx,No_dia);  diam_idx(diam_idx==0) = No_dia;
  layout   = [new_arcs(ceil(l_idx./No_dia),:),...
                                   avail_diams(diam_idx), length(l_idx)];
else
  l_idx    = find(length ~= 0);
  diam_idx = mod(l_idx,No_dia);  diam_idx(diam_idx==0) = No_dia;
  layout   = [new_arcs(ceil(l_idx./No_dia),:), avail_diams(diam_idx)];
end

% Output of computational results
for fp = [1, out_file, wds_file]
fprintf(fp,'------------------------------------------------------------\n');
fprintf(fp,'Bounds for the optimal value of the distribution network:\n\n');
if (verification == 0)
  fprintf(fp,'                          [%13.4f , %13.4f]\n\n\n',...
                                               global_lb , global_ub);
else
  fprintf(fp,'                          [%14s , %14s]\n\n\n',...
                                   s_inf (global_lb), s_sup (global_ub));
end
% [...]
if (verification == 0)
  fprintf(fp,'------------------------------------------------------------\n');
  fprintf(fp,'Maximal deviation of solution to nonlinear constraints \n');
  fprintf(fp,'Delta:  %4.10f\n', norm(delta, inf));
  fprintf(fp,'------------------------------------------------------------\n');
  fprintf(fp,'Accuracy of optimal solution: %4.10f\n', global_ub - global_lb);
  fprintf(fp,'------------------------------------------------------------\n');
else
  fprintf(fp,'------------------------------------------------------------\n');
  fprintf(fp,'Maximal diameter of solution intervals \n');
  fprintf(fp,'Length:  %e\n',     max(diam(length)));
  fprintf(fp,'Flow:    %e\n',     max(diam(flow)));
  fprintf(fp,'Head:    %e\n\n',   max(diam(head)));
  fprintf(fp,'Accuracy of optimal solution: %4.10f\n',...
                                   diam(infsup(global_lb,global_ub)));
  fprintf(fp,'------------------------------------------------------------\n');
end
fprintf(fp,'\n\n\n');
end % for (output files)
fclose(out_file); fclose(wds_file);
```

```
% --------------------------------------------------------------------------
% for LaTeX purposes:
fp = fopen(strcat('./results/',mdate,'.tex'), 'w');
fprintf(fp,'--------------------------------------------------------------\n');
fprintf(fp,'\\begin{table}[!htbp] \n');
% [...]
fprintf(fp,'\\end{table} \n\n\n');
fclose(fp);


% --------------------------------------------------------------------------
% SUBFUNCTIONS
% --------------------------------------------------------------------------
function [x_inf] = s_inf (x, len, prec)
%
% S_INF - Infimum of the interval x as decimal string,
%          formatted as %(len).(prec)f
global wds_error
MAX_ITER = 20;

if ~exist('len')
  len = 13; prec = 4;
end
formatstr = [ '%' sprintf('%d',len) '.' sprintf('%d',prec) 'f' ];
x_inf = sprintf(formatstr, inf_(x));

for i = 1:MAX_ITER
  if (intval(x_inf) <= inf_(x))
    return;
  else
    x_inf = sprintf(formatstr, inf_(x)-i*10^-(prec+1));
  end
end
feval(wds_error, 'S_INF: maximum number of iterations reached.');


% --------------------------------------------------------------------------
function [x_sup] = s_sup (x, len, prec)
%
% S_SUP - Supremum of the interval x as decimal string,
%          formatted as %(len).(prec)f
global wds_error
MAX_ITER = 20;

if ~exist('len')
  len = 13; prec = 4;
end
formatstr = [ '%' sprintf('%d',len) '.' sprintf('%d',prec) 'f' ];
x_sup = sprintf(formatstr, sup(x));

for i = 1:MAX_ITER
  if (intval(x_sup) >= sup(x))
    return;
  else
    x_sup = sprintf(formatstr, sup(x)+i*10^-(prec+1));
  end
end
feval(wds_error, 'S_SUP: maximum number of iterations reached.');
```

Appendix C

Evaluation of Computational Time

Name	Time		Calls	Self time		Location
wds	1,517.98	100.0%	1	0.46	0.0%	
bb	1,517.13	99.9%	1	**517.52**	**34.1%**	
nop/relax	607.54	40.0%	304,919	**238.83**	**15.7%**	
repmat	255.51	16.8%	1,640,999	**255.51**	**16.8%**	r13.1/toolbox/
lp_cplex_mex	167.72	11.0%	17,272	**167.72**	**11.0%**	
kron	107.08	7.1%	89,861	**78.23**	**5.2%**	r13.1/toolbox/
fzero	92.05	6.1%	17,298	**66.00**	**4.3%**	r13.1/toolbox/
nop/adjust_head	69.78	4.6%	9,761	**39.63**	**2.6%**	
nop/adjust_lbeq	52.78	3.5%	9,240	18.53	1.2%	
nop/upper_bounding	39.26	2.6%	8,032	15.13	1.0%	
nop/adjust_flow	32.33	2.1%	9,259	13.75	0.9%	
meshgrid	28.85	1.9%	179,722	28.85	1.9%	r13.1/toolbox/
nop/q_dt_hw	26.71	1.8%	609,838	26.71	1.8%	
nop/select_idx	21.37	1.4%	8,032	7.26	0.5%	
nop/q_fkt_hw	16.41	1.1%	304,919	16.41	1.1%	
fzero/parse_all	15.17	1.0%	17,298	5.53	0.4%	r13.1/toolbox/
nop/rel_eq_bar_hw	7.86	0.5%	179,662	7.86	0.5%	
optimget	6.70	0.4%	34,596	3.00	0.2%	r13.1/toolbox/
nop/rel_eq_tilde_hw	4.05	0.3%	96,888	4.05	0.3%	
optimget/optimgetfast	3.70	0.2%	34,596	3.70	0.2%	r13.1/toolbox/
fcnchk	2.94	0.2%	17,298	2.94	0.2%	r13.1/toolbox/
output	0.17	0.0%	1	0.09	0.0%	
nop	0.12	0.0%	1	0.10	0.0%	
newyork	0.06	0.0%	1	0.05	0.0%	
datestr	0.06	0.0%	1	0.06	0.0%	r13.1/toolbox/
strcat	0.06	0.0%	6	0.03	0.0%	r13.1/toolbox/

Table C.1: Extracts of MATLAB Profile summary report, while calculating New York (12") on sdome with default parameters. 21 M-functions, 13 M-subfunctions and 3 MEX-functions have been called, the clock precision is stated as 0.01 seconds. Self times greater that 2 % are marked bold and add up to approximately 90 %.

Name	Time		Calls	Self time		Location
wds	37,468.02	100.0%	1	0.72	0.0%	
vbb	37,453.55	100.0%	1	508.21	1.4%	
vnop/relax	29,585.85	79.0%	368,346	365.63	1.0%	
intval/power	18,391.95	49.1%	2,790,808	**2,631.58**	**7.0%**	INTLAB
vnop/q_dt_hw	12,337.70	32.9%	736,692	208.79	0.6%	
vnop/sgn_pow	11,975.74	32.0%	1,328,813	250.60	0.7%	
vnop/q_fkt_hw	6,855.86	18.3%	368,346	60.46	0.2%	
intval/times	6,710.03	17.9%	8,919,717	**5,751.54**	**15.4%**	INTLAB
intval/exp	4,833.67	12.9%	2,790,808	**1,372.48**	**3.7%**	INTLAB
exp_rnd	3,272.24	8.7%	5,581,616	**3,159.73**	**8.4%**	INTLAB
intval/subsref	3,151.29	8.4%	7,777,411	**3,151.29**	**8.4%**	INTLAB
intval/subsasgn	2,751.88	7.3%	6,811,852	**2,741.76**	**7.3%**	INTLAB
intval/intval	2,725.08	7.3%	6,287,346	**2,657.26**	**7.1%**	INTLAB
vnop/rel_eq_bar_hw	2,650.44	7.1%	88,952	34.54	0.1%	
intval/minus	2,474.60	6.6%	5,611,881	**2,318.62**	**6.2%**	INTLAB
vnop/adjust_head	2,424.10	6.5%	11,685	200.67	0.5%	
repmat	2,405.22	6.4%	3,532,988	730.01	1.9%	r13.1/toolbox/
vnop/bisect_vlb	2,360.26	6.3%	10,274	8.97	0.0%	
intval/log	2,266.94	6.1%	914,777	357.23	1.0%	INTLAB
vnop/rel_eq_tilde_hw	1,905.71	5.1%	66,544	22.86	0.1%	
log_rnd	1,847.86	4.9%	1,829,554	**1,785.32**	**4.8%**	INTLAB
vnop/bisect_vub	1,610.15	4.3%	7,568	6.49	0.0%	
intval/rdivide	1,589.83	4.2%	1,922,959	**1,491.62**	**4.0%**	INTLAB
intval/mpower	1,248.74	3.3%	155,496	23.11	0.1%	INTLAB
intval/plus	1,189.09	3.2%	2,716,441	**1,113.92**	**3.0%**	INTLAB
intval/mrdivide	1,118.76	3.0%	1,186,264	109.10	0.3%	INTLAB
intval/issparse	984.10	2.6%	17,062,758	**984.10**	**2.6%**	INTLAB
setround	887.70	2.4%	93,040,404	**887.70**	**2.4%**	INTLAB
vnop/adjust_flow	851.64	2.3%	11,290	26.73	0.1%	
infsup	818.10	2.2%	880,943	307.95	0.8%	INTLAB
intval/mtimes	793.58	2.1%	626,365	272.75	0.7%	INTLAB
vnop/adjust_lbeq	672.07	1.8%	11,162	16.95	0.0%	
kron	521.00	1.4%	89,169	48.49	0.1%	r13.1/toolbox/
intval/sum	512.37	1.4%	1,190,700	478.40	1.3%	INTLAB
intval/size	509.36	1.4%	4,485,208	509.36	1.4%	INTLAB
intval/inf_	408.90	1.1%	3,593,798	408.90	1.1%	INTLAB
intval/abs	353.86	0.9%	892,194	344.95	0.9%	INTLAB
intval/length	310.22	0.8%	5,372,665	310.22	0.8%	INTLAB
intval/isnan	310.00	0.8%	3,728,817	310.00	0.8%	INTLAB
intval/mid	304.75	0.8%	71,872	302.88	0.8%	INTLAB
vnop/verify_lb	280.90	0.7%	11,162	30.21	0.1%	
intval/reshape	188.63	0.5%	1,667,089	188.63	0.5%	INTLAB
intval/sup	157.06	0.4%	3,378,090	157.06	0.4%	INTLAB
intval/full	149.33	0.4%	1,587,627	149.33	0.4%	INTLAB
lp_cplex_mex	144.24	0.4%	11,613	144.24	0.4%	

Table C.2: Extracts of MATLAB Profile report, while calculating New York (12") on `sdome` with default parameters and `verification` = 2, times for determining the floating point solutions are not included. 78 M-functions, 27 M-subfunctions and 4 MEX-functions have been called, the clock precision is stated as 0.01 seconds. Self times greater that 2 % are marked bold and add up to approximately 80 %.

lambda					
	Verification	`relax`	`sgn_pow`	`lp_cplex_mex`	CPU Time (s)
Two-loop	0	26 % (1,141)		24 % (318)	4
Two-loop	1	83 % (1,155)	32 % (6,258)	1 % (328)	74
New York (12")	0	34 % (327,096)		15 % (18,463)	522
New York (12")	2[1]	80 % (377,388)	32 % (1,365,141)	1 % (11,927)	15,737
Hanoi	0	28 % (122,499)		34 % (9,074)	220
Hanoi	1	82 % (96,579)	31 % (324,910)	1 % (7,156)	3,745
		`adjust_lbeq`	$\psi(f)$	`lp_cplex_mex`	
Two-loop	0	50 % (317)	22 % (4,006,071)	1 % (623)	292
Two-loop	1	93 % (367)	81 % (79,746)	1 % (727)	2,580
Two-loop Exp.	0	47 % (158)	21 % (2,120,801)	1 % (301)	157
Two-loop Exp.	1	78 % (162)	77 % (40,451)	1 % (323)	1,370

sdome					
	Verification	`relax`	`sgn_pow`	`lp_cplex_mex`	CPU Time (s)
Two-loop	0	39 % (1,141)		21 % (316)	10
Two-loop	1	83 % (1,155)	33 % (6,258)	1 % (329)	183
New York (12")	0	40 % (304,919)		11 % (17,272)	1,518
New York (12")	2[1]	79 % (368,346)	32 % (1,328,813)	1 % (11,613)	37,468
Hanoi	0	36 % (122,499)		25 % (9,074)	625
Hanoi	1	80 % (96,579)	31 % (324,910)	1 % (7,156)	9,210
		`adjust_lbeq`	$\psi(f)$	`lp_cplex_mex`	
Two-loop	0	51 % (317)	39 % (4,006,071)	1 % (623)	1,835
Two-loop	1	86 % (367)	67 % (79,746)	1 % (727)	7,540
Two-loop Exp.	0	49 % (137)	40 % (1,795,563)	1 % (256)	841
Two-loop Exp.	1	73 % (177)	65 % (44,051)	1 % (353)	4,305

Table C.3: Amount of time spent on selected functions, the time including calls of subfunctions is provided in percent of the total computational time, which is part of the last column. The amount of calls of these functions is given in brackets. The second part is in both cases based on the Darcy-Weisbach problem, $\psi(f)$ is used for the local subfunction `psi_fkt` for unverified and `verify_psi` for the verified calculation.

[1] Only the second part of the calculation excluding the determination of the floating point solution is considered.

Appendix D

Computational Results for the Water Distribution Networks

D.1 Computational Results of NOP based on Hazen-Williams Formula

D.1.1 Optimal Design for the Test Networks

Arc	Diameter	Length
(1, 2)	18	1000.0000
(2, 3)	10	795.4085
(2, 3)	12	204.5915
(2, 4)	16	1000.0000
(4, 5)	1	1000.0000
(4, 6)	14	310.3542
(4, 6)	16	689.6458
(6, 7)	8	11.1392
(6, 7)	10	988.8608
(3, 5)	8	98.4793
(3, 5)	10	901.5207
(5, 7)	1	999.9771
(5, 7)	2	0.0229

Arc	Flow
(1, 2)	1120.0000
(2, 3)	368.3316
(2, 4)	651.6684
(4, 5)	0.9757
(4, 6)	530.6927
(6, 7)	200.6927
(3, 5)	268.3316
(5, 7)	-0.6927

Node	Head
(1)	0.0000
(2)	53.2508
(3)	30.0000
(4)	43.8574
(5)	30.0002
(6)	30.0000
(7)	30.0000

Parameters	
lp_solver	cplex
verification	0
accuracy	10^{-6}
no_relax	6
branching	1
max_iter	2000

Table D.1: Computational results of the two-loop network calculated with Hazen-Williams formula and $C_{HW} = 130$, obtained after a CPU time of **4 seconds** with bounds for the optimal value of **403,385.2701** and **403,385.5249.**

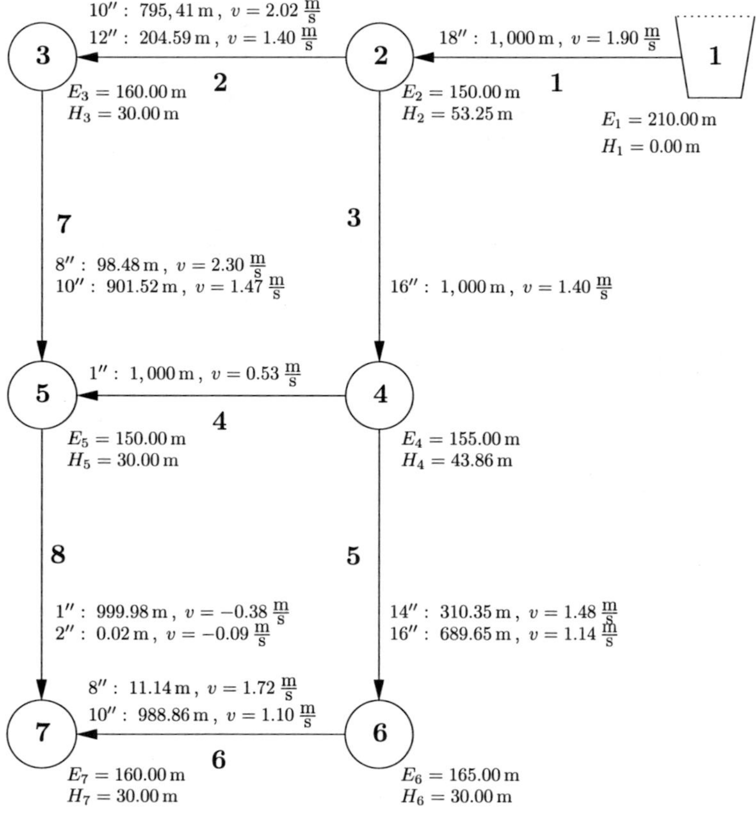

Figure D.1: Illustration of solution for two-loop network as part of Table D.1, where the pipeline flow rate q is substituted by the average fluid velocity v.

Arc	Flow
(1, 2)	1120.0000
(2, 3)	368.3316
(2, 4)	651.6684
(4, 5)	0.9757
(4, 6)	530.6927
(6, 7)	200.6927
(3, 5)	268.3316
(5, 7)	-0.6927

Arc	Diameter	Length
(6, 7)	8	11.1391
(6, 7)	10	988.8609
(3, 5)	8	98.4943
(3, 5)	10	901.5057
(5, 7)	1	999.9866
(5, 7)	2	0.0134

Node	Head
(1)	0.0000
(2)	53.2508
(3)	30.0000
(4)	43.8574
(5)	30.0000
(6)	30.0000
(7)	30.0000

Parameters	
lp_solver	cplex
verification	0
accuracy	10^{-6}
no_relax	6
branching	5
max_iter	1000

	no_relax	bran-ching	CPU Time (s)	# Sub-Problems	Lower Bound (US$)	Upper Bound (US$)
lambda						
	6	1	1	79	65,013.2711	65,013.2992
	6	2	1	77	65,013.2919	65,013.3135
	6	5	1	9	65,013.2990	65,013.3393
	6	6	1	67	65,013.2931	65,013.3243
	4	1	1	73	65,013.2858	65,013.3000
	4	2	2	87	65,013.2945	65,013.2996
	4	5	1	9	65,013.2990	65,013.3393
	4	6	1	67	65,013.2931	65,013.3243
sdome						
	6	1	2	83		infeasible
	6	2	3	77	65,013.2919	65,013.3135
	6	5	1	9	Inf	65,013.5180
	6	6	2	67	65,013.2931	65,013.3243
	4	1	2	73		infeasible
	4	2	3	101		infeasible
	4	5	1	37	65,013.2990	65,013.3243
	4	6	2	67	65,013.2931	65,013.3243

Table D.2: Computational results of the two-loop expansion network calculated with Hazen-Williams formula and $C_{HW} = 130$, obtained after a CPU time of **1 second** with bounds for the optimal value of **65,013.2990** and **65,013.3393.**
When using the linear solver `linprog` and an accuracy of 10^{-6} the problem is infeasible, regardless of selecting 4 or 6 tangents as relaxations and the branching strategy and regardless of calculating on `lambda` or `sdome`. When selecting branching strategy 3 or 4 it is infeasible as well.

Arc	Diameter	Length
(7, 8)	120	1939.1604
(7, 8)	132	7660.8396
(10,17)	96	26399.9965
(10,17)	108	0.0035
(12,18)	96	30934.3632
(12,18)	108	265.6368
(18,19)	84	23999.9985
(18,19)	96	0.0015
(11,20)	72	14247.4294
(11,20)	84	152.5706
(9,16)	72	26399.9988
(9,16)	84	0.0012

Arc	Flow
(1, 2)	881.6190
(2, 3)	789.2190
(3, 4)	696.8190
(4, 5)	608.6190
(5, 6)	520.4190
(6, 7)	432.2190
(7, 8)	166.7710
(8, 9)	255.8190
(9,10)	58.5000
(11, 9)	134.6435
(12,11)	482.6810
(13,12)	833.9810
(14,13)	951.0810
(15,14)	1043.4810
(1,15)	1135.8810
(10,17)	39.1442
(12,18)	159.5383
(18,19)	82.9080
(11,20)	110.1076
(20,16)	8.0375
(9,16)	80.9812
(7,21)	177.2480
(21, 8)	177.2480
(10,22)	18.3558
(22,17)	18.3558
(9,23)	80.9812
(23,16)	80.9812
(20,24)	-67.9299
(24,11)	-67.9299
(12,25)	74.6617
(25,18)	74.6617
(18,26)	34.1920
(26,19)	34.1920

Node	Head
(1)	300.0000
(2)	294.2425
(3)	286.2351
(4)	283.8904
(5)	281.8149
(6)	280.2051
(7)	277.6694
(8)	276.5599
(9)	273.7117
(10)	273.6802
(11)	273.8051
(12)	275.0887
(13)	278.0589
(14)	285.5407
(15)	293.3168
(16)	260.0000
(17)	272.8000
(18)	261.1455
(19)	255.0000
(20)	260.6751
(21)	277.1147
(22)	273.2401
(23)	266.8558
(24)	267.2401
(25)	268.1171
(26)	258.0728

Table D.3: Computational results of the New York (12") network calculated with Hazen-Williams formula, $C_{HW} = 100$ and default parameters. After a CPU time of **420 seconds** bounds for the optimal value of **38,049,839.4065** and **38,049,864.8659** are obtained.

Arc	Diameter	Length
(7, 8)	120	2025.2330
(7, 8)	132	7574.7670
(10,17)	96	26294.7509
(10,17)	108	105.2490
(12,18)	96	30928.8363
(12,18)	108	271.1637
(18,19)	84	23998.6544
(18,19)	96	1.3456
(11,20)	72	14352.3337
(11,20)	84	47.6663
(9,16)	72	26266.1622
(9,16)	84	133.8378

Arc	Flow
(1, 2)	881.6017
(2, 3)	789.2017
(3, 4)	696.8017
(4, 5)	608.6017
(5, 6)	520.4017
(6, 7)	432.2017
(7, 8)	166.5435
(8, 9)	255.8017
(9,10)	58.5000
(11, 9)	134.7716
(12,11)	482.6983
(13,12)	833.9983
(14,13)	951.0983
(15,14)	1043.4983
(1,15)	1135.8983
(10,17)	39.1559
(12,18)	159.5404
(18,19)	82.9084
(11,20)	109.9515
(20,16)	7.9268
(9,16)	81.0953
(7,21)	177.4582
(21, 8)	177.4582
(10,22)	18.3441
(22,17)	18.3441
(9,23)	80.9779
(23,16)	80.9779
(20,24)	-67.9753
(24,11)	-67.9753
(12,25)	74.6596
(25,18)	74.6596
(18,26)	34.1916
(26,19)	34.1916

Node	Head
(1)	300.0000
(2)	294.2427
(3)	286.2357
(4)	283.8910
(5)	281.8156
(6)	280.2059
(7)	277.6705
(8)	276.5585
(9)	273.7106
(10)	273.6791
(11)	273.8042
(12)	275.0879
(13)	278.0582
(14)	285.5402
(15)	293.3166
(16)	260.0000
(17)	272.8000
(18)	261.1455
(19)	255.0001
(20)	260.6580
(21)	277.1145
(22)	273.2396
(23)	266.8553
(24)	267.2311
(25)	268.1167
(26)	258.0728

Table D.4: Computational results of the New York $(12'')$ network calculated with Hazen-Williams formula, $C_{HW} = 100$ and the same parameters, as in Table D.3, except the linear solver `linprog`. After a CPU time of **3,410 seconds** bounds for the optimal value of **38,052,280.8570** and **38,052,280.8570** are obtained.

Arc	Diameter	Length
(7, 8)	112	0.0001
(7, 8)	116	9599.9999
(10,17)	96	1269.6330
(10,17)	100	25130.3670
(12,18)	96	9473.8384
(12,18)	100	21726.1616
(18,19)	80	23999.9994
(18,19)	84	0.0006
(11,20)	72	11730.9479
(11,20)	76	2669.0521
(9,16)	72	26399.9976
(9,16)	76	0.0024

Arc	Flow
(1, 2)	879.5380
(2, 3)	787.1380
(3, 4)	694.7380
(4, 5)	606.5380
(5, 6)	518.3380
(6, 7)	430.1380
(7, 8)	142.1693
(8, 9)	253.7380
(9,10)	58.5000
(11, 9)	136.0370
(12,11)	484.7620
(13,12)	836.0620
(14,13)	953.1620
(15,14)	1045.5620
(1,15)	1137.9620
(10,17)	40.3919
(12,18)	163.0774
(18,19)	79.7179
(11,20)	111.4025
(20,16)	8.7250
(9,16)	80.6375
(7,21)	199.7687
(21, 8)	199.7687
(10,22)	17.1081
(22,17)	17.1081
(9,23)	80.6375
(23,16)	80.6375
(20,24)	-67.3225
(24,11)	-67.3225
(12,25)	71.1226
(25,18)	71.1226
(18,26)	37.3821
(26,19)	37.3821

Node	Head
(1)	300.0000
(2)	294.2676
(3)	286.2993
(4)	283.9674
(5)	281.9050
(6)	280.3072
(7)	277.7940
(8)	276.4097
(9)	273.6042
(10)	273.5727
(11)	273.6994
(12)	274.9933
(13)	277.9772
(14)	285.4893
(15)	293.2942
(16)	260.0000
(17)	272.8000
(18)	262.2481
(19)	255.0000
(20)	260.7858
(21)	277.1019
(22)	273.1863
(23)	266.8021
(24)	267.2426
(25)	268.6207
(26)	258.6241

Table D.5: Computational results of the New York (4") network calculated with Hazen-Williams formula, $C_{HW} = 100$ and the default parameters. After a CPU time of **1,173 seconds** bounds for the optimal value of **37,896,865.4104** and **37,896,903.2702** are obtained.

Arc	Diameter	Length
(7, 8)	108	0.0001
(7, 8)	112	0.0001
(7, 8)	116	9577.9819
(7, 8)	120	22.0164
(7, 8)	124	0.0006
(7, 8)	128	0.0002
(7, 8)	132	0.0001
(10,17)	88	0.0002
(10,17)	92	0.0006
(10,17)	96	1286.2279
(10,17)	100	25113.7699
(10,17)	104	0.0005
(10,17)	108	0.0002
(10,17)	112	0.0001
(12,18)	88	0.0001
(12,18)	92	0.0004
(12,18)	96	9447.2687
(12,18)	100	21752.7293
(12,18)	104	0.0006
(12,18)	108	0.0002
(12,18)	112	0.0001
(18,19)	72	0.0003
(18,19)	76	27.8501
(18,19)	80	23972.1479
(18,19)	84	0.0007
(18,19)	88	0.0002
(18,19)	92	0.0001
(11,20)	64	0.0001
(11,20)	68	0.0003
(11,20)	72	11733.1651
(11,20)	76	2666.8332
(11,20)	80	0.0005
(11,20)	84	0.0002
(9,16)	68	0.0003
(9,16)	72	26399.2670
(9,16)	76	0.7314
(9,16)	80	0.0004
(9,16)	84	0.0002

Arc	Flow
(1, 2)	879.5394
(2, 3)	787.1394
(3, 4)	694.7394
(4, 5)	606.5394
(5, 6)	518.3394
(6, 7)	430.1394
(7, 8)	142.1856
(8, 9)	253.7394
(9,10)	58.5000
(11, 9)	136.0364
(12,11)	484.7606
(13,12)	836.0606
(14,13)	953.1606
(15,14)	1045.5606
(1,15)	1137.9606
(10,17)	40.3910
(12,18)	163.0821
(18,19)	79.7134
(11,20)	111.4011
(20,16)	8.7242
(9,16)	80.6380
(7,21)	199.7538
(21, 8)	199.7538
(10,22)	17.1090
(22,17)	17.1090
(9,23)	80.6378
(23,16)	80.6378
(20,24)	-67.3231
(24,11)	-67.3231
(12,25)	71.1179
(25,18)	71.1179
(18,26)	37.3866
(26,19)	37.3866

Node	Head
(1)	300.0000
(2)	294.2676
(3)	286.2992
(4)	283.9674
(5)	281.9050
(6)	280.3071
(7)	277.7940
(8)	276.4098
(9)	273.6043
(10)	273.5728
(11)	273.6995
(12)	274.9933
(13)	277.9772
(14)	285.4893
(15)	293.2942
(16)	260.0000
(17)	272.8000
(18)	262.2498
(19)	255.0000
(20)	260.7857
(21)	277.1019
(22)	273.1864
(23)	266.8021
(24)	267.2426
(25)	268.6215
(26)	258.6249

Table D.6: Computational results of the New York (4") network calculated with Hazen-Williams formula, $C_{HW} = 100$ and the default parameters, except the linear solver `linprog`. After a CPU time of **10,533 seconds** bounds for the optimal value of **37,896,984.1314** and **37,896,989.6332** are obtained.

Arc	Diameter	Length
(1, 2)	40	100.0000
(2, 3)	40	1350.0000
(3, 4)	40	900.0000
(4, 5)	40	1150.0000
(5, 6)	40	1450.0000
(6, 7)	40	450.0000
(7, 8)	40	850.0000
(8, 9)	40	850.0000
(9,10)	30	72.6662
(9,10)	40	727.3338
(10,11)	30	950.0000
(11,12)	24	1200.0000
(12,13)	24	3500.0000
(10,14)	16	250.5635
(10,14)	20	549.4365
(14,15)	16	500.0000
(15,16)	12	550.0000
(16,17)	12	2730.0000
(17,18)	16	1750.0000
(18,19)	20	427.4603
(18,19)	24	372.5397
(3,19)	24	400.0000
(3,20)	40	2200.0000
(20,21)	16	491.3604
(20,21)	20	1008.6396
(21,22)	12	500.0000
(20,23)	40	2650.0000
(23,24)	30	1230.0000
(24,25)	30	1300.0000
(25,26)	20	850.0000
(26,27)	12	300.0000
(16,27)	12	750.0000
(23,28)	16	1500.0000
(28,29)	12	2000.0000
(29,30)	12	1600.0000
(30,31)	16	150.0000
(31,32)	16	748.1671
(31,32)	20	111.8329
(25,32)	24	950.0000

Arc	Flow
(1, 2)	19940.0000
(2, 3)	19050.0000
(3, 4)	7966.1358
(4, 5)	7836.1358
(5, 6)	7111.1358
(6, 7)	6106.1358
(7, 8)	4756.1358
(8, 9)	4206.1358
(9,10)	3681.1358
(10,11)	2000.0000
(11,12)	1500.0000
(12,13)	940.0000
(10,14)	1156.1358
(14,15)	541.1358
(15,16)	261.1358
(16,17)	-132.1014
(17,18)	-997.1014
(18,19)	-2342.1014
(3,19)	2402.1014
(3,20)	7831.7628
(20,21)	1415.0000
(21,22)	485.0000
(20,23)	5141.7628
(23,24)	3501.0703
(24,25)	2681.0703
(25,26)	1186.7628
(26,27)	286.7628
(16,27)	83.2372
(23,28)	595.6925
(28,29)	305.6925
(29,30)	-54.3075
(30,31)	-414.3075
(31,32)	-519.3075
(25,32)	1324.3075

Node	Head
(1)	100.0000
(2)	97.1402
(3)	61.6631
(4)	56.9577
(5)	51.1256
(6)	44.9822
(7)	43.5444
(8)	41.8346
(9)	40.4729
(10)	39.1934
(11)	37.6341
(12)	34.2068
(13)	30.0000
(14)	33.6561
(15)	32.0992
(16)	30.2959
(17)	32.8296
(18)	49.7309
(19)	58.9306
(20)	50.5177
(21)	35.1598
(22)	30.0000
(23)	44.3592
(24)	38.6646
(25)	34.9928
(26)	31.1698
(27)	30.0000
(28)	38.7790
(29)	30.0000
(30)	30.2863
(31)	30.5711
(32)	32.8386

Table D.7: Computational results of the Hanoi network calculated with Hazen-Williams formula, $C_{HW} = 130$ and the same parameters as in Table D.1, except a maximum number of $20,000$ iterations. After a CPU time of **185 seconds** bounds for the optimal value of **6,055,536.3162** and **6,055,542.3685** are obtained.

D.1.2 Verified Optimal Design for the Networks

Arc	Diameter	Length	
(1, 2)	16	[0.0000,	0.0001]
(1, 2)	18	[999.9999,	1000.0000]
(2, 3)	10	[795.4084,	795.4085]
(2, 3)	12	[204.5915,	204.5916]
(2, 4)	14	[0.0000,	0.0001]
(2, 4)	16	[999.9999,	1000.0000]
(4, 5)	1	[999.9999,	1000.0000]
(4, 5)	2	[0.0000,	0.0001]
(4, 6)	14	[310.3541,	310.3542]
(4, 6)	16	[689.6458,	689.6459]
(6, 7)	8	[11.1391,	11.1392]
(6, 7)	10	[988.8608,	988.8609]
(3, 5)	8	[98.4792,	98.4793]
(3, 5)	10	[901.5207,	901.5208]
(5, 7)	1	[999.9770,	999.9771]
(5, 7)	2	[0.0229,	0.0230]

Arc	Flow	
(1, 2)	[1119.9999,	1120.0001]
(2, 3)	[368.3315,	368.3316]
(2, 4)	[651.6684,	651.6685]
(4, 5)	[0.9756,	0.9757]
(4, 6)	[530.6927,	530.6928]
(6, 7)	[200.6927,	200.6928]
(3, 5)	[268.3315,	268.3316]
(5, 7)	[-0.6928,	-0.6927]

Node	Head	
(1)	[0.0000,	0.0000]
(2)	[53.2507,	53.2508]
(3)	[30.0000,	30.0000]
(4)	[43.8574,	43.8575]
(5)	[30.0002,	30.0003]
(6)	[30.0000,	30.0000]
(7)	[30.0000,	30.0000]

Parameters	
`lp_solver`	`cplex`
`verification`	1
`accuracy`	10^{-6}
`no_relax`	6
`branching`	1
`max_iter`	2,000

Table D.8: Verified results of the two-loop network calculated with Hazen-Williams formula and $C_{HW} = 130$, obtained after a CPU time of **52 seconds** with bounds for the optimal value of **403,385.2700** and **403,385.5250.**

Arc		Flow
(1, 2)	[1119.9999,	1120.0001]
(2, 3)	[368.3315,	368.3316]
(2, 4)	[651.6684,	651.6685]
(4, 5)	[0.9756,	0.9757]
(4, 6)	[530.6927,	530.6928]
(6, 7)	[200.6927,	200.6928]
(3, 5)	[268.3315,	268.3316]
(5, 7)	[-0.6928,	-0.6927]

Node		Head
(1)	[0.0000,	0.0000]
(2)	[53.2507,	53.2508]
(3)	[30.0000,	30.0001]
(4)	[43.8574,	43.8575]
(5)	[30.0000,	30.0000]
(6)	[30.0000,	30.0000]
(7)	[30.0000,	30.0000]

Arc	Diameter		Length
(6, 7)	8	[11.1391,	11.1392]
(6, 7)	10	[988.8608,	988.8609]
(3, 5)	8	[98.4942,	98.4943]
(3, 5)	10	[901.5057,	901.5058]
(5, 7)	1	[999.9866,	999.9867]
(5, 7)	2	[0.0133,	0.0134]

Parameters	
`lp_solver`	`cplex`
`verification`	1
`accuracy`	10^{-6}
`no_relax`	6
`branching`	3
`max_iter`	1,000

Table D.9: Verified results of the two-loop expansion network calculated with Hazen-Williams formula and $C_{HW} = 130$, obtained after a CPU time of **3 seconds** with bounds for the optimal value of **65,013.2989** and **65,013.3394.**

Arc	Flow	
(1, 2)	[881.6189,	881.6190]
(2, 3)	[789.2189,	789.2190]
(3, 4)	[696.8189,	696.8190]
(4, 5)	[608.6189,	608.6190]
(5, 6)	[520.4189,	520.4190]
(6, 7)	[432.2189,	432.2190]
(7, 8)	[166.7710,	166.7711]
(8, 9)	[255.8189,	255.8190]
(9,10)	[58.4999,	58.5001]
(11, 9)	[134.6435,	134.6436]
(12,11)	[482.6810,	482.6811]
(13,12)	[833.9810,	833.9811]
(14,13)	[951.0810,	951.0811]
(15,14)	[1043.4810,	1043.4811]
(1,15)	[1135.8810,	1135.8811]
(10,17)	[39.1441,	39.1442]
(12,18)	[159.5382,	159.5383]
(18,19)	[82.9080,	82.9081]
(11,20)	[110.1075,	110.1076]
(20,16)	[8.0375,	8.0376]
(9,16)	[80.9812,	80.9813]
(7,21)	[177.2479,	177.2480]
(21, 8)	[177.2479,	177.2480]
(10,22)	[18.3558,	18.3559]
(22,17)	[18.3558,	18.3559]
(9,23)	[80.9812,	80.9813]
(23,16)	[80.9812,	80.9813]
(20,24)	[-67.9300,	-67.9299]
(24,11)	[-67.9300,	-67.9299]
(12,25)	[74.6617,	74.6618]
(25,18)	[74.6617,	74.6618]
(18,26)	[34.1919,	34.1920]
(26,19)	[34.1919,	34.1920]

Node	Head	
(1)	[300.0000,	300.0000]
(2)	[294.2424,	294.2425]
(3)	[286.2351,	286.2352]
(4)	[283.8903,	283.8904]
(5)	[281.8148,	281.8149]
(6)	[280.2051,	280.2052]
(7)	[277.6694,	277.6695]
(8)	[276.5599,	276.5600]
(9)	[273.7116,	273.7117]
(10)	[273.6801,	273.6802]
(11)	[273.8051,	273.8052]
(12)	[275.0886,	275.0887]
(13)	[278.0588,	278.0589]
(14)	[285.5406,	285.5407]
(15)	[293.3168,	293.3169]
(16)	[260.0000,	260.0000]
(17)	[272.7999,	272.8000]
(18)	[261.1455,	261.1456]
(19)	[255.0000,	255.0000]
(20)	[260.6751,	260.6752]
(21)	[277.1146,	277.1147]
(22)	[273.2400,	273.2401]
(23)	[266.8558,	266.8559]
(24)	[267.2401,	267.2402]
(25)	[268.1171,	268.1172]
(26)	[258.0727,	258.0728]

Arc	Diameter	Length	
(7, 8)	120	[1939.1470,	1939.1471]
(7, 8)	132	[7660.8529,	7660.8530]
(10,17)	84	[0.0040,	0.0041]
(10,17)	96	[26399.9959,	26399.9960]
(12,18)	96	[30934.3696,	30934.3697]
(12,18)	108	[265.6303,	265.6304]
(18,19)	84	[23999.9880,	23999.9881]
(18,19)	96	[0.0119,	0.0120]
(11,20)	72	[14247.4291,	14247.4292]
(11,20)	84	[152.5708,	152.5709]
(9,16)	72	[26399.9968,	26399.9969]
(9,16)	84	[0.0031,	0.0032]

Parameters	
`lp_solver`	`cplex`
`verification`	2
`accuracy`	10^{-6}
`no_relax`	6
`branching`	3
`max_iter`	40,000

Table D.10: Verified results of the New York (12") network calculated with Hazen-Williams formula and $C_{HW} = 100$, obtained after a CPU time of **11,779 seconds** with bounds for the optimal value of **38,049,834.3497** and **38,049,865.4756.**

Arc	Flow	
(1, 2)	[879.5379,	879.5380]
(2, 3)	[787.1379,	787.1380]
(3, 4)	[694.7379,	694.7380]
(4, 5)	[606.5379,	606.5380]
(5, 6)	[518.3379,	518.3380]
(6, 7)	[430.1379,	430.1380]
(7, 8)	[142.1693,	142.1694]
(8, 9)	[253.7379,	253.7380]
(9,10)	[58.4999,	58.5001]
(11, 9)	[136.0370,	136.0371]
(12,11)	[484.7620,	484.7621]
(13,12)	[836.0620,	836.0621]
(14,13)	[953.1620,	953.1621]
(15,14)	[1045.5620,	1045.5621]
(1,15)	[1137.9620,	1137.9621]
(10,17)	[40.3918,	40.3919]
(12,18)	[163.0773,	163.0774]
(18,19)	[79.7179,	79.7180]
(11,20)	[111.4024,	111.4025]
(20,16)	[8.7249,	8.7250]
(9,16)	[80.6375,	80.6376]
(7,21)	[199.7686,	199.7687]
(21, 8)	[199.7686,	199.7687]
(10,22)	[17.1081,	17.1082]
(22,17)	[17.1081,	17.1082]
(9,23)	[80.6375,	80.6376]
(23,16)	[80.6375,	80.6376]
(20,24)	[-67.3226,	-67.3225]
(24,11)	[-67.3226,	-67.3225]
(12,25)	[71.1226,	71.1227]
(25,18)	[71.1226,	71.1227]
(18,26)	[37.3820,	37.3821]
(26,19)	[37.3820,	37.3821]

Node	Head	
(1)	[300.0000,	300.0000]
(2)	[294.2675,	294.2676]
(3)	[286.2992,	286.2993]
(4)	[283.9674,	283.9675]
(5)	[281.9050,	281.9051]
(6)	[280.3071,	280.3072]
(7)	[277.7940,	277.7941]
(8)	[276.4097,	276.4098]
(9)	[273.6042,	273.6043]
(10)	[273.5726,	273.5727]
(11)	[273.6994,	273.6995]
(12)	[274.9932,	274.9933]
(13)	[277.9771,	277.9772]
(14)	[285.4892,	285.4893]
(15)	[293.2941,	293.2942]
(16)	[260.0000,	260.0000]
(17)	[272.7999,	272.8000]
(18)	[262.2481,	262.2482]
(19)	[255.0000,	255.0000]
(20)	[260.7858,	260.7859]
(21)	[277.1018,	277.1019]
(22)	[273.1863,	273.1864]
(23)	[266.8021,	266.8022]
(24)	[267.2426,	267.2427]
(25)	[268.6207,	268.6208]
(26)	[258.6240,	258.6241]

Arc	Diameter	Length	
(7, 8)	116	[9599.9680,	9599.9681]
(7, 8)	120	[0.0319,	0.0320]
(10,17)	96	[1269.6495,	1269.6496]
(10,17)	100	[25130.3504,	25130.3505]
(12,18)	96	[9473.8413,	9473.8414]
(12,18)	100	[21726.1586,	21726.1587]
(18,19)	80	[23999.9968,	23999.9969]
(18,19)	84	[0.0031,	0.0032]
(11,20)	72	[11730.9459,	11730.9460]
(11,20)	76	[2669.0540,	2669.0541]
(9,16)	72	[26399.9955,	26399.9956]
(9,16)	76	[0.0044,	0.0045]

Parameters	
`lp_solver`	`cplex`
`verification`	2
`accuracy`	10^{-6}
`no_relax`	6
`branching`	3
`max_iter`	80,000

Table D.11: Verified results of the New York (4") network calculated with Hazen-Williams formula and $C_{HW} = 100$, obtained after a CPU time of **20,993 seconds** with bounds for the optimal value of **37,896,865.7786** and **37,896,903.6035.**

Arc	Diameter	Length
(1, 2)	30	[0.0000, 0.0001]
(1, 2)	40	[99.9999, 100.0000]
(2, 3)	30	[0.0000, 0.0001]
(2, 3)	40	[1349.9999, 1350.0000]
(3, 4)	30	[0.0000, 0.0001]
(3, 4)	40	[899.9999, 900.0000]
(4, 5)	30	[0.0000, 0.0001]
(4, 5)	40	[1149.9999, 1150.0000]
(5, 6)	30	[0.0000, 0.0001]
(5, 6)	40	[1449.9999, 1450.0000]
(6, 7)	30	[0.0000, 0.0001]
(6, 7)	40	[449.9999, 450.0000]
(7, 8)	30	[0.0000, 0.0001]
(7, 8)	40	[849.9999, 850.0000]
(8, 9)	30	[0.0000, 0.0001]
(8, 9)	40	[849.9999, 850.0000]
(9,10)	30	[72.6662, 72.6663]
(9,10)	40	[727.3337, 727.3338]
(10,11)	24	[0.0000, 0.0001]
(10,11)	30	[949.9999, 950.0000]
(11,12)	20	[0.0000, 0.0001]
(11,12)	24	[1199.9999, 1200.0000]
(12,13)	20	[0.0000, 0.0001]
(12,13)	24	[3499.9999, 3500.0000]
(10,14)	16	[250.5634, 250.5635]
(10,14)	20	[549.4365, 549.4366]
(14,15)	12	[0.0000, 0.0001]
(14,15)	16	[499.9999, 500.0000]
(15,16)	12	[549.9999, 550.0000]
(15,16)	16	[0.0000, 0.0001]
(16,17)	12	[2729.9999, 2730.0000]
(16,17)	16	[0.0000, 0.0001]
(17,18)	12	[0.0000, 0.0001]
(17,18)	16	[1749.9999, 1750.0000]
(18,19)	20	[427.4603, 427.4604]
(18,19)	24	[372.5396, 372.5397]
(3,19)	20	[0.0000, 0.0001]
(3,19)	24	[399.9999, 400.0000]
(3,20)	30	[0.0000, 0.0001]
(3,20)	40	[2199.9999, 2200.0000]
(20,21)	16	[491.3603, 491.3604]
(20,21)	20	[1008.6396, 1008.6397]
(21,22)	12	[499.9999, 500.0000]
(21,22)	16	[0.0000, 0.0001]
(20,23)	30	[0.0000, 0.0001]
(20,23)	40	[2649.9999, 2650.0000]
(23,24)	24	[0.0000, 0.0001]
(23,24)	30	[1229.9999, 1230.0000]
(24,25)	24	[0.0000, 0.0001]
(24,25)	30	[1299.9999, 1300.0000]
(25,26)	20	[849.9999, 850.0000]
(25,26)	24	[0.0000, 0.0001]
(26,27)	12	[299.9999, 300.0000]
(26,27)	16	[0.0000, 0.0001]
(16,27)	12	[749.9999, 750.0000]
(16,27)	16	[0.0000, 0.0001]
(23,28)	12	[0.0000, 0.0001]
(23,28)	16	[1499.9999, 1500.0000]
(28,29)	12	[1999.9999, 2000.0000]
(28,29)	16	[0.0000, 0.0001]
(29,30)	12	[1599.9999, 1600.0000]
(29,30)	16	[0.0000, 0.0001]
(30,31)	12	[0.0000, 0.0001]
(30,31)	16	[149.9999, 150.0000]
(31,32)	16	[748.1670, 748.1671]
(31,32)	20	[111.8329, 111.8330]
(25,32)	20	[0.0000, 0.0001]
(25,32)	24	[949.9999, 950.0000]

Arc	Flow
(1, 2)	[19940.0000, 19940.0000]
(2, 3)	[19050.0000, 19050.0000]
(3, 4)	[7966.1357, 7966.1358]
(4, 5)	[7836.1357, 7836.1358]
(5, 6)	[7111.1357, 7111.1358]
(6, 7)	[6106.1357, 6106.1358]
(7, 8)	[4756.1357, 4756.1358]
(8, 9)	[4206.1357, 4206.1358]
(9,10)	[3681.1357, 3681.1358]
(10,11)	[2000.0000, 2000.0000]
(11,12)	[1500.0000, 1500.0000]
(12,13)	[940.0000, 940.0000]
(10,14)	[1156.1357, 1156.1358]
(14,15)	[541.1357, 541.1358]
(15,16)	[261.1357, 261.1358]
(16,17)	[-132.1015, -132.1014]
(17,18)	[-997.1015, -997.1014]
(18,19)	[-2342.1015, -2342.1014]
(3,19)	[2402.1014, 2402.1015]
(3,20)	[7831.7627, 7831.7628]
(20,21)	[1415.0000, 1415.0000]
(21,22)	[485.0000, 485.0000]
(20,23)	[5141.7627, 5141.7628]
(23,24)	[3501.0703, 3501.0704]
(24,25)	[2681.0703, 2681.0704]
(25,26)	[1186.7627, 1186.7628]
(26,27)	[286.7627, 286.7628]
(16,27)	[83.2372, 83.2373]
(23,28)	[595.6924, 595.6925]
(28,29)	[305.6924, 305.6925]
(29,30)	[-54.3076, -54.3075]
(30,31)	[-414.3076, -414.3075]
(31,32)	[-519.3076, -519.3075]
(25,32)	[1324.3075, 1324.3076]

Node	Head
(1)	[100.0000, 100.0000]
(2)	[97.1401, 97.1402]
(3)	[61.6631, 61.6632]
(4)	[56.9576, 56.9577]
(5)	[51.1256, 51.1257]
(6)	[44.9822, 44.9823]
(7)	[43.5444, 43.5445]
(8)	[41.8346, 41.8347]
(9)	[40.4728, 40.4729]
(10)	[39.1934, 39.1935]
(11)	[37.6341, 37.6342]
(12)	[34.2068, 34.2069]
(13)	[30.0000, 30.0000]
(14)	[33.6561, 33.6562]
(15)	[32.0992, 32.0993]
(16)	[30.2959, 30.2960]
(17)	[32.8295, 32.8296]
(18)	[49.7308, 49.7309]
(19)	[58.9305, 58.9306]
(20)	[50.5176, 50.5177]
(21)	[35.1597, 35.1598]
(22)	[30.0000, 30.0000]
(23)	[44.3591, 44.3592]
(24)	[38.6645, 38.6646]
(25)	[34.9928, 34.9929]
(26)	[31.1698, 31.1699]
(27)	[30.0000, 30.0000]
(28)	[38.7789, 38.7790]
(29)	[30.0000, 30.0000]
(30)	[30.2862, 30.2863]
(31)	[30.5710, 30.5711]
(32)	[32.8385, 32.8386]

Table D.12: Verified results of the Hanoi network calculated with Hazen-Williams formula, $C_{HW} = 130$ and the same parameters as in Table D.8, except a maximum number of $20,000$ iterations. After a CPU time of **2,730 seconds** bounds for the optimal value of **6,055,536.3256** and **6,055,542.3706** are obtained.

D.1.3 Comparison of Calculated Results

Floating Point Calculation					
Test Network	Verification	CPU Time (s)	# Sub-Problems	Lower Bound (US$)	Upper Bound (US$)
Two-loop	0	4	163	403,385.2701	403,385.5249
Two-loop Expansion	0	1	9	65,013.2990	65,013.3393
New York (12")	0	420	10,929	38,049,839.4065	38,049,864.8659
New York (4")	0	1,173	19,387	37,896,865.4104	37,896,903.2702
Hanoi	0	185	4,537	6,055,536.3162	6,055,542.3685

Verified Calculation					
Test Network	Verification	CPU Time (s)	# Sub-Problems	Lower Bound (US$)	Upper Bound (US$)
Two-loop	1	52	165	403,385.2700	403,385.5250
Two-loop Expansion	1	3	7	65,013.2989	65,013.3394
New York (12")	2	11,779	23,058	38,049,834.3497	38,049,865.4756
New York (4")	2	20,993	39,412	37,896,865.7786	37,896,903.6035
Hanoi	1	2,730	3,577	6,055,536.3256	6,055,542.3706

Table D.13: Comparison of computational results on `lambda` which are based on calculations with Hazen-Williams formula and default parameters.

Floating Point Calculation					
Test Network	Verifi-cation	CPU Time (s)	# Sub-Problems	Lower Bound (US$)	Upper Bound (US$)
Two-loop	0	6	163	403,385.6092	403,385.7645
Two-loop Expansion	0	1	9	Inf	65,013.5180
New York (12")	0	1,005	9,913	38,049,834.7706	38,049,864.9960
New York(4")	0	2,691	17,473	37,896,865.5709	37,896,903.3067
Hanoi	0	430	4,537	6,055,536.3161	6,055,542.3685

Verified Calculation					
Test Network	Verifi-cation	CPU Time (s)	# Sub-Problems	Lower Bound (US$)	Upper Bound (US$)
Two-loop	1	77	165	403,385.2700	403,385.5250
Two-loop Expansion	1	5	7	65,013.2989	65,013.3394
New York (12")	2	17,640	21,820	38,049,827.5943	38,049,865.3111
New York (4")	2	36,978	38,358	37,896,865.7815	37,896,903.6171
Hanoi	1	4,225	3,577	6,055,536.3256	6,055,542.3706

Table D.14: Comparison of computational results on `sdome` which are based on calculations with Hazen-Williams formula and default parameters.

D.2 Computational Results of DNOP based on Darcy-Weisbach Formula

D.2.1 Verified Optimal Design for the Test Networks

Arc	Flow	
(1, 2)	[1120.0000,	1120.0000]
(2, 3)	[368.2681,	368.2682]
(2, 4)	[651.7318,	651.7319]
(4, 5)	[1.0286,	1.0287]
(4, 6)	[530.7032,	530.7033]
(6, 7)	[200.7032,	200.7033]
(3, 5)	[268.2681,	268.2682]
(5, 7)	[-0.7033,	-0.7032]

Arc	Diameter	Length	
(1, 2)	16	[0.0000,	0.0001]
(1, 2)	18	[999.9999,	1000.0000]
(2, 3)	8	[0.0000,	0.0001]
(2, 3)	10	[999.9999,	1000.0000]
(2, 4)	14	[754.3114,	754.3115]
(2, 4)	16	[245.6885,	245.6886]
(4, 5)	1	[999.9999,	1000.0000]
(4, 5)	2	[0.0000,	0.0001]
(4, 6)	12	[0.0000,	0.0001]
(4, 6)	14	[999.9999,	1000.0000]
(6, 7)	8	[168.2625,	168.2626]
(6, 7)	10	[831.7374,	831.7375]
(3, 5)	8	[581.2022,	581.2023]
(3, 5)	10	[418.7977,	418.7978]
(5, 7)	1	[999.9984,	999.9985]
(5, 7)	2	[0.0015,	0.0016]

Node	Head	
(1)	[0.0000,	0.0000]
(2)	[54.9469,	54.9470]
(3)	[33.5792,	33.5793]
(4)	[44.3538,	44.3539]
(5)	[30.0000,	30.0001]
(6)	[30.0000,	30.0000]
(7)	[30.0000,	30.0000]

Parameters	
`k_value`	0.0003
`viscosity`	$1.306 \cdot 10^{-6}$
`lp_solver`	`cplex`
`verification`	1
`accuracy`	10^{-6}
`no_relax`	0
`branching`	1
`max_iter`	2,000

Table D.15: Verified results of the two-loop network calculated with Darcy-Weisbach formula and an equivalent sand roughness of 0.003 mm obtained after a CPU time of **1,815 seconds** with bounds for the optimal value of **350,625.4386** and **350,625.4779.**

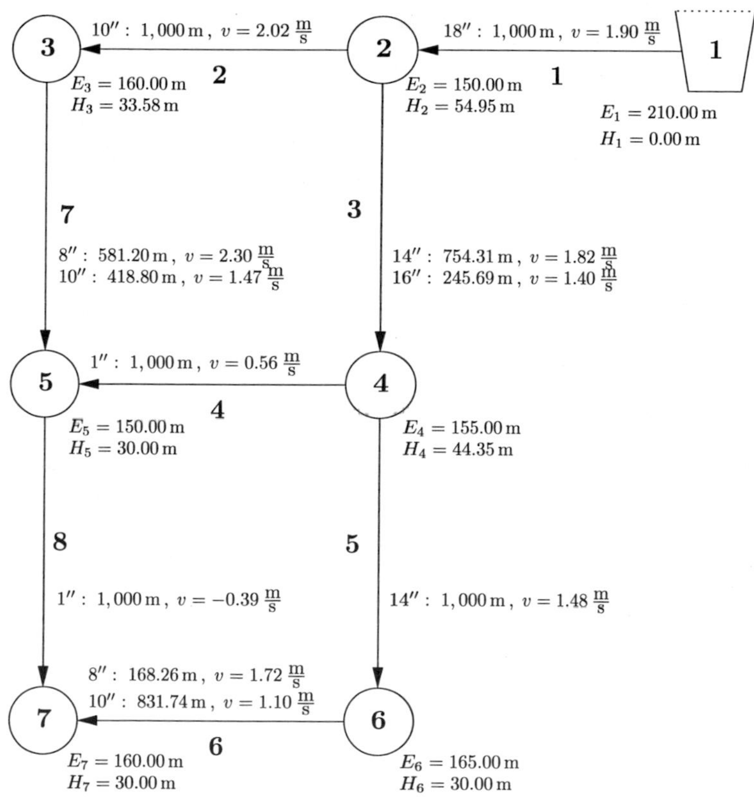

Figure D.2: Illustration of solution for two-loop network as part of Table D.15, where the pipeline flow rate q is substituted by the average fluid velocity v.

Arc		Flow
(1, 2)	[1119.9999,	1120.0001]
(2, 3)	[368.2005,	368.2006]
(2, 4)	[651.7994,	651.7995]
(4, 5)	[1.0962,	1.0963]
(4, 6)	[530.7032,	530.7033]
(6, 7)	[200.7032,	200.7033]
(3, 5)	[268.2005,	268.2006]
(5, 7)	[-0.7033,	-0.7032]

Node		Head
(1)	[0.0000,	0.0000]
(2)	[54.9469,	54.9470]
(3)	[34.9442,	34.9443]
(4)	[46.6245,	46.6246]
(5)	[30.0000,	30.0000]
(6)	[33.6972,	33.6973]
(7)	[30.0000,	30.0000]

Arc	Diameter		Length
(6, 7)	8	[675.3244,	675.3245]
(6, 7)	10	[324.6755,	324.6756]
(3, 5)	8	[692.0473,	692.0474]
(3, 5)	10	[307.9526,	307.9527]
(5, 7)	1	[999.9845,	999.9846]
(5, 7)	2	[0.0154,	0.0155]

Parameters	
k_value	0.0003
viscosity	$1.306\ 10^{-6}$
lp_solver	cplex
verification	1
accuracy	10^{-6}
no_relax	6
branching	3
max_iter	1,000

Table D.16: Verified results of the two-loop expansion network calculated with Darcy-Weisbach and an equivalent sand roughness of 0.003 mm obtained after a CPU time of **966 seconds** with bounds for the optimal value of **53,693.6472** and **53,693.7005.**

Arc		Flow
(1, 2)	[866.9158,	866.9159]
(2, 3)	[774.5158,	774.5159]
(3, 4)	[682.1158,	682.1159]
(4, 5)	[593.9158,	593.9159]
(5, 6)	[505.7158,	505.7159]
(6, 7)	[417.5158,	417.5159]
(7, 8)	[10.4247,	10.4248]
(8, 9)	[241.1158,	241.1159]
(9,10)	[58.4999,	58.5001]
(11, 9)	[163.4012,	163.4013]
(12,11)	[497.3841,	497.3842]
(13,12)	[848.6841,	848.6842]
(14,13)	[965.7841,	965.7842]
(15,14)	[1058.1841,	1058.1842]
(1,15)	[1150.5841,	1150.5842]
(10,17)	[7.9275,	7.9276]
(12,18)	[105.9203,	105.9204]
(18,19)	[41.8661,	41.8662]
(11,20)	[34.0701,	34.0702]
(20,16)	[-6.0171,	-6.0170]
(9,16)	[24.5686,	24.5687]
(7,21)	[318.8910,	318.8911]
(21, 8)	[318.8910,	318.8911]
(10,22)	[49.5724,	49.5725]
(22,17)	[49.5724,	49.5725]
(9,23)	[151.4484,	151.4485]
(23,16)	[151.4484,	151.4485]
(20,24)	[-129.9128,	-129.9127]
(24,11)	[-129.9128,	-129.9127]
(12,25)	[128.2796,	128.2797]
(25,18)	[128.2796,	128.2797]
(18,26)	[75.2338,	75.2339]
(26,19)	[75.2338,	75.2339]

Node		Head
(1)	[300.0000,	300.0000]
(2)	[297.1920,	297.1921]
(3)	[293.3247,	293.3248]
(4)	[292.2045,	292.2046]
(5)	[291.2247,	291.2248]
(6)	[290.4752,	290.4753]
(7)	[289.3145,	289.3146]
(8)	[287.6916,	287.6917]
(9)	[286.4437,	286.4438]
(10)	[286.4283,	286.4284]
(11)	[286.5082,	286.5083]
(12)	[287.1693,	287.1694]
(13)	[288.6929,	288.6930]
(14)	[292.5395,	292.5396]
(15)	[296.5468,	296.5469]
(16)	[264.3854,	264.3855]
(17)	[283.7324,	283.7325]
(18)	[268.1606,	268.1607]
(19)	[255.0000,	255.0001]
(20)	[264.1850,	264.1851]
(21)	[288.5031,	288.5032]
(22)	[285.0804,	285.0805]
(23)	[275.4146,	275.4147]
(24)	[275.3466,	275.3467]
(25)	[277.6650,	277.6651]
(26)	[261.5803,	261.5804]

Arc	Diameter		Length
(7, 8)	36	[9599.9986,	9599.9987]
(7, 8)	48	[0.0013,	0.0014]
(10,17)	36	[26399.9987,	26399.9988]
(10,17)	48	[0.0012,	0.0013]
(12,18)	60	[9244.0334,	9244.0335]
(12,18)	72	[21955.9665,	21955.9666]
(18,19)	36	[0.0003,	0.0004]
(18,19)	48	[23999.9996,	23999.9997]
(11,20)	36	[14399.9877,	14399.9879]
(11,20)	48	[0.0121,	0.0123]
(9,16)	36	[26399.7212,	26399.7226]
(9,16)	48	[0.2774,	0.2788]

Parameters	
`k_value`	$2.4 \cdot 10^{-3}$
`viscosity`	$1.407 \cdot 10^{-5}$
`lp_solver`	`cplex`
`verification`	3
`accuracy`	10^{-6}
`no_relax`	6
`branching`	3
`max_iter`	40,000

Table D.17: Verified results of the New York (12") network calculated with Darcy-Weisbach formula and an significantly reduced equivalent sand roughness of 0.0002 ft. After a CPU time of **192,899 seconds** bounds for the optimal value of **16,879,362.3978** and **16,879,375.8937** are obtained. Using the default equivalent sand roughness of 0.02 ft the problem can be verified to be infeasible.

Arc	Diameter	Length	
(1, 2)	30	[0.0000,	0.0001]
(1, 2)	40	[99.9999,	100.0000]
(2, 3)	30	[0.0000,	0.0001]
(2, 3)	40	[1349.9999,	1350.0000]
(3, 4)	30	[0.0000,	0.0001]
(3, 4)	40	[899.9999,	900.0000]
(4, 5)	30	[0.0000,	0.0001]
(4, 5)	40	[1149.9999,	1150.0000]
(5, 6)	30	[0.0000,	0.0001]
(5, 6)	40	[1449.9999,	1450.0000]
(6, 7)	30	[0.0000,	0.0001]
(6, 7)	40	[449.9999,	450.0000]
(7, 8)	30	[0.0000,	0.0001]
(7, 8)	40	[849.9999,	850.0000]
(8, 9)	30	[0.0000,	0.0001]
(8, 9)	40	[849.9999,	850.0000]
(9,10)	30	[0.0000,	0.0001]
(9,10)	40	[799.9999,	800.0000]
(10,11)	24	[0.0000,	0.0001]
(10,11)	30	[949.9999,	950.0000]
(11,12)	24	[0.0000,	0.0001]
(11,12)	30	[1199.9999,	1200.0000]
(12,13)	24	[3406.1249,	3406.1250]
(12,13)	30	[93.8750,	93.8751]
(10,14)	12	[793.8926,	793.8927]
(10,14)	16	[6.1073,	6.1074]
(14,15)	12	[499.9999,	500.0000]
(14,15)	16	[0.0000,	0.0001]
(15,16)	12	[24.8924,	24.8925]
(15,16)	16	[525.1075,	525.1076]
(16,17)	24	[2711.6952,	2711.6953]
(16,17)	30	[18.3047,	18.3048]
(17,18)	24	[0.0000,	0.0001]
(17,18)	30	[1749.9999,	1750.0000]
(18,19)	24	[0.0000,	0.0001]
(18,19)	30	[799.9999,	800.0000]
(3,19)	24	[0.0000,	0.0001]
(3,19)	30	[399.9999,	400.0000]
(3,20)	30	[0.0000,	0.0001]
(3,20)	40	[2199.9999,	2200.0000]
(20,21)	16	[6.7874,	6.7875]
(20,21)	20	[1493.2125,	1493.2126]
(21,22)	12	[499.9999,	500.0000]
(21,22)	16	[0.0000,	0.0001]
(20,23)	30	[0.0000,	0.0001]
(20,23)	40	[2649.9999,	2650.0000]
(23,24)	24	[0.0000,	0.0001]
(23,24)	30	[1229.9999,	1230.0000]
(24,25)	24	[1092.7763,	1092.7764]
(24,25)	30	[207.2236,	207.2237]
(25,26)	12	[640.8524,	640.8525]
(25,26)	16	[209.1475,	209.1476]
(26,27)	16	[0.0000,	0.0001]
(26,27)	20	[299.9999,	300.0000]
(16,27)	16	[0.0000,	0.0001]
(16,27)	20	[749.9999,	750.0000]
(23,28)	16	[1380.6082,	1380.6083]
(23,28)	20	[119.3917,	119.3918]
(28,29)	12	[0.0000,	0.0001]
(28,29)	16	[1999.9999,	2000.0000]
(29,30)	12	[1599.9999,	1600.0000]
(29,30)	16	[0.0000,	0.0001]
(30,31)	12	[0.0000,	0.0001]
(30,31)	16	[149.9999,	150.0000]
(31,32)	12	[0.0000,	0.0001]
(31,32)	16	[859.9999,	860.0000]
(25,32)	20	[0.0000,	0.0001]
(25,32)	24	[949.9999,	950.0000]

Arc	Flow	
(1, 2)	[19940.0000,	19940.0000]
(2, 3)	[19050.0000,	19050.0000]
(3, 4)	[7227.5000,	7227.5000]
(4, 5)	[7097.5000,	7097.5000]
(5, 6)	[6372.5000,	6372.5000]
(6, 7)	[5367.5000,	5367.5000]
(7, 8)	[4017.5000,	4017.5000]
(8, 9)	[3467.5000,	3467.5000]
(9,10)	[2942.5000,	2942.5000]
(10,11)	[2000.0000,	2000.0000]
(11,12)	[1500.0000,	1500.0000]
(12,13)	[940.0000,	940.0000]
(10,14)	[417.5000,	417.5000]
(14,15)	[-197.5000,	-197.5000]
(15,16)	[-477.5000,	-477.5000]
(16,17)	[-1726.8751,	-1726.8749]
(17,18)	[-2591.8751,	-2591.8749]
(18,19)	[-3936.8751,	-3936.8749]
(3,19)	[3996.8749,	3996.8751]
(3,20)	[6975.6249,	6975.6251]
(20,21)	[1415.0000,	1415.0000]
(21,22)	[485.0000,	485.0000]
(20,23)	[4285.6249,	4285.6251]
(23,24)	[2543.1249,	2543.1251]
(24,25)	[1723.1249,	1723.1251]
(25,26)	[330.6249,	330.6251]
(26,27)	[-569.3751,	-569.3749]
(16,27)	[939.3749,	939.3751]
(23,28)	[697.5000,	697.5000]
(28,29)	[407.5000,	407.5000]
(29,30)	[47.5000,	47.5000]
(30,31)	[-312.5000,	-312.5000]
(31,32)	[-417.5000,	-417.5000]
(25,32)	[1222.5000,	1222.5000]

Node	Head	
(1)	[100.0000,	100.0000]
(2)	[96.6065,	96.6066]
(3)	[54.7710,	54.7711]
(4)	[50.6800,	50.6801]
(5)	[45.6363,	45.6364]
(6)	[40.4926,	40.4927]
(7)	[39.3534,	39.3535]
(8)	[38.1331,	38.1332]
(9)	[37.2175,	37.2176]
(10)	[36.5913,	36.5914]
(11)	[35.0685,	35.0686]
(12)	[33.9686,	33.9687]
(13)	[30.0000,	30.0000]
(14)	[30.0000,	30.0000]
(15)	[30.9682,	30.9683]
(16)	[32.5280,	32.5281]
(17)	[42.7890,	42.7891]
(18)	[47.4435,	47.4436]
(19)	[52.2797,	52.2798]
(20)	[45.4459,	45.4460]
(21)	[35.5587,	35.5588]
(22)	[30.0000,	30.0000]
(23)	[41.1296,	41.1297]
(24)	[37.9776,	37.9777]
(25)	[33.6204,	33.6205]
(26)	[30.0000,	30.0000]
(27)	[30.3325,	30.3326]
(28)	[33.8294,	33.8295]
(29)	[30.2110,	30.2111]
(30)	[30.0000,	30.0000]
(31)	[30.1627,	30.1628]
(32)	[31.7933,	31.7934]

Table D.18: Verified results of the Hanoi network calculated with Darcy-Weisbach formula an equivalent sand roughness of 0.25 mm, `verification` = 3 and default parameters except a reduced accuracy of 10^{-2}. After a CPU time of **61,170 seconds** bounds for the optimal value of **6,548,469.8583** and **6,614,612.1670** are obtained.

D.2.2 Comparison of Calculated Results

Floating Point Calculation – Darcy-Weisbach					
Test Network	Verification	CPU Time (s)	# Sub-Problems	Lower Bound (US$)	Upper Bound (US$)
Two-loop	0	182	317	350,625.4396	350,625.7877
Two-loop Expansion	0	98	201	53,693.6397	53,693.6591
New York (12")	0	285	169		infeasible
New York (12")[1]	0	19,279	8,091	16,879,367.0544	16,879,376.6967
New York (4")	0	481	105		infeasible
Hanoi	0	6,159	5,699	6,548,469.8584	6,614,612.1645

Verified Calculation					
Test Network	Verification	CPU Time (s)	# Sub-Problems	Lower Bound (US$)	Upper Bound (US$)
Two-loop	1	1,815	367	350,625.4386	350,625.4779
Two-loop Expansion	1	966	203	53,693.6472	53,693.7005
New York (12")	1	2,212	97		infeasible
New York (12")[1]	3	192,899	7,589	16,879,362.3978	16,879,375.8937
New York (4")	1	4,866	97		infeasible
Hanoi	3	61,170	5,699	6,548,469.8583	6,614,612.1670

Table D.19: Comparison of computational results on `lambda` which are based on calculations with Darcy-Weisbach formula and default parameters, except for the verified calculation of Hanoi, where an accuracy of 10^{-2} is used.

[1] As the New York (12") network is infeasible, this is recalculated with a significantly reduced equivalent sand roughness of 0.0002 ft instead of 0.02 ft.

D.3 Computational Results of NOP based on Adjusted Hazen-Williams

D.3.1 Verified Optimal Design for the Test Networks

Arc	Flow	
(1, 2)	[1119.9999,	1120.0001]
(2, 3)	[368.4502,	368.4503]
(2, 4)	[651.5497,	651.5498]
(4, 5)	[0.9134,	0.9135]
(4, 6)	[530.6362,	530.6363]
(6, 7)	[200.6362,	200.6363]
(3, 5)	[268.4502,	268.4503]
(5, 7)	[-0.6363,	-0.6362]

Arc	Diameter	Length	
(1, 2)	16	[0.0000,	0.0001]
(1, 2)	18	[999.9999,	1000.0000]
(2, 3)	8	[0.0000,	0.0001]
(2, 3)	10	[999.9999,	1000.0000]
(2, 4)	14	[747.0954,	747.0955]
(2, 4)	16	[252.9045,	252.9046]
(4, 5)	1	[999.9999,	1000.0000]
(4, 5)	2	[0.0000,	0.0001]
(4, 6)	12	[0.0000,	0.0001]
(4, 6)	14	[999.9999,	1000.0000]
(6, 7)	8	[167.5862,	167.5863]
(6, 7)	10	[832.4137,	832.4138]
(3, 5)	8	[566.2369,	566.2370]
(3, 5)	10	[433.7630,	433.7631]
(5, 7)	1	[999.9820,	999.9821]
(5, 7)	2	[0.0179,	0.0180]

Node	Head	
(1)	[0.0000,	0.0000]
(2)	[54.9361,	54.9362]
(3)	[33.5070,	33.5071]
(4)	[44.3583,	44.3584]
(5)	[30.0002,	30.0003]
(6)	[30.0000,	30.0000]
(7)	[30.0000,	30.0000]

Parameters	
k_value	0.0003
viscosity	$1.306\ 10^{-6}$
lp_solver	cplex
verification	1
accuracy	10^{-6}
no_relax	6
branching	1
max_iter	2,000

Table D.20: Verified results of the two-loop network calculated with adjusted Hazen-Williams and an equivalent sand roughness of 0.003 mm obtained after a CPU time of **52 seconds** with bounds for the optimal value of **350,982.4592** and **350,982.7817.**

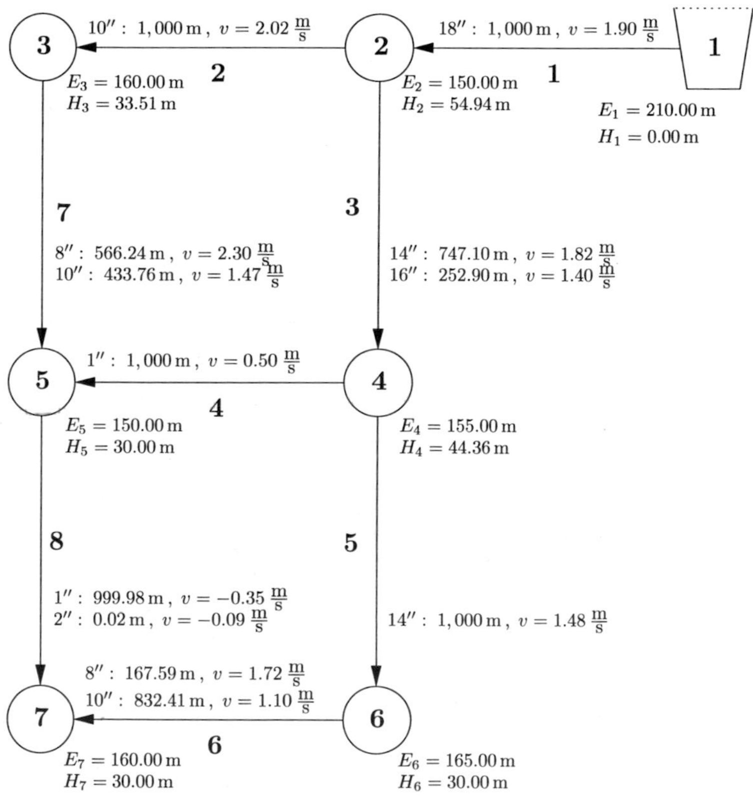

Figure D.3: Illustration of solution for two-loop network as part of Table D.20, where the pipeline flow rate q is substituted by the average fluid velocity v .

Arc		Flow
(1, 2)	[1119.9999,	1120.0001]
(2, 3)	[368.3934,	368.3935]
(2, 4)	[651.6065,	651.6066]
(4, 5)	[0.9702,	0.9703]
(4, 6)	[530.6362,	530.6363]
(6, 7)	[200.6362,	200.6363]
(3, 5)	[268.3934,	268.3935]
(5, 7)	[-0.6363,	-0.6362]

Node		Head
(1)	[0.0000,	0.0000]
(2)	[54.9361,	54.9362]
(3)	[34.8820,	34.8821]
(4)	[46.6127,	46.6128]
(5)	[30.0000,	30.0000]
(6)	[33.6849,	33.6850]
(7)	[30.0000,	30.0000]

Arc	Diameter		Length
(6, 7)	8	[667.9907,	667.9908]
(6, 7)	10	[332.0092,	332.0093]
(3, 5)	8	[676.4179,	676.4180]
(3, 5)	10	[323.5820,	323.5821]
(5, 7)	1	[999.9998,	999.9999]
(5, 7)	2	[0.0001,	0.0002]

Parameters	
`k_value`	0.0003
`viscosity`	$1.306\ 10^{-6}$
`lp_solver`	`cplex`
`verification`	1
`accuracy`	10^{-6}
`no_relax`	6
`branching`	3
`max_iter`	1,000

Table D.21: Verified results of the two-loop expansion network calculated with adjusted Hazen-Williams and an equivalent sand roughness of 0.003 mm obtained after a CPU time of **23 seconds** with bounds for the optimal value of **53,900.2742** and **53,900.3219.**

Arc	Flow
(1, 2)	[869.2811, 869.2812]
(2, 3)	[776.8811, 776.8812]
(3, 4)	[684.4811, 684.4812]
(4, 5)	[596.2811, 596.2812]
(5, 6)	[508.0811, 508.0812]
(6, 7)	[419.8811, 419.8812]
(7, 8)	[10.3702, 10.3703]
(8, 9)	[243.4811, 243.4812]
(9,10)	[58.4999, 58.5001]
(11, 9)	[161.9707, 161.9708]
(12,11)	[495.0188, 495.0189]
(13,12)	[846.3188, 846.3189]
(14,13)	[963.4188, 963.4189]
(15,14)	[1055.8188, 1055.8189]
(1,15)	[1148.2188, 1148.2189]
(10,17)	[7.8651, 7.8652]
(12,18)	[107.8448, 107.8449]
(18,19)	[41.6834, 41.6835]
(11,20)	[33.3099, 33.3100]
(20,16)	[-6.9520, -6.9519]
(9,16)	[24.2043, 24.2044]
(7,21)	[321.3108, 321.3109]
(21, 8)	[321.3108, 321.3109]
(10,22)	[49.6348, 49.6349]
(22,17)	[49.6348, 49.6349]
(9,23)	[152.7475, 152.7476]
(23,16)	[152.7475, 152.7476]
(20,24)	[-129.7382, -129.7381]
(24,11)	[-129.7382, -129.7381]
(12,25)	[126.3551, 126.3552]
(25,18)	[126.3551, 126.3552]
(18,26)	[75.4165, 75.4166]
(26,19)	[75.4165, 75.4166]

Node	Head
(1)	[300.0000, 300.0000]
(2)	[297.1882, 297.1883]
(3)	[293.3269, 293.3270]
(4)	[292.2126, 292.2127]
(5)	[291.2425, 291.2426]
(6)	[290.5051, 290.5052]
(7)	[289.3727, 289.3728]
(8)	[287.7597, 287.7598]
(9)	[286.5318, 286.5319]
(10)	[286.5193, 286.5194]
(11)	[286.5882, 286.5883]
(12)	[287.2223, 287.2224]
(13)	[288.7283, 288.7284]
(14)	[292.5512, 292.5513]
(15)	[296.5471, 296.5472]
(16)	[263.6632, 263.6633]
(17)	[283.9216, 283.9217]
(18)	[268.4991, 268.4992]
(19)	[255.0000, 255.0001]
(20)	[263.4489, 263.4490]
(21)	[288.5662, 288.5663]
(22)	[285.2205, 285.2206]
(23)	[275.0975, 275.0976]
(24)	[275.0186, 275.0187]
(25)	[277.8607, 277.8608]
(26)	[261.7495, 261.7496]

Arc	Diameter	Length
(7, 8)	36	[9599.8862, 9599.8863]
(7, 8)	48	[0.1137, 0.1138]
(10,17)	36	[26399.9394, 26399.9395]
(10,17)	48	[0.0605, 0.0606]
(12,18)	60	[7246.8590, 7246.8591]
(12,18)	72	[23953.1409, 23953.1410]
(18,19)	36	[0.0022, 0.0023]
(18,19)	48	[23999.9977, 23999.9978]
(11,20)	36	[14399.9960, 14399.9961]
(11,20)	48	[0.0039, 0.0040]
(9,16)	36	[26399.8979, 26399.8993]
(9,16)	48	[0.1007, 0.1021]

Parameters	
`lp_solver`	`cplex`
`verification`	3
`accuracy`	10^{-6}
`no_relax`	6
`branching`	3
`max_iter`	40000

Table D.22: Verified results of the New York (12") network calculated with Adjusted Hazen-Williams formula and an equivalent sand roughness of 0.0002 ft. After a CPU time of **8,561 seconds** bounds for the optimal value of **16,968,690.9573** and **16,968,704.3609** are obtained. Using the default equivalent sand roughness of 0.02 ft the problem can be verified to be infeasible.

Arc	Flow	
(1, 2)	[869.2864,	869.2865]
(2, 3)	[776.8864,	776.8865]
(3, 4)	[684.4864,	684.4865]
(4, 5)	[596.2864,	596.2865]
(5, 6)	[508.0864,	508.0865]
(6, 7)	[419.8864,	419.8865]
(7, 8)	[10.3711,	10.3712]
(8, 9)	[243.4864,	243.4865]
(9,10)	[58.4999,	58.5001]
(11, 9)	[161.9652,	161.9653]
(12,11)	[495.0135,	495.0136]
(13,12)	[846.3135,	846.3136]
(14,13)	[963.4135,	963.4136]
(15,14)	[1055.8135,	1055.8136]
(1,15)	[1148.2135,	1148.2136]
(10,17)	[7.8654,	7.8655]
(12,18)	[107.8314,	107.8315]
(18,19)	[41.6838,	41.6839]
(11,20)	[33.3108,	33.3109]
(20,16)	[-6.9517,	-6.9516]
(9,16)	[24.2052,	24.2053]
(7,21)	[321.3152,	321.3153]
(21, 8)	[321.3152,	321.3153]
(10,22)	[49.6345,	49.6346]
(22,17)	[49.6345,	49.6346]
(9,23)	[152.7464,	152.7465]
(23,16)	[152.7464,	152.7465]
(20,24)	[-129.7376,	-129.7375]
(24,11)	[-129.7376,	-129.7375]
(12,25)	[126.3685,	126.3686]
(25,18)	[126.3685,	126.3686]
(18,26)	[75.4161,	75.4162]
(26,19)	[75.4161,	75.4162]

Node	Head	
(1)	[300.0000,	300.0000]
(2)	[297.1882,	297.1883]
(3)	[293.3269,	293.3270]
(4)	[292.2126,	292.2127]
(5)	[291.2425,	291.2426]
(6)	[290.5051,	290.5052]
(7)	[289.3727,	289.3728]
(8)	[287.7598,	287.7599]
(9)	[286.5320,	286.5321]
(10)	[286.5195,	286.5196]
(11)	[286.5884,	286.5885]
(12)	[287.2224,	287.2225]
(13)	[288.7284,	288.7285]
(14)	[292.5512,	292.5513]
(15)	[296.5472,	296.5473]
(16)	[263.6658,	263.6659]
(17)	[283.9222,	283.9223]
(18)	[268.4972,	268.4973]
(19)	[255.0000,	255.0001]
(20)	[263.4516,	263.4517]
(21)	[288.5663,	288.5664]
(22)	[285.2208,	285.2209]
(23)	[275.0989,	275.0990]
(24)	[275.0200,	275.0201]
(25)	[277.8598,	277.8599]
(26)	[261.7486,	261.7487]

Arc	Diameter	Length	
(7, 8)	36	[9599.8946,	9599.8947]
(7, 8)	40	[0.1053,	0.1054]
(10,17)	36	[26399.9937,	26399.9938]
(10,17)	40	[0.0062,	0.0063]
(12,18)	64	[1271.9508,	1271.9509]
(12,18)	68	[29928.0491,	29928.0492]
(18,19)	48	[23999.9626,	23999.9627]
(18,19)	52	[0.0373,	0.0374]
(11,20)	36	[14399.9869,	14399.9871]
(11,20)	40	[0.0129,	0.0131]
(9,16)	36	[26399.7261,	26399.7286]
(9,16)	40	[0.2714,	0.2739]

Parameters	
`lp_solver`	`cplex`
`verification`	3
`accuracy`	10^{-6}
`no_relax`	6
`branching`	3
`max_iter`	80,000

Table D.23: Verified results of the New York (4") network calculated with adjusted Hazen-Williams formula and a reduced equivalent sand roughness of 0.0002 ft. After a CPU time of **4,617 seconds** bounds for the optimal value of **16,801,968.8573** and **16,801,975.4907** are obtained. Using the default equivalent sand roughness of 0.02 ft the problem can be verified to be infeasible.

Arc	Diameter	Length	
(1, 2)	30	[0.0000,	0.0001]
(1, 2)	40	[99.9999,	100.0000]
(2, 3)	30	[0.0000,	0.0001]
(2, 3)	40	[1349.9999,	1350.0000]
(3, 4)	30	[0.0000,	0.0001]
(3, 4)	40	[899.9999,	900.0000]
(4, 5)	30	[0.0000,	0.0001]
(4, 5)	40	[1149.9999,	1150.0000]
(5, 6)	30	[0.0000,	0.0001]
(5, 6)	40	[1449.9999,	1450.0000]
(6, 7)	30	[0.0000,	0.0001]
(6, 7)	40	[449.9999,	450.0000]
(7, 8)	30	[0.0000,	0.0001]
(7, 8)	40	[849.9999,	850.0000]
(8, 9)	30	[0.0000,	0.0001]
(8, 9)	40	[849.9999,	850.0000]
(9,10)	30	[0.0000,	0.0001]
(9,10)	40	[799.9999,	800.0000]
(10,11)	24	[0.0000,	0.0001]
(10,11)	30	[949.9999,	950.0000]
(11,12)	30	[1199.9999,	1200.0000]
(11,12)	40	[0.0000,	0.0001]
(12,13)	20	[0.0000,	0.0001]
(12,13)	24	[3499.9999,	3500.0000]
(10,14)	12	[658.6195,	658.6196]
(10,14)	16	[141.3804,	141.3805]
(14,15)	12	[499.9999,	500.0000]
(14,15)	16	[0.0000,	0.0001]
(15,16)	12	[104.0538,	104.0539]
(15,16)	16	[445.9461,	445.9462]
(16,17)	20	[0.0000,	0.0001]
(16,17)	24	[2729.9999,	2730.0000]
(17,18)	30	[1749.9999,	1750.0000]
(17,18)	40	[0.0000,	0.0001]
(18,19)	24	[0.0000,	0.0001]
(18,19)	30	[799.9999,	800.0000]
(3,19)	24	[0.0000,	0.0001]
(3,19)	30	[399.9999,	400.0000]
(3,20)	30	[0.0000,	0.0001]
(3,20)	40	[2199.9999,	2200.0000]
(20,21)	16	[14.2756,	14.2757]
(20,21)	20	[1485.7243,	1485.7244]
(21,22)	12	[499.9999,	500.0000]
(21,22)	16	[0.0000,	0.0001]
(20,23)	30	[0.0000,	0.0001]
(20,23)	40	[2649.9999,	2650.0000]
(23,24)	24	[0.0000,	0.0001]
(23,24)	30	[1229.9999,	1230.0000]
(24,25)	20	[0.0000,	0.0001]
(24,25)	24	[1299.9999,	1300.0000]
(25,26)	12	[814.3721,	814.3722]
(25,26)	16	[35.6278,	35.6279]
(26,27)	16	[0.0000,	0.0001]
(26,27)	20	[299.9999,	300.0000]
(16,27)	16	[0.0000,	0.0001]
(16,27)	20	[749.9999,	750.0000]
(23,28)	12	[0.0000,	0.0001]
(23,28)	16	[1499.9999,	1500.0000]
(28,29)	16	[1999.9999,	2000.0000]
(28,29)	20	[0.0000,	0.0001]
(29,30)	12	[1599.9999,	1600.0000]
(29,30)	16	[0.0000,	0.0001]
(30,31)	12	[0.0000,	0.0001]
(30,31)	16	[149.9999,	150.0000]
(31,32)	12	[0.0000,	0.0001]
(31,32)	16	[859.9999,	860.0000]
(25,32)	20	[14.4787,	14.4788]
(25,32)	24	[935.5212,	935.5213]

Arc	Flow	
(1, 2)	[19940.0000,	19940.0000]
(2, 3)	[19050.0000,	19050.0000]
(3, 4)	[7254.7157,	7254.7158]
(4, 5)	[7124.7157,	7124.7158]
(5, 6)	[6399.7157,	6399.7158]
(6, 7)	[5394.7157,	5394.7158]
(7, 8)	[4044.7157,	4044.7158]
(8, 9)	[3494.7157,	3494.7158]
(9,10)	[2969.7157,	2969.7158]
(10,11)	[2000.0000,	2000.0000]
(11,12)	[1500.0000,	1500.0000]
(12,13)	[940.0000,	940.0000]
(10,14)	[444.7157,	444.7158]
(14,15)	[-170.2843,	-170.2842]
(15,16)	[-450.2843,	-450.2842]
(16,17)	[-1728.0729,	-1728.0728]
(17,18)	[-2593.0729,	-2593.0728]
(18,19)	[-3938.0729,	-3938.0728]
(3,19)	[3998.0728,	3998.0729]
(3,20)	[6947.2113,	6947.2114]
(20,21)	[1415.0000,	1415.0000]
(21,22)	[485.0000,	485.0000]
(20,23)	[4257.2113,	4257.2114]
(23,24)	[2515.0338,	2515.0339]
(24,25)	[1695.0338,	1695.0339]
(25,26)	[302.2113,	302.2114]
(26,27)	[-597.7887,	-597.7886]
(16,27)	[967.7886,	967.7887]
(23,28)	[697.1775,	697.1776]
(28,29)	[407.1775,	407.1776]
(29,30)	[47.1775,	47.1776]
(30,31)	[-312.8225,	-312.8224]
(31,32)	[-417.8225,	-417.8224]
(25,32)	[1222.8224,	1222.8225]

Node	Head	
(1)	[100.0000,	100.0000]
(2)	[96.5910,	96.5911]
(3)	[54.5679,	54.5680]
(4)	[50.4668,	50.4669]
(5)	[45.4117,	45.4118]
(6)	[40.2638,	40.2639]
(7)	[39.1267,	39.1268]
(8)	[37.9159,	37.9160]
(9)	[37.0107,	37.0108]
(10)	[36.3946,	36.3947]
(11)	[34.9174,	34.9175]
(12)	[33.8649,	33.8650]
(13)	[30.0000,	30.0000]
(14)	[30.0000,	30.0000]
(15)	[30.6856,	30.6857]
(16)	[32.6160,	32.6161]
(17)	[42.7446,	42.7447]
(18)	[47.3074,	47.3075]
(19)	[52.0989,	52.0990]
(20)	[45.3710,	45.3711]
(21)	[35.5062,	35.5063]
(22)	[30.0000,	30.0000]
(23)	[41.1912,	41.1913]
(24)	[38.1734,	38.1735]
(25)	[33.5321,	33.5322]
(26)	[30.0000,	30.0000]
(27)	[30.3477,	30.3478]
(28)	[33.6281,	33.6282]
(29)	[30.1705,	30.1706]
(30)	[30.0000,	30.0000]
(31)	[30.1534,	30.1535]
(32)	[31.7185,	31.7186]

Table D.24: Verified results of the Hanoi network calculated with adjusted Hazen-Williams formula, $\epsilon = 0.25$ mm, default parameters and `verification` $= 1$. After a CPU time of **1,402 seconds** bounds for the optimal value of **6,591,289.3765** and **6,591,295.8450** are obtained.

Two-Loop	
d_k	C_{HW_k}
1	128.46
2	140.65
3	145.70
4	147.88
6	149.24
8	149.34
10	149.09
12	148.75
14	148.41
16	148.08
18	147.64
20	147.22
22	146.81
24	146.42
c_d	1.8265

New York	**(12")**	**(4")**	**(12")**	**(4")**
	$\epsilon = 0.02$ ft		$\epsilon = 0.0002$ ft	
d_k	C_{HW_k}	C_{HW_k}	C_{HW_k}	C_{HW_k}
36	80.69	80.69	130.11	130.11
40		82.60		132.22
44		84.35		134.13
48	85.96	85.96	135.78	135.79
52		87.45		137.38
56		88.85		138.83
60	90.15	90.15	140.10	140.11
64		91.39		141.30
68		92.55		142.45
72	93.66	93.66	143.47	143.48
76		94.71		144.43
80		95.72		145.33
84	96.68	96.68	146.16	146.17
88		97.60		146.96
92		98.48		147.69
96	99.33	99.33	148.37	148.38
100		100.15		149.04
104		100.94		149.66
108	101.70	101.70	150.23	150.24
112		102.44		150.79
116		103.15		151.32
120	103.85	103.85	151.82	151.83
124		104.52		152.32
128		105.17		152.78
132	105.81	105.81	153.21	153.22
136		106.43		153.65
140		107.03		154.07
144	107.61	107.61	154.45	154.46
148		108.19		154.85
152		108.75		155.22
156	109.29	109.29	155.57	155.58
160		109.82		155.93
164		110.35		156.20
168	110.85	110.85	156.45	156.46
172		111.35		156.70
176		111.84		156.93
180	112.32	112.32	157.14	157.14
184		112.78		157.35
188		113.24		157.54
192	113.69	113.69	157.72	157.73
196		114.13		157.91
200		114.56		158.08
204	114.99	114.99	158.23	158.24
c_d	1.9989	1.9989	1.9350	1.9351

Hanoi	
d_k	C_{HW_k}
12	137.88
16	145.17
20	150.91
24	155.65
30	161.53
40	168.86
c_d	1.9903

Table D.25: Adjusted Hazen-Williams coefficients C_{HW_k} and adjusted discard coefficients c_d used in Tables D.20 to D.24 and in Table D.26.

D.3.2 Comparison of Calculated Results

Floating Point Calculation – Adjusted Hazen-Williams					
Test Network	Verification	CPU Time (s)	# Sub-Problems	Lower Bound (US$)	Upper Bound (US$)
Two-loop	0	3	165	350,982.4396	350,982.7077
Two-loop Expansion	0	2	107	53,901.0008	53,901.0089
New York (12")	0	7	113		infeasible
New York (12")[2]	0	78	2,169	16,968,676.8426	16,968,691.9980
New York (4")	0	11	143		infeasible
New York (4")[2]	0	133	2,389	16,801,969.5385	16,801,970.2876
Hanoi	0	72	1,643	6,591,289.3766	6,591,295.8396

Verified Calculation					
Test Network	Verification	CPU Time (s)	# Sub-Problems	Lower Bound (US$)	Upper Bound (US$)
Two-loop	1	52	163	350,982.4592	350,982.7817
Two-loop Expansion	1	23	93	53,900.2742	53,900.3219
New York (12")	1	410	143		infeasible
New York (12")[2]	3	8,561	9,235	16,968,690.9573	16,968,704.3609
New York (4")	1	398	131		infeasible
New York (4")[2]	3	4,617	3,961	16,801,968.8573	16,801,975.4907
Hanoi	1	1,402	1,645	6,591,289.3765	6,591,295.8450

Table D.26: Comparison of computational results on `lambda` which are based on adjusted Hazen-Williams formula and default parameters.

[2] As the New York networks are infeasible, these are recalculated with a significantly reduced equivalent sand roughness of 0.0002 ft instead of 0.02 ft.

D.4 Comparison of Branch and Bound Parameters

D.4.1 Comparison of Linear Solvers

CPLEX				
Test Network	CPU Time (s)	# Sub-Problems	Lower Bound (US$)	Upper Bound (US$)
Two-loop	4	163	403,385.2701	403,385.5249
Two-loop Exp.	1	9	65,013.2990	65,013.3393
New York (12")	420	10,929	38,049,839.4065	38,049,864.8659
New York (4")	1,173	19,387	37,896,865.4104	37,896,903.2702
Hanoi	185	4,537	6,055,536.3162	6,055,542.3685

LINPROG				
Test Network	CPU Time (s)	# Sub-Problems	Lower Bound (US$)	Upper Bound (US$)
Two-loop	45	165	403,385.2626	403,385.5105
Two-loop Exp.	1	5		infeasible
New York (12")	3,410	12,541	38,052,280.8570	38,052,280.8570
New York (4")	10,533	23,467	37,896.984.1314	37,896,989.6332
Hanoi	1,544	4,141	6,055,536.2597	6,055,542.3063

Table D.27: Comparison of the influence of the linear solver "cplex" and "linprog" on `lambda`. Results are based on calculations with Hazen-Williams equation and default parameters.

CPLEX				
Test Network	CPU Time (s)	# Sub-Problems	Lower Bound (US$)	Upper Bound (US$)
Two-loop	6	163	403,385.6092	403,385.7645
Two-loop Exp.	1	9	Inf	65,013.5180
New York (12")	1,005	9,913	38,049,834.7706	38,049,864.9960
New York (4")	2,691	17,473	37,896,865.5709	37,896,903.3067
Hanoi	430	4,537	6,055,536.3161	6,055,542.3685

LINPROG				
Test Network	CPU Time (s)	# Sub-Problems	Lower Bound (US$)	Upper Bound (US$)
Two-loop	101	149	403,473.9334	403,406.1124
Two-loop Exp.	2	3		infeasible
New York (12")	11,314	15,057	38,056,282.3414	38,056,310.5509
New York (4")	50,282	35,611	37,898,725.8909	37,898,725.8909
Hanoi	4,166	4,243	6,055,536.2644	6,055,542.3180

Table D.28: Comparison of the influence of the linear solver "cplex" and "linprog" on `sdome`. Results are based on calculations with Hazen-Williams equation, and the same parameters as in Table D.27.

D.4.2 Comparison of Verification Strategies and Accuracies

Verification Strategy 1				
Test Network	CPU Time (s)	# Sub-Problems	Lower Bound (US$)	Upper Bound (US$)
Two-loop	52	165	403,385.2700	403,385.5250
Two-loop (DW)	1,815	367	350,625.4386	350,625.4779
Two-loop Exp.	3	7	65,013.2989	65,013.3394
New York (12")				38,049,865.3380[3]
New York (4")				37,896,903.6699[3]
Hanoi	2,730	3,577	6,055,536.3256	6,055,542.3706

Verification Strategy 2				
Two-loop	54	322	403,385.1223	403,385.5250
Two-loop (DW)	1,945	682	350,625.4372	350,625.7877
Two-loop Exp.	5	84	65,013.2989	65,013.3136
New York (12")	11,779	23,058	38,049,834.3497	38,049,865.4756
New York (4")	20,993	39,412	37,896,865.7786	37,896,903.6035
Hanoi	2,913	8,114	6,055,536.3256	6,055,542.3706

Verification Strategy 3				
Two-loop	58	165	403,385.2700	403,385.5250
Two-loop (DW)	1,899	367	350,625.4386	350,625.4779
Two-loop Exp.	3	7	65,013.2989	65,013.3394
New York (12")	11,750	12,129	38,049,834.3497	38,049,865.2094
New York (4")	20,946	20,023	37,896,865.6158	37,896,903.5093
Hanoi	2,885	3,577	6,055,536.3256	6,055,542.3698

Verification Strategy 4				
Two-loop	54	322	403,385.1223	403,385.5250
Two-loop (DW)	2,019	682	350,625.4372	350,625.7877
Two-loop Exp.	5	84	65,013.2989	65,013.3136
New York (12")	11,783	23,058	38,049,834.3497	38,049,865.4756
New York (4")	21,032	39,412	37,896,865.7786	37,896,903.6035
Hanoi	2,909	8,114	6,055,536.3256	6,055,542.3706

Table D.29: Comparison of different verification strategies on `lambda`, if not otherwise specified calculations are based on Hazen-William formula default parameters and the same maximum number of iterations as in Table D.30.

[3]The accuracy reduction used for pruning subproblems has not been large enough, so a recalculation with reduced value for the constant `BBUB` would be necessary.

Test Network	CPU Time (s)	# Sub-Problems	Lower Bound (US$)	Upper Bound (US$)
Verification Strategy 1				
Two-loop	77	165	403,385.2700	403,385.5250
Two-loop (DW)	2,811	367	350,625.4386	350,625.4779
Two-loop Exp.	5	7	65,013.2989	65,013.3394
New York (12")				38,049,865.4423[4]
New York (4")				37,896,903.6768[4]
Hanoi	4,225	3,577	6,055,536.3256	6,055,542.3706
Verification Strategy 2				
Two-loop[5]				
Two-loop (DW)	3,233	682	350.625.4372	350.625.7877
Two-loop Exp.	7	84	65,013.2989	65,013.3136
New York (12")	17,640	21,820	38,049,827.5943	38,049,865.3111
New York (4")	36,978	38,358	37,896,865.7815	37,896,903.6171
Hanoi	4,620	8,114	6.055,536.3256	6.055,542.3706
Verification Strategy 3				
Two-loop	92	165	403,385.2700	403,385.5250
Two-loop (DW)	2,931	367	350,625.4386	350,625.4779
Two-loop Exp.	5	7	65,013.2989	65,013.3394
New York (12")	17,411	11,907	38,049,827.5943	38,049,865.1806
New York (4")	36,889	20,883	37,896,865.6159	37,896,903.5116
Hanoi	4,621	3,577	6.055,536.3256	6.055,542.3698
Verification Strategy 4				
Two-loop	91	362	403,385.3212	403,385.3221
Two-loop (DW)	3,360	682	350,625.4372	350,625.7877
Two-loop Exp.	7	84	65,013.2989	65,013.3136
New York (12")	17,541	21,820	38,049,827.5943	38,049,865.3111
New York (4")	36,505	38,358	37,896,865.7815	37,896,903.6171
Hanoi	4,609	8,114	6.055,536.3256	6.055,542.3706

Table D.30: Comparison of different verification strategies on `sdome`, if not otherwise specified calculations are based on Hazen-William formula, default parameters and a reduced number of iterations for `verification` = 4.[6]

[4]The accuracy reduction used for pruning subproblems has not been large enough, so a recalculation with reduced value for the constant `BBUB` would be necessary.

[5]The maximum number of iterations has been reached.

[6]A value of 200 has been selected for the parameter `max_iter` for the two-loop network, 100 for two-loop expansion, 15,000 for New York (12"), 25,000 for New York (4") and 6,000 for Hanoi.

Two-loop Test Network: Floating Point				
accuracy	CPU Time (s)	# Sub-Problems	Lower Bound (US$)	Upper Bound (US$)
10^{-6}	6	163	403,385.6092	403,385.7645
10^{-5}	6	147	403,384.2993	403,388.1347
10^{-4}	5	141	403,381.5498	403,393.4277
10^{-3}	5	133	403,373.6256	403,762.2280
10^{-2}	5	121	403,306.0722	405,319.3277
Two-loop Test Network: Verified Calculation (1)				
10^{-6}	77	165	403,385.2700	403,385.5250
10^{-5}	72	149	403,384.2992	403,388.1347
10^{-4}	70	143	403,381.5498	403,393.4277
10^{-3}	68	135	403,373.6255	403,762.2280
10^{-2}	65	123	403,305.9957	405,319.3278
Two-loop Test Network: Verified Calculation (3)				
10^{-6}	92	165	403,385.2700	403,385.5250
10^{-5}	86	149	403,384.2992	403,388.1347
10^{-4}	84	143	403,381.5498	403,393.4277
10^{-3}	81	135	403,373.6255	403,762.2280
10^{-2}	76	123	403,305.9957	405,319.3278
Two-loop Test Network: Verified Two Phase B&B				
10^{-5}	77	308	403,383.7165	403,385.7645
10^{-4}	72	290	403,354.4229	403,385.7645
10^{-3}	69	280	403,241.0804	403,385.7645
10^{-2}	55	244	399,908.4092	403,385.7645

Table D.31: Comparison of results with verified and floating point calculations for different accuracies. Results are based on calculations on `sdome` with Hazen-Williams formula and default parameters.

Two-loop Expansion Network: Floating Point				
accuracy	CPU Time (s)	# Sub-Problems	Lower Bound (US$)	Upper Bound (US$)
10^{-6}	2	67	65,013.2931	65,013.3243
10^{-5}	1	35	65,012.9575	65,013.3393
10^{-4}	1	27	65,012.6493	65,013.3393
10^{-3}	1	19	64,954.3951	65,013.3393
10^{-2}	1	9	64,721.9075	65,013.3393
Two-loop Expansion Network: Verified Calculation (1)				
10^{-6}	15	67	65,013.2934	65,013.3245
10^{-5}	10	35	65,012.9558	65,013.3394
10^{-4}	8	27	65,012.6487	65,013.3394
10^{-3}	7	19	64,954.3944	65,013.3394
10^{-2}	5	9	64,721.9066	65,013.3394
Two-loop Expansion Network: Verified Calculation (3)				
10^{-6}	18	67	65,013.2934	65,013.3245
10^{-5}	10	35	65,012.9558	65,013.3394
10^{-4}	9	27	65,012.6487	65,013.3394
10^{-3}	7	19	64,954.3944	65,013.3394
10^{-2}	5	9	64,721.9066	65,013.3394
Two-loop Expansion Network: Verified Two Phase B&B				
10^{-6}	17	142	65,013.2505	65,013.3136
10^{-5}	12	112	65,012.9558	65,013.3136
10^{-4}	11	104	65,012.6487	65,013.3136
10^{-3}	9	96	64,954.3944	65,013.3136
10^{-2}	7	86	64,721.9066	65,013.3136

Table D.32: Comparison of results with verified and floating point calculations for different accuracies. Results are based on calculations on `sdome` with Hazen-Williams formula, $C_{HW} = 130$ and default parameters.

New York Test Network (12"): Floating Point				
accuracy	CPU Time (s)	# Sub-Problems	Lower Bound (US$)	Upper Bound (US$)
10^{-6}	1,005	9,913	38,049,834.7706	38,049,864.9960
10^{-4}	984	9,841	38,049,643.1351	38,049,932.2901
10^{-2}	983	9,841	38,049,643.1351	38,049,932.2901

New York Test Network: Verified Calculation (3)				
10^{-6}	17,411	11,907	38,049,827.5943	38,049,865.1806
10^{-4}	17,322	11,821	38,048,674.7712	38,049,865.4588
10^{-2}	17,033	11,671	37,682,664.9037	38,055,941.7422

New York Test Network: Verified Two Phase B&B				
10^{-6}	17,640	21,820	38,049,827.5943	38,049,865.3111
10^{-4}	16,811	21,258	38,046,082.7313	38,049,865.3111
10^{-2}	12,642	17,492	37,669,469.1727	38,049,865.3111

Table D.33: Comparison of results with verified and floating point calculations for different accuracies. Results are based on calculations on `sdome` with Hazen-Williams formula, $C_{HW} = 100$ and default parameters. The accuracy for the upper bound is not large enough for the verified result, when using `verification` = 1.

New York Test Network (4"): Floating Point				
accuracy	CPU Time (s)	# Sub-Problems	Lower Bound (US$)	Upper Bound (US$)
10^{-6}	2,691	17,473	37,896,865.5709	37,896,903.3067
10^{-4}	2,255	14,205	37,896,145.2176	37,897,411.9185
10^{-2}	2,229	14,205	37,896,145.2176	37,897,411.9185

New York Test Network: Verified Calculation (3)				
10^{-6}	36,889	20,883	37,896,865.6159	37,896,903.5116
10^{-4}	29,775	16,517	37,893,754.3773	37,897,447.1510
10^{-2}	28,634	16,051	37,531,831.8878	37,897,447.1510

New York Test Network: Verified Two Phase B&B				
10^{-6}	36,978	38,358	37,896,865.7815	37,896,903.6171
10^{-4}	25,002	31,020	37,893,122.2537	37,896,903.6171
10^{-2}	15,119	24,706	37,518,759.5986	37,896,903.6171

Table D.34: Comparison of results with verified and floating point calculations for different accuracies. Results are based on calculations on `sdome` with Hazen-Williams formula, $C_{HW} = 100$ and default parameters.
When using `verification` $= 1$ the accuracy for the upper bound is not large enough for the verified result with an `accuracy` of 10^{-6}, whereas with an `accuracy` of 10^{-4} or 10^{-2} no inclusion for the floating point solution could be determined.

Hanoi Test Network: Floating Point				
accuracy	CPU Time (s)	# Sub-Problems	Lower Bound (US$)	Upper Bound (US$)
10^{-6}	430	4,537	6,055,536.3161	6,055,542.3685
10^{-4}	85	839	6,054,945.7335	6,055,542.6150
10^{-2}	15	133	5,999,393.2437	6,059,058.7748

Hanoi Test Network: Verified Calculation (1)				
10^{-6}	4,225	3,577	6,055,536.3256	6,055,542.3706
10^{-4}	1,240	849	6,054,947.1940	6,055,542.4190
10^{-2}	258	129	5,998,381.7577	6,058,793.2643

Hanoi Test Network: Verified Calculation (3)				
10^{-6}	4,621	3,577	6,055,536.3256	6,055,542.3698
10^{-4}	1,354	849	6,054,947.1940	6,055,542.4190
10^{-2}	276	129	5,998,381.7577	6,058,793.2643

Hanoi Test Network: Verified Two Phase B&B				
10^{-6}	4,620	8,114	6,055,536.3256	6,055,542.3706
10^{-4}	1,675	5,386	6,054,947.1940	6,055,542.3706
10^{-2}	699	4,662	5,995,196.7083	6,055,542.3706

Table D.35: Comparison of results with verified and floating point calculations for different accuracies. Results are based on calculations on `sdome` with Hazen-Williams formula, $C_{HW} = 130$ and default parameters.

D.4.3 Comparison of Different Numbers of Relaxation

Test Network	Verification	CPU Time (s)	# Sub-Problems	Lower Bound (US$)	Upper Bound (US$)
Without Relaxation					
Two-loop	0	4	331	403,385.3063	403,385.5205
Two-loop	1	20	453	403,385.2925	403,385.5655
Two-loop (AHW)	0	4	343	350,982.5272	350,982.7937
Two-loop Exp.	0	1	61	65,013.2988	65,013.3390
Two-loop Exp.	1	4	63	65,013.2976	65,013.3391
Two-loop Exp. (DW)	0	115	231	53,693.6522	53,693.7021
New York (12")[7]	0				
New York (4")[7]	0				
Hanoi[7]	0				
4 Tangents for Relaxation					
Two-loop	0	3	173	403,385.2278	403,385.5931
Two-loop	1	55	207	403,385.2278	403,385.5931
Two-loop (AHW)	0	4	189	350,982.5105	350,982.7308
Two-loop Exp.	0	1	9	65,013.2990	65,013.3393
Two-loop Exp.	1	3	7	65,013.2989	65,013.3394
Two-loop Exp. (DW)	0	102	205	53,693.6462	53,693.6809
New York (12")	0	343	9,545	38,049,828.0508	38,049,863.8945
New York (4")	0	1,169	20,037	37,896,865.2726	37,896,903.1257
Hanoi	0	131	3,597	6,055,536.3315	6,055,542.3685
6 Tangents for Relaxation					
Two-loop	0	4	163	403,385.2701	403,385.5249
Two-loop	1	52	165	403,385.2700	403,385.5250
Two-loop (AHW)	0	3	165	350,982.4396	350,982.7077
Two-loop Exp.	0	1	9	65,013.2990	65,013.3393
Two-loop Exp.	1	3	7	65,013.2989	65,013.3394
Two-loop Exp. (DW)	0	98	201	53,693.6397	53,693.6591
New York (12")	0	420	10,929	38,049,839.4065	38,049,864.8659
New York (4")	0	1,173	19,387	37,896,865.4104	37,896,903.2702
Hanoi	0	185	4,537	6,055,536.3162	6,055,542.3685

Table D.36: Comparison of different numbers of relaxation, if not otherwise specified calculations are based on Hazen-Williams equation and default parameters.

[7]The maximum number of iterations has been reached.

D.4.4 Comparison of Different Branching Strategies

Test Network	Verifi-cation	CPU Time (s)	# Sub-Problems	Lower Bound (US$)	Upper Bound (US$)
Branching Strategy 1					
Two-loop	0	4	163	403,385.2701	403,385.5249
Two-loop	1	52	165	403,385.2700	403,385.5250
Two-loop (DW)	0	182	317	350,625.4396	350,625.7877
Hanoi	0	185	4,537	6,055,536.3162	6,055,542.3685
Branching Strategy 3					
Two-loop	0	4	163	403,385.2701	403,385.5249
Two-loop	1	60	165	403,385.2700	403,385.5250
Two-loop (DW)	0	188	317	350,625.4396	350,625.7877
Hanoi	0	196	4,537	6,055,536.3162	6,055,542.3685
Branching Strategy 5					
Two-loop	0	4	163	403,385.2701	403,385.5249
Two-loop (DW)	0	189	317	350,625.4396	350,625.7877
Hanoi	0	196	4,537	6,055,536.3162	6,055,542.3685
Branching Strategy 2					
Two-loop	0	3	125	403,385.1188	403,385.3444
Two-loop	1	36	125	403,385.1187	403,385.3444
Two-loop (DW)	0	maximum number of iterations reached			
Hanoi	0	301	7,493	6,055,536.3201	6,055,542.3684
Branching Strategy 4					
Two-loop	0	3	125	403,385.1188	403,385.3444
Two-loop	1	42	125	403,385.1187	403,385.3444
Two-loop (DW)	0	maximum number of iterations reached			
Hanoi	0	319	7,499	6,055,536.3201	6,055,542.3684
Branching Strategy 6					
Two-loop	0	3	125	403,385.1188	403,385.3444
Two-loop (DW)	0	maximum number of iterations reached			
Hanoi	0	317	7,499	6,055,536.3201	6,055,542.3684

Table D.37: Comparison of different branching strategies for non-expansion networks and default parameters.

Test Network	Verification	CPU Time (s)	# Sub-Problems	Lower Bound (US$)	Upper Bound (US$)
Branching Strategy 1					
Two-loop Exp.	0	1	79	65,013.2711	65,013.2992
Two-loop Exp.	1	21	67	65,013.2390	65,013.3010
Two-loop Exp. (DW)	0	140	259	53,693.6530	53,693.7025
New York (12")	0	541	14,523	38,049,833.1417	38,049,864.9609
New York (12")	3	22,846	21,265	38,049,828.1671	38,049,865.2382
New York (4")	0	2,078	34,671	37,896,865.3229	37,896,903.2120
New York (4")	3	41,234	37,709	37,896,865.7367	37,896,903.5088
Branching Strategy 3					
Two-loop Exp.	0	1	1		infeasible
Two-loop Exp.	1	3	7	65,013.2989	65,013.3394
Two-loop Exp. (DW)	0	1	1		infeasible
New York (12")	0	418	10,929	38,049,839.4065	38,049,864.8659
New York (12")	3	11,750	12,129	38,049,834.3497	38,049,865.2094
New York (4")	0	1,175	19,387	37,896,865.4104	37,896,903.2702
New York (4")	3	20,946	20,023	37,896,865.6158	37,896,903.5093
Branching Strategy 5					
Two-loop Exp.	0	1	9	65,013.2990	65,013.3393
Two-loop Exp. (DW)	0	98	201	53,693.6397	53,693.6591
New York (12")	0	420	10,929	38,049,839.4065	38,049,864.8659
New York (4")	0	1,173	19,387	37,896,865.4104	37,896,903.2702
Branching Strategy 2					
Two-loop Exp.	0	1	77	65,013.2919	65,013.3135
Two-loop Exp.	1	23	77	65,013.2919	65,013.3136
Two-loop Exp. (DW)	0	143	269	53,694.0380	53,694.0658
New York [8]					
Branching Strategy 4					
Two-loop Exp.	0	1	1		infeasible
Two-loop Exp.	1	11	67	65,013.2934	65,013.3245
Two-loop Exp. (DW)	0	1	1		infeasible
New York [8]					
Branching Strategy 6					
Two-loop Exp.	0	1	67	65,013.2931	65,013.3243
Two-loop Exp. (DW)	0	118	243	53,693.8435	53,693.8165
New York [8]					

Table D.38: Comparison of different branching strategies for expansion networks and default parameters.

[8]The maximum number of iterations has been reached for the New York networks with floating point calculations for branching strategies 2, 4 and 6 and for New York (12") calculated with `verification` = 3 for the branching strategies 2 and 4.

D.5 Comparison of Hydraulic Parameters

D.5.1 Comparison of Head-Loss Formulae

Hazen-Williams				
Test Network	CPU Time (s)	# Sub-Problems	Lower Bound (US$)	Upper Bound (US$)
Two-loop	52	165	403,385.2700	403,385.5250
Two-loop Exp.	3	7	65,013.2989	65,013.3394
New York (12")	11,779	23,058	38,049,834.3497	38,049,865.4756
New York (4")	20,993	39,412	37,896,865.7786	37,896,903.6035
Hanoi	2,730	3,577	6,055,536.3256	6,055,542.3706

Darcy-Weisbach				
Two-loop	1,815	367	350,625.4386	350,625.4779
Two-loop Exp.	966	203	53,693.6472	53,693.7005
New York (12")	2,212	97		infeasible
New York (4")	4,866	97		infeasible
Hanoi	61,170	5,699	6,548,469.8583	6,614,612.1670

Adjusted Hazen-Williams				
Two-loop	52	163	350,982.4592	350,982.7817
Two-loop Exp.	23	93	53,900.2742	53,900.3219
New York (12")	410	143		infeasible
New York (4")	398	131		infeasible
Hanoi	1,402	1,645	6,591,289.3765	6,591,295.8450

Table D.39: Comparison of head-loss formulae, results are calculated with default parameters and the same verification strategy and accuracies as used in Tables D.13, D.19 and D.26.

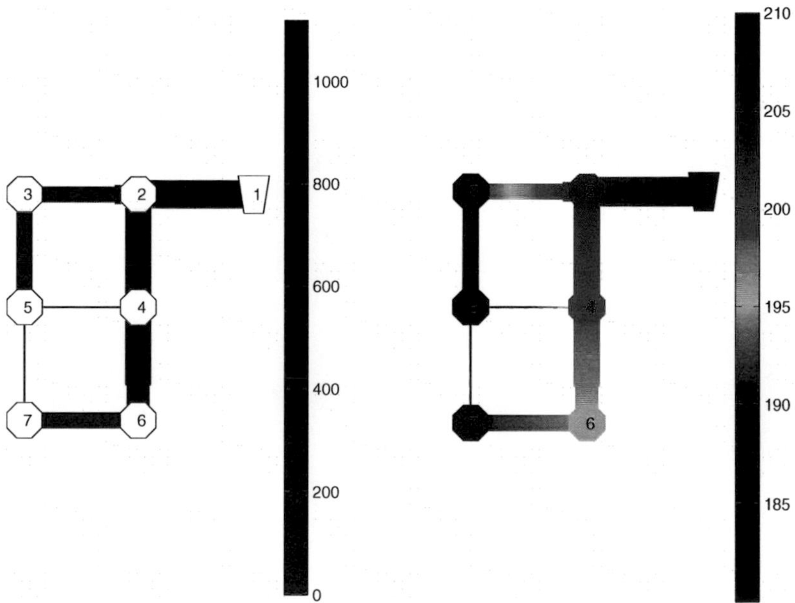

Figure D.4: Illustration of optimal solution of the two-loop network, as part of Table D.8, calculated with Hazen-Williams formula, flow on the left in cubic-meter per second and head (sum of elevation and head established above) in meter on the right side. The diameter of the pipes is proportional to the thickness of the lines.

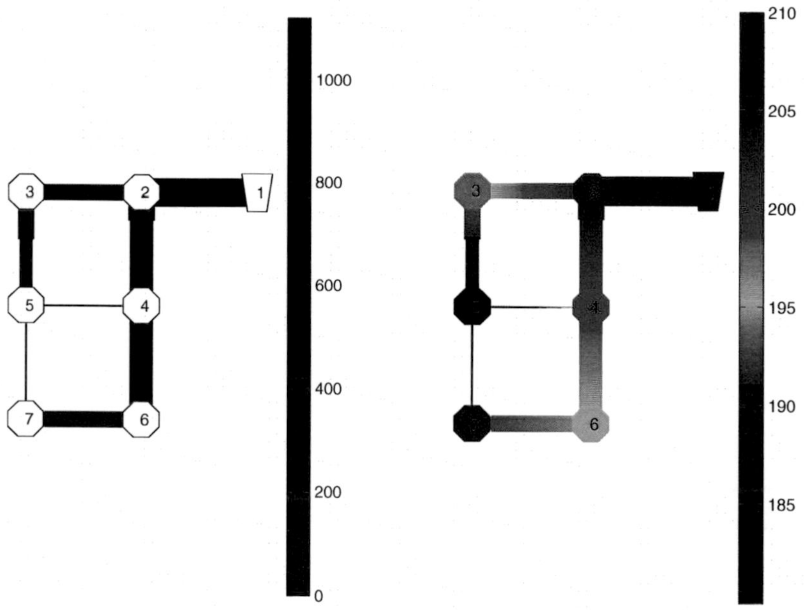

Figure D.5: Illustration of optimal solution of the two-loop network, as part of Table D.15, calculated with Darcy-Weisbach formula, flow on the left in cubic-meter per second and head (sum of elevation and head established above) in meter on the right side. The diameter of the pipes is proportional to the thickness of the lines.

		HW		DW		AHW	
Pipe	Diameter	Length	Velocity	Length	Velocity	Length	Velocity
(1, 2)	18	1000.00	1.8950	1000.00	1.8950	1000.00	1.8950
(2, 3)	10	795.41	2.0192	1000.00	2.0189	1000.00	2.0198
(2, 3)	12	204.59	1.4022				
(2, 4)	14			754.31	1.8229	747.10	1.8223
(2, 4)	16	1000.00	1.3955	245.69	1.3956	252.90	1.3952
(4, 5)	1	1000.00	0.5349	1000.00	0.5639	1000.00	0.5008
(4, 6)	14	310.35	1.4843	1000.00	1.4843	1000.00	1.4842
(4, 6)	16	689.65	1.1364				
(6, 7)	8	11.14	1.7191	168.26	1.7192	167.59	1.7186
(6, 7)	10	988.86	1.1002	831.74	1.1003	832.41	1.0999
(3, 5)	8	98.49	2.2984	581.20	2.2979	566.26	2.2994
(3, 5)	10	901.51	1.4710	418.80	1.4706	433.74	1.4716
(5, 7)	1	1000.00	-0.3798	1000.00	-0.3855	1000.00	-0.3488
Node			Head		Head		Head
(1)			0.00		0.00		0.00
(2)			53.25		54.95		54.94
(3)			30.00		33.58		33.51
(4)			43.86		44.35		44.36
(5)			30.00		30.00		30.00
(6)			30.00		30.00		30.00
(7)			30.00		30.00		30.00
CPU time (s)			54		1,838		53
# Subproblems			173		371		173
Lower bound (US$)			403,385.3080		350,625.4432		350,982.5190
Upper bound (US$)			403,385.3286		350,625.4779		350,982.5452

Table D.40: Comparison of solutions for the two-loop network using a Hazen-Williams coefficient of 130 and an equivalent sand roughness of $\epsilon = 0.003$ mm. Calculations are based on default parameters except `verification` = 1 and an `accuracy` or 10^{-7}.

	Hazen-Williams			adjusted Hazen-Williams		
Newly to be designed arcs						
Pipe	Diameter	Length	Velocity	Diameter	Length	Velocity
(7, 8)	120	591.05	0.6472	36	2,926.08	0.4498
(7, 8)	132	2,335.03	0.5349			
(10,17)				48	1,537.68	0.4757
(10,17)	96	8,046.72	0.2374	60	6,509.04	0.3044
(12,18)	96	9,428.80	0.9674			
(12,18)	108	80.96	0.7644	96	9,509.76	0.9665
(18,19)	84	7,315.20	0.6566	72	7,127.23	0.7820
(18,19)				84	187.97	0.5745
(11,20)	72	4,342.62	1.1870	72	1,606.93	1.3972
(11,20)	84	46.50	0.8721	84	2,782.19	1.0265
(9,16)	72	8,046.72	0.8730	60	8,046.72	0.8654
Fixed arcs						
Pipe	Diameter	Length	Velocity	Diameter	Length	Velocity
(1, 2)	180	11,600.00	1.5206	180	11,600.00	1.4958
(2, 3)	180	19,800.00	1.3613	180	19,800.00	1.3364
(3, 4)	180	7,300.00	1.2019	180	7,300.00	1.1771
(4, 5)	180	8,300.00	1.0498	180	8,300.00	1.0249
(5, 6)	180	8,600.00	0.8976	180	8,600.00	0.8728
(6, 7)	180	19,100.00	0.7455	180	19,100.00	0.7207
(8, 9)	132	12,500.00	0.8205	132	12,500.00	0.7743
(9,10)	180	9,600.00	0.1009	180	9,600.00	0.1009
(11, 9)	204	11,200.00	0.1808	204	11,200.00	0.1785
(12,11)	204	14,500.00	0.6482	204	14,500.00	0.6675
(13,12)	204	12,200.00	1.1199	204	12,200.00	1.1392
(14,13)	204	24,100.00	1.2772	204	24,100.00	1.2965
(15,14)	204	21,100.00	1.4012	204	21,100.00	1.4206
(1,15)	204	15,500.00	1.5253	204	15,500.00	1.5446
(20,16)	60	38,400.00	0.1248	60	38,400.00	0.3747
(7,21)	132	4,800.00	0.5685	132	4,800.00	1.0238
(21, 8)	132	4,800.00	0.5685	132	4,800.00	1.0238
(10,22)	72	13,200.00	0.1979	72	13,200.00	0.4084
(22,17)	72	13,200.00	0.1979	72	13,200.00	0.4084
(9,23)	72	13,200.00	0.8730	72	13,200.00	0.9714
(23,16)	72	13,200.00	0.8730	72	13,200.00	0.9714
(20,24)	60	7,200.00	-1.0545	60	7,200.00	-1.0018
(24,11)	60	7,200.00	-1.0545	60	7,200.00	-1.0018
(12,25)	72	15,600.00	0.8049	72	15,600.00	0.8065
(25,18)	72	15,600.00	0.8049	72	15,600.00	0.8065
(18,26)	60	12,000.00	0.5308	60	12,000.00	0.6917
(26,19)	60	12,000.00	0.5308	60	12,000.00	0.6917

Table D.41: Comparison of solutions of New York (12") with the parameters of Table D.42, diameters are measured in inch, length in meter and velocities in m/s.

	HW	**AHW**
Node	Head	Head
(1)	91.4400	91.4400
(2)	89.6851	89.8624
(3)	87.2445	87.7122
(4)	86.5298	87.0971
(5)	85.8972	86.5667
(6)	85.4065	86.1679
(7)	84.6336	85.5640
(8)	84.2955	84.6659
(9)	83.4273	83.9966
(10)	83.4177	83.9906
(11)	83.4558	84.0153
(12)	83.8470	84.3528
(13)	84.7523	85.1788
(14)	87.0328	87.2916
(15)	89.4030	89.5119
(16)	79.2480	79.2480
(17)	83.1494	83.1494
(18)	79.5972	80.4831
(19)	77.7240	77.7240
(20)	79.4538	80.5456
(21)	84.4646	85.1150
(22)	83.2836	83.5700
(23)	81.3377	81.6223
(24)	81.4548	82.2804
(25)	81.7221	82.4179
(26)	78.6606	79.1036
CPU time (s)	11,779	10,943
# Subproblems	11,259	23,058
Lower bound (US$)	38,049,834.3497	28,787,866.1283
Upper bound (US$)	38,049,865.4756	28,787,894.4094

Table D.42: Comparison of optimal head for the New York (12") test network, measured in meter.
Calculations are based on default parameters, `verification` $= 2$ for Hazen-Williams and `verification` $= 3$ for adjusted Hazen-Williams and for the latter an equivalent sand roughness of $\epsilon = 0.01$ ft is used to avoid infeasibility.

	HW			DW			AHW		
Pipe	Diam.	Length	Velocity	Diam.	Length	Velocity	Diam.	Length	Velocity
(1, 2)	40	100.00	6.8320	40	100.00	6.8320	40	100.00	6.8320
(2, 3)	40	1350.00	6.5270	40	1350.00	6.5270	40	1350.00	6.5270
(3, 4)	40	900.00	2.7294	40	900.00	2.4763	40	900.00	2.4857
(4, 5)	40	1150.00	2.6849	40	1150.00	2.4318	40	1150.00	2.4411
(5, 6)	40	1450.00	2.4365	40	1450.00	2.1834	40	1450.00	2.1927
(6, 7)	40	450.00	2.0921	40	450.00	1.8390	40	450.00	1.8484
(7, 8)	40	850.00	1.6296	40	850.00	1.3765	40	850.00	1.3858
(8, 9)	40	850.00	1.4411	40	850.00	1.1881	40	850.00	1.1974
(9,10)	30	72.59	2.2423						
(9,10)	40	727.41	1.2613	40	800.00	1.0082	40	800.00	1.0175
(10,11)	30	950.00	1.2182	30	950.00	1.2182	30	950.00	1.2182
(11,12)	24	1200.00	1.4276	30	1200.00	0.9137	30	1200.00	0.9137
(12,13)	24	3500.00	0.8946	24	3406.12	0.8946	24	3500.00	0.8946
(12,13)				30	93.88	0.5726			
(10,14)	16	250.42	2.4759	12	793.89	1.5894	12	658.62	1.6930
(10,14)	20	549.58	1.5845	16	6.11	0.8940	16	141.38	0.9523
(14,15)	16	500.00	1.1589	12	500.00	-0.7519	12	500.00	-0.6483
(15,16)	12	550.00	0.9943	12	24.89	-1.8178	12	104.05	-1.7142
(15,16)				16	525.11	-1.0225	16	445.95	-0.9642
(16,17)	12	2730.00	-0.5027	24	2711.70	-1.6435	24	2730.00	-1.6447
(16,17)				30	18.30	-1.0519			
(17,18)	16	1750.00	-2.1351	30	1750.00	-1.5787	30	1750.00	-1.5795
(18,19)	20	427.82	-3.2098						
(18,19)	24	372.18	-2.2290	30	800.00	-2.3980	30	800.00	-2.3987
(3,19)	24	400.00	2.2861	30	400.00	2.4345	30	400.00	2.4353
(3,20)	40	2200.00	2.6834	40	2200.00	2.3900	40	2200.00	2.3803
(20,21)	16	491.36	3.0301	16	6.79	3.0301	16	14.28	3.0301
(20,21)	20	1008.64	1.9393	20	1493.21	1.9393	20	1485.72	1.9393
(21,22)	12	500.00	1.8464	12	500.00	1.8464	12	500.00	1.8464
(20,23)	40	2650.00	1.7617	40	2650.00	1.4684	40	2650.00	1.4586
(23,24)	30	1230.00	2.1325	30	1230.00	1.5490	30	1230.00	1.5319
(24,25)				24	1092.78	1.6400	24	1300.00	1.6132
(24,25)	30	1300.00	1.6331	30	207.22	1.0496			
(25,26)	16			12	640.85	1.2587	12	814.37	1.1505
(25,26)	20	850.00	1.6265	16	209.15	0.7080	16	35.63	0.6472
(26,27)	12	300.00	1.0917	20	300.00	-0.7803	20	300.00	-0.8193
(16,27)	12	750.00	0.3169	20	750.00	1.2874	20	750.00	1.3264
(23,28)				16	1380.61	1.4936			
(23,28)	16	1500.00	1.2756	20	119.39	0.9559	16	1500.00	1.4929
(28,29)	12	2000.00	1.1638	16	2000.00	0.8726	16	2000.00	0.8719
(29,30)	12	1600.00	-0.2067	12	1600.00	0.1808	12	1600.00	0.1796
(30,31)	16	150.00	-0.8872	16	150.00	-0.6692	16	150.00	-0.6699
(31,32)	16	748.17	-1.1121	16	860.00	-0.8940	16	860.00	-0.8947
(31,32)	20	111.83	-0.7117						
(25,32)							20	14.48	1.6759
(25,32)	24	950.00	1.2604	24	950.00	1.1635	24	935.52	1.1638

Table D.43: Comparison of solutions for Hanoi, for parameters used see Table D.44.

	HW	**DW**	**AHW**
Node	Head	Head	Head
(1)	100.0000	100.0000	100.0000
(2)	97.1402	96.6066	96.5910
(3)	61.6631	54.7711	54.5679
(4)	56.9576	50.6800	50.4668
(5)	51.1255	45.6363	45.4118
(6)	44.9821	40.4927	40.2639
(7)	43.5442	39.3534	39.1267
(8)	41.8344	38.1332	37.9160
(9)	40.4726	37.2176	37.0108
(10)	39.1934	36.5914	36.3946
(11)	37.6341	35.0686	34.9174
(12)	34.2068	33.9686	33.8649
(13)	30.0000	30.0000	30.0000
(14)	33.6569	30.0000	30.0000
(15)	32.0998	30.9682	30.6857
(16)	30.2959	32.5281	32.6160
(17)	32.8280	42.7891	42.7446
(18)	49.7279	47.4435	47.3075
(19)	58.9307	52.2798	52.0990
(20)	50.5177	45.4459	45.3710
(21)	35.1598	35.5587	35.5063
(22)	30.0000	30.0000	30.0000
(23)	44.3592	41.1297	41.1913
(24)	38.6646	37.9776	38.1735
(25)	34.9928	33.6204	33.5321
(26)	31.1698	30.0000	30.0000
(27)	30.0000	30.3325	30.3478
(28)	38.7790	33.8295	33.6281
(29)	30.0000	30.2111	30.1705
(30)	30.2863	30.0000	30.0000
(31)	30.5711	30.1627	30.1535
(32)	32.8386	31.7933	31.7186
CPU time (s)	2,885	61,170	1,474
# Subproblems	3,577	5,669	1,645
Lower bound (US$)	6,055,536.3256	6,548,469.8583	6,591,289.3765
Upper bound (US$)	6,055,542.3698	6,614,612.1670	6,591,295.8422

Table D.44: Comparison of optimal head for the Hanoi test network, measured in meter. Calculations are based on default parameters and `verification` = 3, except an reduced accuracy of 10^{-2} for Darcy-Weisbach.

k_value	CPU Time (s)	# Sub-Problems	Upper Bound (US$)
150	56	173	353,855.4352
148	55	173	358,103.6617
146	58	181	362,303.4108
144	57	179	366,454.5916
142	56	173	370,557.5339
140	56	175	374,611.7128
138	56	173	379,120.4527
136	55	163	384,709.9222
134	53	161	391,014.9248
132	53	161	397,240.1536
130	52	165	403,385.5250
128	54	171	409,631.1144
126	50	159	416,355.8586
124	50	155	423,725.7256
122	58	177	431,015.7991
120	58	175	438,204.2910
118	52	163	445,336.2401
116	60	193	452,710.1223
114	52	163	459,976.4079
112	55	173	467,551.6563
110	54	171	477,637.5611
108	60	233	487,569.2719
106	67	267	497,346.0813
104	69	271	506,967.8025
102	66	261	516,433.4676
100	70	281	528,723.5172
98	72	291	541,838.6721
96	73	297	554,727.8135
94	76	311	567,390.4465
92	82	335	582,729.0777
90	73	297	598,138.7446
88	63	247	613,510.1778
86	57	215	632,915.1822
84	54	201	653,232.8227
82	53	197	676,575.7127
80	51	187	702,949.6969

Table D.45: Optimal value depending on C_{HW} calculated for the two-loop network with Hazen-Williams formula, default parameters and `verification` = 1.

	DW			AHW		
k_value	CPU Time (s)	# Sub-Problems	Upper Bound (US$)	CPU Time (s)	# Sub-Problems	Upper Bound (US$)
0.00015	1,858	375	349,579.7492	52	163	350,119.5371
0.0010	1,829	369	354,029.1627	56	171	354,692.6466
0.0100	1,800	357	380,841.7509	53	157	382,537.9886
0.0200	1,937	391	404,216.4445	57	175	405,395.8605
0.0300	1,898	381	420,536.9537	57	179	421,519.4931
0.0400	2,011	403	434,059.8770	56	177	434,840.8210
0.0500	2,092	419	445,082.6102	55	171	445,765.6912
0.0600	2,122	425	454,174.3044	55	169	454,752.9531
0.0700	1,998	401	461,819.9352	59	187	462,320.1565
0.0800	2,141	429	468,396.3888	50	157	468,836.3943
0.0900	2,180	437	476,331.4393	54	169	476,930.7976
0.1000	2,270	455	483,749.3947	53	169	484,289.8511
0.1100	2,419	485	490,401.4582	55	179	490,893.4571
0.1200	2,498	501	496,422.0307	58	187	496,872.7830
0.1300	2,579	515	501,912.8383	57	181	502,328.6687
0.1400	2,631	521	506,953.8212	56	179	507,339.6714
0.1500	2,731	547	511,829.6526	58	189	512,220.7976
0.1600	2,577	517	516,536.4336	59	195	516,901.7362
0.1700	2,576	517	520,923.7908	57	181	521,266.6663
0.1800	2,665	535	526,089.0760	56	181	526,551.3364
0.1900	2,828	567	531,328.4347	56	177	531,766.3017
0.2000	2,705	543	536,264.5892	65	209	536,679.7643
0.2100	2,849	571	540,927.4834	66	211	541,322.8322
0.2200	2,988	599	545,343.6215	70	227	545,720.9363
0.2300	3,037	609	549,536.4063	70	227	549,896.6478
0.2400	3,097	621	553,524.8961	69	223	553,869.8710
0.2500	2,882	579	557,327.0672	69	221	557,658.0591
0.2600	2,883	579	560,957.9452	65	209	561,275.6356
0.2700	2,894	581	564,430.6878	69	225	564,736.6778
0.2800	2,803	563	567,758.2711	68	221	568,055.3607
0.2900	2,755	553	571,024.2968	68	223	571,312.9367
0.3000	2,755	553	574,479.5444	70	231	574,846.8488

Table D.46: Optimal value depending on `k_value` calculated for the two-loop network for Darcy-Weisbach and adjusted Hazen-Williams formula calculated with default parameters and `verification` = 1.

New York (12")				
Hazen-Williams				
k_value	CPU Time (s)	# Sub-Problems	Lower Bound (US$)	Upper Bound (US$)
150	3,695	3,531	16,395,983.4581	16,395,993.4448
140	3,800	3,529	17,343,386.0715	17,343,394.6941
130	3,415	3,173	19,040,776.8786	19,040,792.4196
120	4,492	4,463	21,414,278.9871	21,414,289.3557
110	19,703	21,451	25,464,046.4937	25,464,071.9137
100	11,750	12,129	38,049,834.3497	38,049,865.2094
90	263	85		infeasible
80	184	69		infeasible
Adjusted Hazen-Williams				
0.0024	8,561	9,235	16,968,690.9573	16,968,704.3609
0.0400	2,860	2,599	21,790,649.4554	21,790,651.5658
0.0800	4,588	4,505	24,973,717.3043	24,973,738.3893
0.1200	10,943	11,259	28,787,866.1283	28,787,894.4094
0.1600	7,060	7,097	33,128,238.6150	33,128,269.0534
0.2000	22,079	23,889	39,618,578.5178	39,618,588.9937
0.2400	409	143		infeasible
0.2800	410	141		infeasible

New York (4")				
Hazen-Williams				
150	6,784	6,303	16,254,689.9018	16,254,697.9389
140	5,418	4,689	17,294,792.8203	17,294,798.3049
130	3,067	2,431	18,915,476.1506	18,915,478.6617
120	3,339	2,739	21,278,182.9476	21,278,197.4292
110	14,749	14,547	25,041,869.7621	25,041,887.6851
100	20,946	20,023	37,896,865.6158	37,896,903.5093
90	250	75		infeasible
80	193	69		infeasible
Adjusted Hazen-Williams				
0.0024	4,617	3,961	16,801,968.8573	16,801,975.4907
0.0400	4,391	3,879	21,642,765.1924	21,642,769.6721
0.0800	12,382	11,561	24,694,104.0304	24,694,111.5048
0.1200	9,935	9,117	28,527,609.9734	28,527,638.0059
0.1600	9,612	8,819	32,947,034.9800	32,947,051.2758
0.2000	14,790	14,245	39,295,657.1103	39,295,682.8690
0.2400	399	131		infeasible
0.2800	428	139		infeasible

Table D.47: Optimal value depending on C_{HW} and ϵ for the New York networks, calculated with `verification` = 3 and default parameters.

Hazen-Williams				
k_value	CPU Time (s)	# Sub-Problems	Lower Bound (US$)	Upper Bound (US$)
200	431	407	4,266,981.3590	4,266,985.5307
190	583	525	4,436,208.5955	4,436,212.6127
180	828	815	4,626,429.4206	4,626,433.6224
170	494	457	4,807,457.4244	4,807,461.5408
160	429	417	5,003,113.5245	5,003,117.7483
150	857	915	5,247,960.1188	5,247,965.3610
140	1,288	1,379	5,613,516.7583	5,613,522.3191
130	2,885	3,577	6,055,536.3256	6,055,542.3698
120	1,083	1,197	6,897,244.7728	6,897,251.6242
110	24	15		infeasible
100	2	1		infeasible

Adjusted Hazen-Williams				
0.0050	1,160	1,219	5,572,841.2686	5,572,846.7258
0.0100	439	395	5,849,439.2839	5,849,444.9572
0.0150	804	811	6,098,736.2974	6,098,741.3433
0.0200	1,990	2,059	6,358,079.8831	6,358,086.2363
0.0250	1,474	1,645	6,591,289.3765	6,591,295.8422
0.0300	1,390	1,515	6,860,275.3340	6,860,281.9638
0.0350	1,363	1,463	7,122,347.7561	7,122,351.7520
0.0400	819	855	7,360,229.6484	7,360,236.9152
0.0450	593	593	7,622,944.8858	7,622,952.3780

Table D.48: Optimal value depending on C_{HW} and ϵ for Hanoi] for the Hanoi network, calculated with default parameters and `verification` = 3.

D.5.2 Comparison of Different Viscosity

Two-Loop					
	Viscosity	CPU Time (s)	# Sub-Problems	Lower Bound (US$)	Upper Bound (US$)
Darcy-Weisbach					
$4°C$	$1.586 \cdot 10^{-6}$	1,899	383	356,608.1925	356,608.3764
$10°C$	$1.306 \cdot 10^{-6}$	1,815	367	350,625.4386	350,625.4779
$20°C$	$1.003 \cdot 10^{-6}$	2,013	405	342,410.1713	342,410.4939
Adjusted Hazen-Williams					
$4°C$	$1.586 \cdot 10^{-6}$	57	177	356,924.5342	356,924.5779
$10°C$	$1.306 \cdot 10^{-6}$	52	163	350,982.4592	350,982.7817
$20°C$	$1.003 \cdot 10^{-6}$	49	179	342,831.1733	342,831.4723

New York (12")					
Darcy-Weisbach					
$4°C$	$1.687 \cdot 10^{-5}$	3,690	169		infeasible
$10°C$	$1.407 \cdot 10^{-5}$	2,212	97		infeasible
$20°C$	$1.008 \cdot 10^{-5}$	2,031	89		infeasible
Adjusted Hazen-Williams					
$4°C$	$1.687 \cdot 10^{-5}$	409	143		infeasible
$10°C$	$1.407 \cdot 10^{-5}$	409	143		infeasible
$20°C$	$1.008 \cdot 10^{-5}$	410	143		infeasible
Adjusted Hazen-Williams with reduced $\epsilon = 0.01$ ft					
$4°C$	$1.687 \cdot 10^{-5}$	10,430	10.749	28,841,675.0126	28,841,703.4200
$10°C$	$1.407 \cdot 10^{-5}$	10.943	11,259	28,787,866.1283	28,787,894.4094
$20°C$	$1.008 \cdot 10^{-5}$	11,615	12,043	28,725,112.9474	28,725,141.6189

Hanoi					
Darcy-Weisbach					
$4°C$	$1.586 \cdot 10^{-6}$	67,828	6,295	6,584,615.0548	6,651,078.6982
$10°C$	$1.306 \cdot 10^{-6}$	61,170	5,699	6,548,469.8583	6,614,612.1670
$20°C$	$1.003 \cdot 10^{-6}$	63,020	5,865	6,516,049.0011	6,581,849.8765
Adjusted Hazen-Williams					
$4°C$	$1.586 \cdot 10^{-6}$	1,399	1,529	6,619,444.7104	6,619,450.9179
$10°C$	$1.306 \cdot 10^{-6}$	1,474	1,645	6,591,289.3765	6,591,295.8422
$20°C$	$1.003 \cdot 10^{-6}$	2,726	3,207	6,566,390.1894	6,566,396.4318

Table D.49: Comparison of different viscosity, calculated with default parameters, except a reduced accuracy of 10^{-2} for Hanoi with Darcy-Weisbach and `verification` = 1 for the two-loop network and `verification` = 3 for New-York and Hanoi.

D.5.3 Comparison of Different Initial Head

Two-Loop				
Initial Head	CPU Time (s)	# Sub-Problems	Lower Bound (US$)	Upper Bound (US$)
180	1	1		infeasible
190	1	1		infeasible
200	57	165	600,494.3411	600,494.7228
210	52	163	350,982.4592	350,982.7817
220	52	191	283,030.9406	283,031.1543
230	51	189	258,081.3468	258,081.5509
240	51	189	252,797.3730	252,797.5770
New York (12")[9]				
280	195	73		infeasible
290	4,094	3,937	20,055,512.2208	20,055,522.8512
300	8,551	9,235	16,968,690.9573	16,968,704.3609
310	5	1		infeasible
320	1	1		infeasible
Hanoi				
80	2	1		infeasible
90	330	393	8,131,472.5765	8,131,480.2319
100	1,406	1,645	6,591,289.3765	6,591,295.8450
110	2,039	2,255	6,320,311.5782	6,320,317.7441
120	2,039	2,255	6,315,290.9013	6,315,297.0671

Table D.50: Optimal value for different initial head using adjusted Hazen-Williams formula, default parameters and `verification` = 1 for the two-loop and Hanoi networks and `verification` = 3 for New York.

[9] A reduced value of $\epsilon = 0.0002$ ft is used.

	HW		**DW**		**AHW**	
Initial Head	# Sub-Problems	Upper Bound (US$)	# Sub-Problems	Upper Bound (US$)	# Sub-Problems	Upper Bound (US$)
180	1	infeasible	1	infeasible	1	infeasible
185	1	infeasible	1	infeasible	1	infeasible
190	1	infeasible	1	infeasible	1	infeasible
195	47	infeasible	63	infeasible	51	infeasible
200	235	757,255.80	411	600,259.85	165	600,494.73
205	181	487,322.96	395	421,004.70	181	421,426.85
210	165	403,385.53	367	350,625.48	163	350,982.79
215	151	354,149.93	355	301,332.72	171	301,508.05
220	179	315,574.25	373	282,900.49	191	283,031.16
225	187	296,240.39	405	266,438.96	199	266,590.33
230	187	282,570.18	317	257,968.05	189	258,081.56
235	191	274,227.19	317	255,339.74	189	255,439.57
240	191	272,279,41	317	252,711.43	189	252,797.58

Table D.51: Optimal value for different initial head for two-loop, calculated with default parameters and `verification` = 1.

D.5.4 Comparison of Different Energy Costs

Energy Costs (US$ per m head)	CPU Time (s)	# Sub-Problems	Lower Bound (US$)	Upper Bound (US$)
Two-Loop				
50	51	163	350,982.4592	350,982.7817
100	51	163	350,982.4592	350,982.7817
150	51	163	350,982.4592	350,982.7817
200	51	163	350,982.4592	350,982.7817
New York (12")				
50	7,854	7,923	38,248,111.3250	38,248,149.2020
100	7,336	7,417	47,659,620.0400	47,659,666.8436
150	7,303	7,393	57,071,127.5854	57,071,184.4876
200	7,268	7,367	66,482,636.6223	66,482,702.1313
Hanoi				
50	122	119	97,269,403.8721	97,269,493.7201
100	123	125	185,592,574.6972	185,592,755.6999
150	122	125	273,769,646.1695	273,769,877.5413
200	98	101	361,853,412.1502	361,853,733.1393

Table D.52: Comparison of optimal value for different energy costs, calculated with adjusted Hazen-Williams formula, default parameters and `verification` = 1 for two-loop and `verification` = 3 for New York and Hanoi.

Energy Costs (US$ per m head)	Pipe Costs (US$)	Energy Costs (US$)	Total Costs (US$)
0.50	259,636.99	9,551.54	269,188.53
1.00	259,636.99	19,103.08	278,740.07
1.50	259,636.99	28,654.62	288,291.61
2.00	259,636.99	38,206.17	297,843.16
2.50	259,637.00	47,757.72	307,394.72
3.00	272,847.31	42,859.43	315,706.74
3.50	302,847.49	18,179.13	321,026.62
4.00	302,847.49	20,776.14	323,623.63
4.50	302,847.49	23,373.16	326,220.65
5.00	302,847.49	25,970.18	328,817.67
5.50	302,847.49	28,567.20	331,414.69
6.00	302,847.49	31,164.22	334,011.71
6.50	302,847.49	33,761.23	336,608.72
7.00	302,847.49	36,358.25	339,205.74
7.50	302,847.55	38,955.27	341,802.82
8.00	302,847.55	41,552.29	344,399.84
8.50	302,847.55	44,149.31	346,996.86
9.00	302,847.55	46,746.33	349,593.88
9.50	342,847.55	8,121.51	350,969.06
10.00	350,982.79	0.00	350,982.79
10.50	350,982.79	0.00	350,982.79
11.00	350,982.79	0.00	350,982.79
11.50	350,982.79	0.00	350,982.79
12.00	350,982.79	0.00	350,982.79

Table D.53: Comparison of pipe and energy costs for two-loop, calculated with adjusted Hazen-Williams formula, default parameters and `verification` = 1. This table contains approximate values, i.e. the pipe costs and total costs part of this table are rounded upwards and the energy costs are determined afterwards.

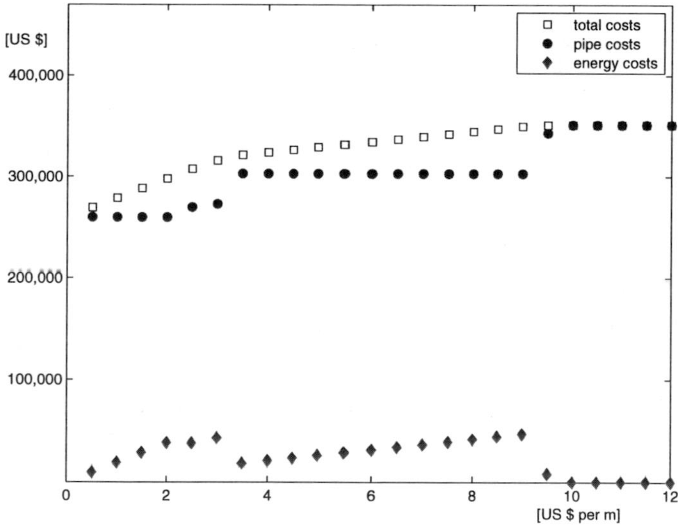

Figure D.6: Illustration of pipe and energy costs of the optimal solution for the two-loop network as part of Table D.53.

New York (12")			
Energy Costs (US$ per m head)	New Pipe Costs (US$)	Energy Costs (US$)	Total Costs (US$)
50	28,836,631.56	9,411,517.64	218,048,699.35
100	28,836,631.56	18,823,035.28	227,460,216.99
150	28,836,631.56	28,234,552.92	236,871,734.63
200	28,836,631.56	37,646,070.57	246,283,252.28

Hanoi			
Energy Costs (US$ per m head)	Pipe Costs (US$)	Energy Costs (US$)	Total Costs (US$)
50	8,929,614.59	88,339,879.14	97,269,493.73
100	8,977,824.51	176,614,931.19	185,592,755.70
150	9,513,399.51	264,256,478.04	273,769,877.55
200	9,525,721.93	352,328,011.21	361,853,733.14

Table D.54: Comparison of pipe and energy costs for New York and Hanoi calculated with default parameters and `verification` = 3. For the New York network the total pipe costs includes the amount of US \$ 179,800,550.15 for the existing pipes, and for all energy costs the optimal head at the source node has a rounded value of 306 ft. Analogous to Table D.53 approximate values are chosen, i.e. the pipe costs and total costs part of this table are rounded upwards and the energy costs are determined afterwards.

Appendix E

Illustration of Darcy-Weisbach and Hazen-Williams Differences

E.1 Head-loss subject to Flow

In this Section in all figures the average fluid velocity of $1\frac{m}{s}$ is marked with a dotted vertical line, of $2.5\frac{m}{s}$ with a dashed line and of $5\frac{m}{s}$ with a solid one.

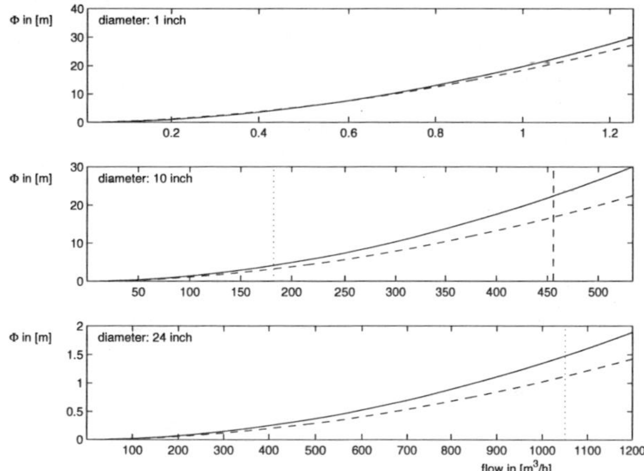

Figure E.1: Two-loop network: Head-loss subject to flow, head-loss according to Hazen-Williams is plotted solid and according to Darcy-Weisbach dashed. Basis are the values of the constants of the two-loop test example and a length of 1,000 meter. The flow range is adjusted to cover typical ranges found in the optimization problem: maximum head-loss of 30 meter with the Hazen-Williams formula or maximum flow of $1,200\,\mathrm{m}^3/\mathrm{h}$, whichever is lower.

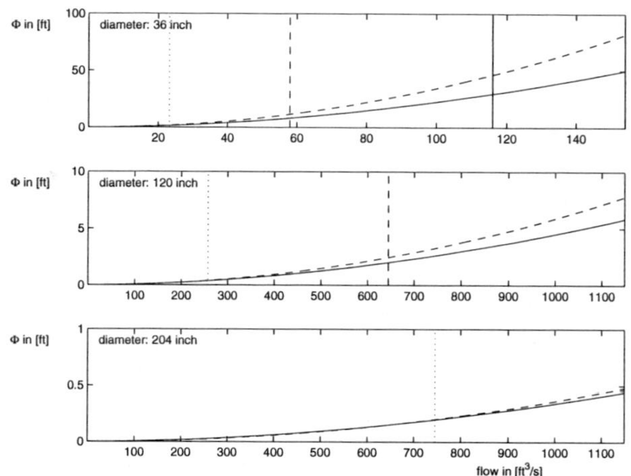

Figure E.2: New-York network: Head-loss subject to flow, based on default values and a length of $1,000$ foot. Head-loss according to Hazen-Williams is plotted solid and according to Darcy-Weisbach dashed. The flow range is adjusted to cover typical ranges found in the optimization problem: maximum head-loss of 50 foot with the Hazen-Williams formula or maximum flow of of $1,150\,\mathrm{ft}^3/\mathrm{s}$, whichever is lower.

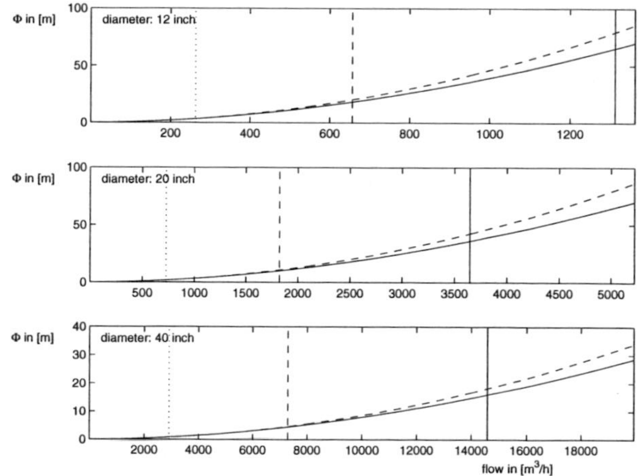

Figure E.3: Hanoi network: Head-loss subject to flow, based on default values and a length of $1,000$ meter. Head-loss according to Hazen-Williams is plotted solid and according to Darcy-Weisbach dashed. The flow range is adjusted to cover typical ranges found in the optimization problem: maximum head-loss of 70 meter with the Hazen-Williams formula or maximum flow of $19,940\,\mathrm{m}^3/\mathrm{h}$, whichever is lower.

E.2 Head-loss Difference subject to Flow

Again, in this Section in all figures the average fluid velocity of $1\frac{m}{s}$ is marked with a dotted vertical line, of $2.5\frac{m}{s}$ with a dashed line and of $5\frac{m}{s}$ with a solid one.

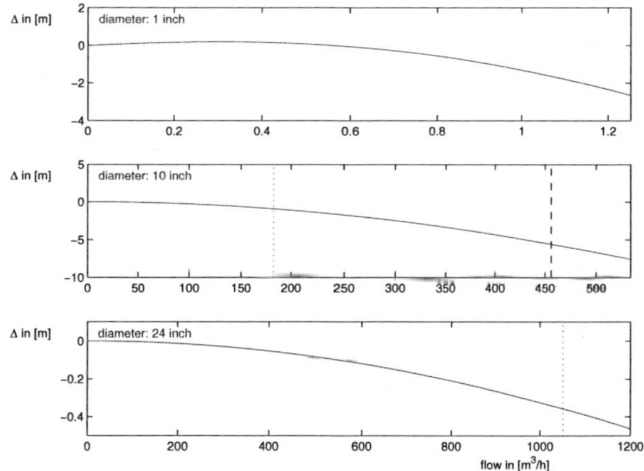

Figure E.4: Two-loop network: Head-loss difference (head-loss according to Darcy-Weisbach minus Hazen-Williams) subject to flow. Basis are the values of the constants of the two-loop test example and a length of $1,000$ meter. The flow range is adjusted to cover typical ranges found in the optimization problem: maximum head-loss of 30 meter with the Hazen-Williams formula or maximum flow of $1,200\,\mathrm{m}^3/\,\mathrm{h}$, whichever is lower.

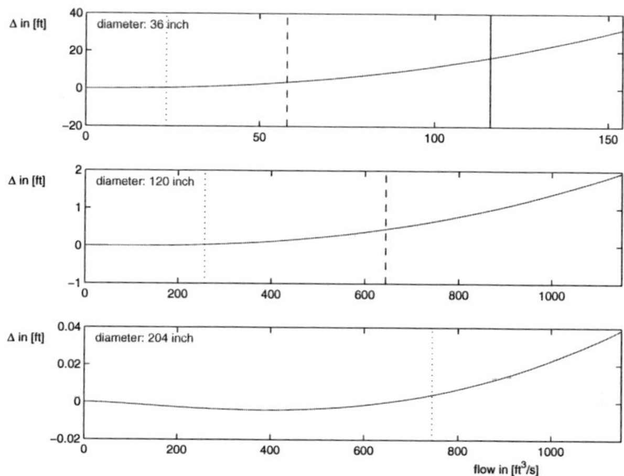

Figure E.5: New-York network: Head-loss difference (head-loss according to Darcy-Weisbach minus Hazen-Williams) subject to flow, based on default values and a length of $1,000$ foot. The flow range is adjusted to cover typical ranges found in the optimization problem: maximum head-loss of 50 foot with the Hazen-Williams formula or maximum flow of of $1,150\,\mathrm{ft}^3/\,\mathrm{s}$, whichever is lower.

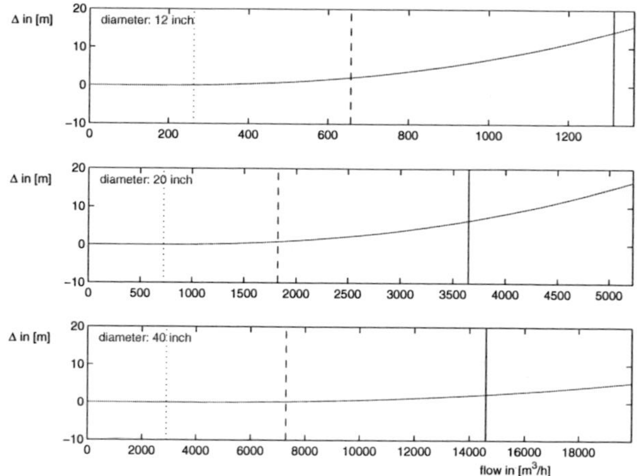

Figure E.6: Hanoi network: Head-loss difference (head-loss according to Darcy-Weisbach minus Hazen-Williams) subject to flow, based on default values and a length of $1,000$ meter. The flow range is adjusted to cover typical ranges found in the optimization problem: maximum head-loss of 70 meter with the Hazen-Williams formula or maximum flow of $19,940\,\mathrm{m}^3/\,\mathrm{h}$, whichever is lower.

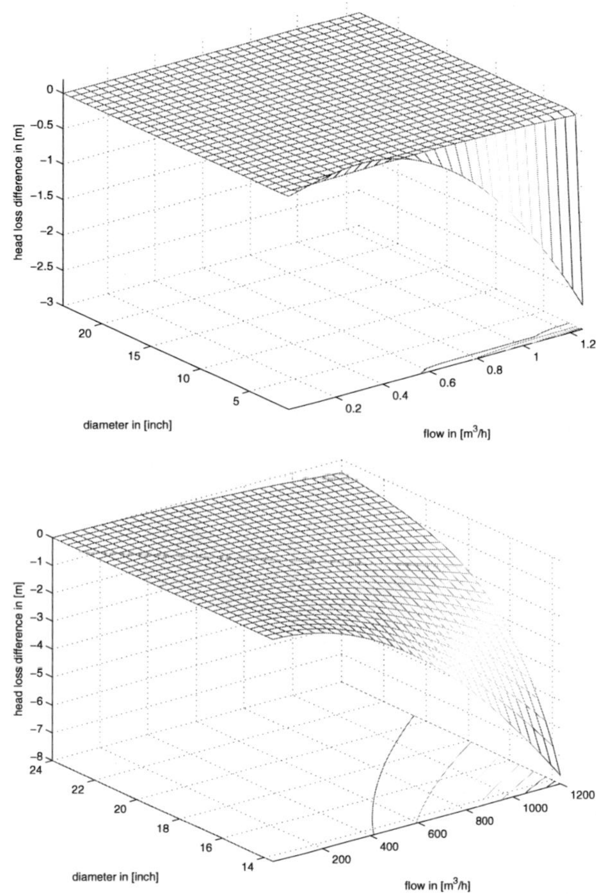

Figure E.7: Two-loop network: Head-loss difference (head-loss according to Darcy-Weisbach minus Hazen-Williams) subject to flow and diameter. Basis are the values of the constants of the two-loop test example and a length of 1,000 meter, first the flow range and second the range of diameters is adjusted to a maximum head-loss of 30 meter with the Hazen-Williams formula.

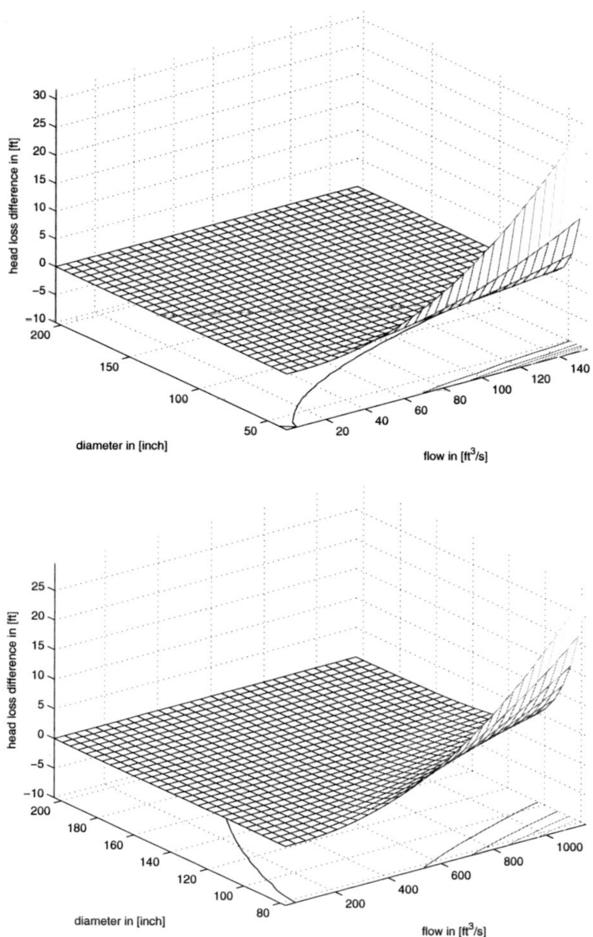

Figure E.8: New-York network: Head-loss difference (head-loss according to Darcy-Weisbach minus Hazen-Williams) subject to flow and diameter. Basis are the values of the constants of the New-York test example and a length of $1,000$ foot, first the flow range and second the range of diameters is adjusted to a maximum head-loss of 50 foot with the Hazen-Williams formula.

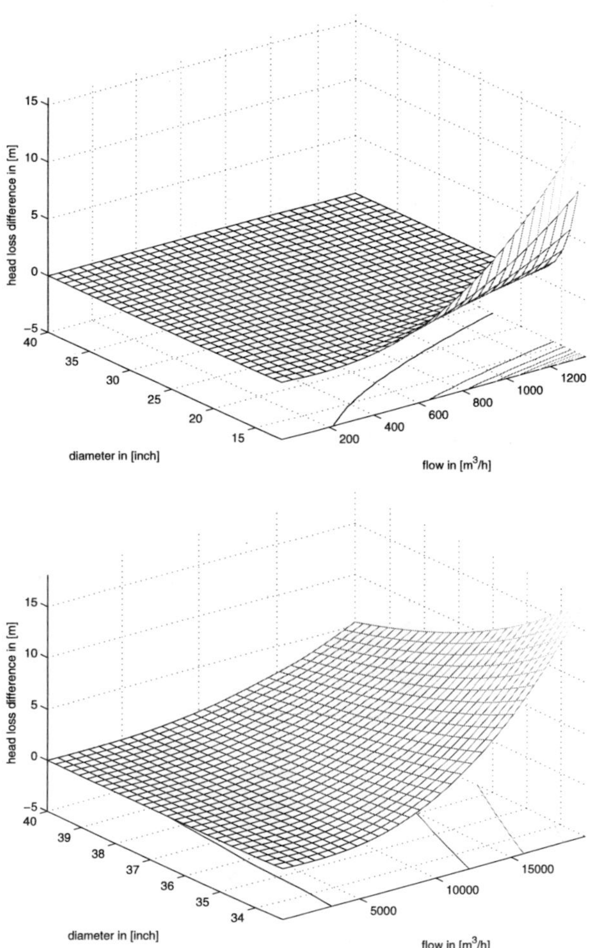

Figure E.9: Hanoi network: Head-loss difference (head-loss according to Darcy-Weisbach minus Hazen-Williams) subject to flow and diameter. Basis are the values of the constants of the Hanoi test example and a length of $1,000$ meter, first the flow range and second the range of diameters is adjusted to a maximum head-loss of 70 meter with the Hazen-Williams formula.

E.3 Relative Difference of Head-loss Formulae

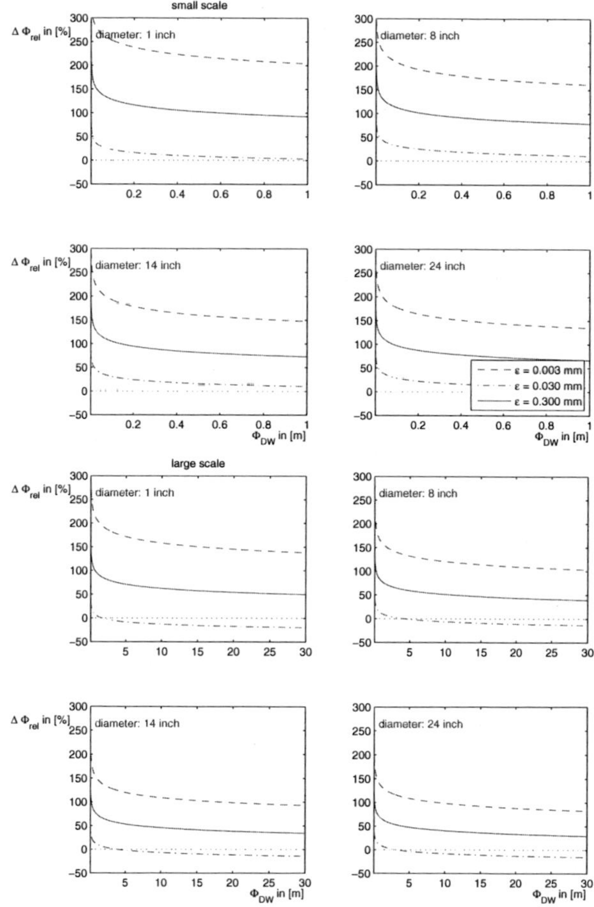

Figure E.10: Two-loop network: Relative difference, i.e. $\Delta\Phi_{rel} = \frac{\Phi_{DW}-\Phi_{HW}}{\Phi_{DW}}$, subject to head-loss according to Darcy-Weisbach. Basis are the values of the constants of the two-loop test example and a length of $1,000$ meter, the flow range is adjusted to a maximum head-loss of 30 meter with the Darcy-Weisbach formula.

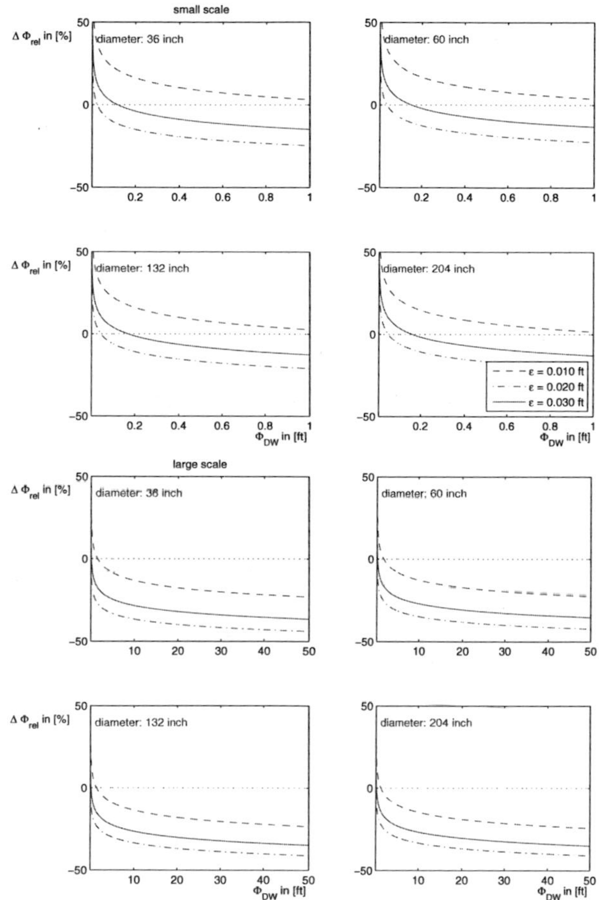

Figure E.11: New-York network: Relative difference, i.e. $\Delta\Phi_{rel} = \frac{\Phi_{DW}-\Phi_{HW}}{\Phi_{DW}}$, subject to head-loss according to Darcy-Weisbach.
Basis are the values of the constants of the two-loop test example and a length of 1,000 feet, the flow range is adjusted to a maximum head-loss of 50 feet with the Darcy-Weisbach formula.

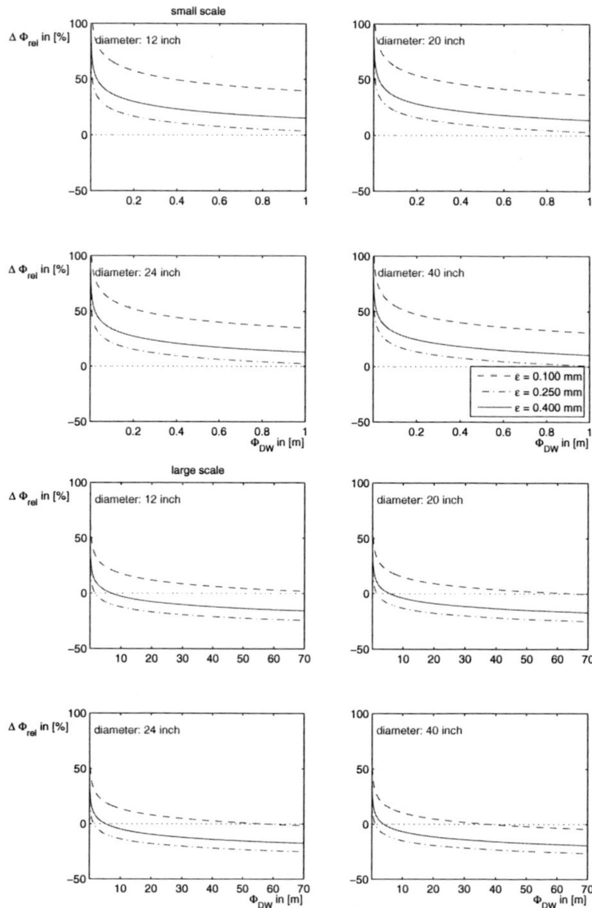

Figure E.12: Hanoi network: Relative difference, i.e. $\Delta\Phi_{rel} = \frac{\Phi_{DW}-\Phi_{HW}}{\Phi_{DW}}$, subject to head-loss according to Darcy-Weisbach. Basis are the values of the constants of the two-loop test example and a length of $1,000$ meter, the flow range is adjusted to a maximum head-loss of 70 meter with the Darcy-Weisbach formula.

Appendix F

Detailed Proofs of Auxiliary Lemmas of Chapter 3

F.1 Proof of Lemma 3.4.10 (ii)

Lemma (cf. p. 97). Considering the continuation $w : I\!R \times I\!R_+ \to I\!R$ of w_k, the auxiliary function $J : I\!R \times I\!R_+ \to I\!R$, defined by

$$J(q,d) := \gamma\, d^{-5}\, w(q,d)$$

is strictly monotone decreasing in d and strictly convex in d for fixed $q > 0$, and strictly monotone increasing in d and strictly concave in d for fixed $q < 0$.

Proof. The monotonicity and convexity is shown by direct differentiation of the function J.

W. l. o. g. it suffices to investigate the case $q > 0$ as J is symmetric in q with respect to the origin, i.e. $J(q,d) = -J(-q,d)$.

Regarding ψ as function in q, d and $f = u(q,d)$, i.e. $\psi : I\!R_+ \setminus \{0\} \times I\!R_+ \setminus \{0\} \times (\frac{1}{4}\left[\log(\frac{\epsilon}{3.71\, d_k})\right]^{-2}, \infty) \to I\!R$ with

$$\psi\,(q,\, d,\, u(q,d)) = \frac{1}{\sqrt{u(q,d)}} + 2\,\log\left(\frac{c_{CW}\, d}{q\sqrt{u(q,d)}} + \frac{\epsilon}{3.71\, d}\right)$$

the implicit function theorem implies

$$\begin{aligned} \frac{\partial u}{\partial d}(q,d) &= -\frac{\frac{\partial \psi}{\partial d}(q,d,\, u(q,d))}{\frac{\partial \psi}{\partial f}(q,d,\, u(q,d))} \\ &= \frac{4\,\left[u\,(q,d)\right]^{\frac{3}{2}}\left(c_{CW}\, d^2 - \frac{\epsilon}{3.71}\, q\sqrt{u\,(q,d)}\right)}{\ln(10)\, c_{CW} d^3 + \ln(10)\, d\,\frac{\epsilon}{3.71}\, q\sqrt{u\,(q,d)} + 2\, c_{CW} d^3\sqrt{u\,(q,d)}} \end{aligned} \quad \text{(F.1.1)}$$

and

$$\frac{\partial^2 u}{\partial d\,\partial d} = 4\,\frac{u\,(q,d)\,(T_1+T_2+T_3+T_4+T_5+T_6+T_7+T_8+T_9+T_{10}+T_{11})}{d^2\left(\ln(10)c_{CW}\,d^2+\ln(10)\frac{\epsilon}{3.71}\,q\sqrt{u\,(q,d)}+2\,c_{CW}\,d^2\sqrt{u\,(q,d)}\right)^3}$$

where

$$T_1 = -8\,c_{CW}{}^2\,d^4\,\tfrac{\epsilon}{3.71}\,q\,\left[u\,(q,d)\right]^2 \qquad T_2 = 4\,c_{CW}{}^3\,d^6\,\left[u\,(q,d)\right]^{\frac{3}{2}}$$

$$T_3 = 3\,\left[\ln(10)\right]^2\,c_{CW}{}^2\,d^4\,\tfrac{\epsilon}{3.71}\,q\,u\,(q,d) \qquad T_4 = 5\,\left[\ln(10)\right]^2\,c_{CW}\,d^2\,\left[\tfrac{\epsilon}{3.71}\right]^2\,q^2\,\left[u\,(q,d)\right]^{\frac{3}{2}}$$

$$T_5 = 6\,\ln(10)\,c_{CW}\,d^2\,\left[\tfrac{\epsilon}{3.71}\right]^2\,q^2\,\left[u\,(q,d)\right]^2 \qquad T_6 = 2\,\ln(10)\,c_{CW}{}^2\,d^4\,\tfrac{\epsilon}{3.71}\,q\,\left[u\,(q,d)\right]^{\frac{3}{2}}$$

$$T_7 = 6\,\ln(10)\,\left[\tfrac{\epsilon}{3.71}\right]^3\,q^3\,\left(u\,(q,d)\right)^{\frac{5}{2}} \qquad T_8 = 12\,c_{CW}\,d^2\,\left[\tfrac{\epsilon}{3.71}\right]^2\,q^2\,\left[u\,(q,d)\right]^{\frac{5}{2}}$$

$$T_9 = -\left[\ln(10)\right]^2\,c_{CW}{}^3\,d^6\,\sqrt{u\,(q,d)} \qquad T_{10} = 2\,\ln(10)\,c_{CW}{}^3\,d^6\,u\,(q,d)$$

$$T_{11} = \left[\ln(10)\right]^2\,\left[\tfrac{\epsilon}{3.71}\right]^3\,q^3\,\left[u\,(q,d)\right]^2\ .$$

Thus, as $q > 0$,

$$J = \gamma\,d^{-5}\,q^2\,u\,(q,d)\ ,$$

and one obtains

$$\frac{\partial J}{\partial d} = -\frac{\gamma\,\left(T_{12}+T_{13}+T_{14}+T_{15}\right)\,q^2\,u\,(q,d)}{d^6\left(\log(10)\,c_{CW}\,d^2+\ln(10)\,\frac{\epsilon}{3.71}\,q\,\sqrt{u\,(q,d)}+2\,c_{CW}\,d^2\,\sqrt{u\,(q,d)}\right)}$$

where

$$T_{12} = 5\,\ln(10)\,c_{CW}\,d^2 \qquad T_{13} = 5\,\ln(10)\,\tfrac{\epsilon}{3.71}\,q\sqrt{u\,(q,d)}$$

$$T_{14} = 6\,c_{CW}\,d^2\,\sqrt{u\,(q,d)} \qquad T_{15} = 4\,\tfrac{\epsilon}{3.71}\,q\,u\,(q,d)\ .$$

As all terms are positive, i.e. $T_{12} > 0$, $\ldots$, $T_{15} > 0$, it follows

$$\frac{\partial J}{\partial d} < 0\,,$$

which implies J to be strictly monotone decreasing in d.

Furthermore one obtains

$$\frac{\partial^2 J}{\partial d\,\partial d} = \frac{2\,\gamma\,q^2\,u\,(q,d)\,\left(T_{16}+T_{17}+T_{18}+\ldots+T_{30}\right)}{d^7\left(\ln(10)c_{CW}\,d^2+\ln(10)\frac{\epsilon}{3.71}\,q\sqrt{u\,(q,d)}+2\,c_{CW}\,d^2\sqrt{u\,(q,d)}\right)^3}\,,$$

where

$$T_{16} = 24\,\left[\tfrac{\epsilon}{3.71}\right]^2 c_{CW}\, d^2\, q^2\, [u\,(q,d)]^{\frac{5}{2}} \qquad T_{17} = 12\,\ln(10)\,\left[\tfrac{\epsilon}{3.71}\right]^3 q^3\, [u\,(q,d)]^{\frac{5}{2}}$$

$$T_{18} = 92\,\ln(10)\,\left[\tfrac{\epsilon}{3.71}\right]^2 c_{CW}\, d^2\, q^2\, [u\,(q,d)]^2 \qquad T_{19} = 15\,[\ln(10)]^3\, {c_{CW}}^3\, d^6$$

$$T_{20} = 184\,\ln(10)\, {c_{CW}}^2\, d^4\, \tfrac{\epsilon}{3.71}\, q\, [u\,(q,d)]^{\frac{3}{2}} \qquad T_{21} = 48\, {c_{CW}}^3\, d^6\, [u\,(q,d)]^{\frac{3}{2}}$$

$$T_{22} = 22\,[\ln(10)]^2\,\left[\tfrac{\epsilon}{3.71}\right]^3 q^3\, [u\,(q,d)]^2 \qquad T_{23} = 45\,[\ln(10)]^3\, c_{CW} d^2 \left[\tfrac{\epsilon}{3.71}\right]^2 q^2\, u\,(q,d)$$

$$T_{24} = 45\,[\ln(10)]^3\, {c_{CW}}^2\, d^4\, \tfrac{\epsilon}{3.71} q\, \sqrt{u\,(q,d)} \qquad T_{25} = 104\,\ln(10)\, {c_{CW}}^3\, d^6\, u\,(q,d)$$

$$T_{26} = 15\,[\ln(10)]^3\,\left[\tfrac{\epsilon}{3.71}\right]^3 q^3\, [u\,(q,d)]^{\frac{3}{2}} \qquad T_{27} = 68\,[\ln(10)]^2\, {c_{CW}}^3\, d^6\, \sqrt{u\,(q,d)}$$

$$T_{28} = 120\,[\ln(10)]^2\, c_{CW} d^2 \left[\tfrac{\epsilon}{3.71}\right]^2 q^2\, [u\,(q,d)]^{\frac{3}{2}} \qquad T_{29} = 166\,[\ln(10)]^2\, \tfrac{\epsilon}{3.71}\, {c_{CW}}^2\, d^4\, q\, u\,(q,d)$$

$$T_{30} = 64\, \tfrac{\epsilon}{3.71}\, {c_{CW}}^2\, d^4\, q\, [u\,(q,d)]^2\ .$$

Obviously all terms T_{16}, $T_{17}, \ldots T_{30}$ are positive, therefore $\frac{\partial^2 J}{\partial d\, \partial d} > 0$ which implies the convexity of J. □

F.2 Proof of Lemma 3.4.14

Lemma (cf. p. 98). The auxiliary function $\widetilde{J}$, defined by $\widetilde{J} : I\!R \times I\!R_+ \to I\!R$ with

$$\widetilde{J}(q,\delta) := \gamma\, \delta^{\frac{5}{3}}\, w(q, \delta^{-\frac{1}{3}}) = \begin{cases} \gamma\, \delta^{\frac{5}{3}}\, \mathrm{sgn}\,(q)\, q^2\, u(|q|, \delta^{-\frac{1}{3}}) & \text{for } q \neq 0 \\ 0 & \text{else} \end{cases}$$

is strictly monotone decreasing and strictly concave in δ for fixed $q < 0$, and strictly monotone increasing and strictly convex in δ for fixed $q > 0$.

Proof. Let $q > 0$ be arbitrary but fixed, then

$$\widetilde{J}(q,\delta) = \gamma\, \delta^{\frac{5}{3}}\, q^2\, u(q, \delta^{-\frac{1}{3}})\ .$$

Thus Equation F.1.1 implies

$$\frac{\partial \widetilde{J}}{\partial \delta}(q,\delta) = \frac{\frac{1}{3}\gamma\, \delta^{\frac{2}{3}}\, q^2\, u(\delta^{-\frac{1}{3}})\ (T_{31} + T_{32} + T_{33} + T_{34})}{\ln(10)\, c_{CW} + \ln(10)\, \frac{\epsilon}{3.71}\, \delta^{\frac{2}{3}}\, q\, \sqrt{u(\delta^{-\frac{1}{3}})} + 2\, c_{CW}\, \sqrt{u(\delta^{-\frac{1}{3}})}}$$

where

$$T_{31} = 5\,\ln(10)\, c_{CW} \qquad T_{32} = 5\,\ln(10)\, \tfrac{\epsilon}{3.71}\, \delta^{\frac{2}{3}}\, q\, \sqrt{u(\delta^{-\frac{1}{3}})}$$

$$T_{33} = 6\, c_{CW}\, \sqrt{u(\delta^{-\frac{1}{3}})} \qquad T_{34} = 4\, \tfrac{\epsilon}{3.71}\, \delta^{\frac{2}{3}}\, q\, u(\delta^{-\frac{1}{3}})\ .$$

Obviously $\frac{\partial \widetilde{J}}{\partial \delta}(q,\delta) > 0$, which proves the first part of the lemma.

The second partial derivative is

$$\frac{\partial^2 \widetilde{J}}{\partial \delta\, \partial \delta}(q,\delta) = \gamma \left[\frac{10}{9}\, \delta^{-\frac{1}{3}}\, q^2\, u(\delta^{-\frac{1}{3}}) - \frac{2}{3}\, \delta^{-\frac{2}{3}}\, q^2\, \frac{\partial u}{\partial d}(\delta^{-\frac{1}{3}}) + \frac{1}{9}\, \delta^{-1}\, q^2\, \frac{\partial^2 u}{\partial d\, \partial d}(\delta^{-\frac{1}{3}}) \right]$$

$$= \frac{\frac{2}{9}\, \gamma\, \delta^{-\frac{1}{3}}\, q^2\, u(\delta^{-\frac{1}{3}})\, (T_{35} + T_{36} + T_{37} + \ldots + T_{48})}{\left[\ln(10)\, c_{CW} + \ln(10)\, \frac{\epsilon}{3.71}\, \delta^{\frac{2}{3}}\, q\, \sqrt{u(\delta^{-\frac{1}{3}})} + 2\, c_{CW}\, \sqrt{u(\delta^{-\frac{1}{3}})} \right]^3}$$

where

$$T_{35} = 5\, [\ln(10)]^3\, {c_{CW}}^3 \qquad T_{36} = 16\, \ln(10)^2\, {c_{CW}}^3\, \sqrt{u(\delta^{-\frac{1}{3}})}$$

$$T_{37} = 16\, \ln(10)\, {c_{CW}}^3\, u(\delta^{-\frac{1}{3}}) \qquad T_{38} = 14\, [\ln(10)]^2\, \left[\tfrac{\epsilon}{3.71}\right]^3\, \delta^2\, q^3\, \left[u(\delta^{\frac{1}{3}})\right]^2$$

$$T_{39} = 15\, [\ln(10)]^3\, \tfrac{\epsilon}{3.71}\, {c_{CW}}^2\, \delta^{\frac{2}{3}}\, q\, \sqrt{u(\delta^{\frac{1}{3}})} \qquad T_{40} = 15\, [\ln(10)]^3\, \left[\tfrac{\epsilon}{3.71}\right]^2 c_{CW}\, \delta^{\frac{4}{3}}\, q^2\, u(\delta^{-\frac{1}{3}})$$

$$T_{41} = 54\, \ln(10)^2\, \tfrac{\epsilon}{3.71}\, {c_{CW}}^2\, q\, \delta^{\frac{2}{3}}\, u(\delta^{-\frac{1}{3}}) \qquad T_{42} = 5\, [\ln(10)]^3\, \left[\tfrac{\epsilon}{3.71}\right]^3\, \delta^2\, q^3\, \left[u(\delta^{-\frac{1}{3}})\right]^{\frac{3}{2}}$$

$$T_{43} = 52\, [\ln(10)]^2\, \left[\tfrac{\epsilon}{3.71}\right]^2 c_{CW}\, \delta^{\frac{4}{3}}\, q^2 \left[u(\delta^{-\frac{1}{3}})\right]^{\frac{3}{2}} \qquad T_{44} = 64\, \ln(10)\, \tfrac{\epsilon}{3.71}\, {c_{CW}}^2\, \delta^{\frac{2}{3}}\, q\, \left[u(\delta^{-\frac{1}{3}})\right]^{\frac{3}{2}}$$

$$T_{45} = 60\, \ln(10)\, \left[\tfrac{\epsilon}{3.71}\right]^2\, c_{CW}\, \delta^{\frac{4}{3}}\, q^2\, \left[u(\delta^{-\frac{1}{3}})\right]^2 \qquad T_{46} = 32\, \tfrac{\epsilon}{3.71}\, {c_{CW}}^2\, \delta^{\frac{2}{3}}\, q\, \left[u(\delta^{-\frac{1}{3}})\right]^2$$

$$T_{47} = 12\, \ln(10)\, \left[\tfrac{\epsilon}{3.71}\right]^3\, \delta^2\, q^3\, \left[u(\delta^{-\frac{1}{3}})\right]^{\frac{5}{2}} \qquad T_{48} = 24\, \left[\tfrac{\epsilon}{3.71}\right]^2\, c_{CW}\, q^2\, \delta^{\frac{4}{3}}\, \left[u(\delta^{-\frac{1}{3}})\right]^{\frac{5}{2}}\,.$$

Clearly $\frac{\partial^2 \widetilde{J}}{\partial \delta\, \partial \delta}(q,\delta) > 0$, which proves the convexity in δ.

The definition of $\widetilde{J}$,

$$\widetilde{J}(q,\delta) = \gamma\, \delta^{\frac{5}{3}}\, w(q, \delta^{-\frac{1}{3}}) = \begin{cases} \gamma\, \delta^{\frac{5}{3}}\, \mathrm{sgn}\,(q)\, u(|q|, \delta^{-\frac{1}{3}}) & \text{for } q \neq 0 \\ 0 & \text{else,} \end{cases}$$

implies symmetry, i.e. $\widetilde{J}(q,\delta) = -\widetilde{J}(-q,\delta)$. Thus for $q < 0$ the function $\widetilde{J}$ is strictly monotone decreasing and strictly concave in δ. □

Bibliography

Optimization and Intervals

[A&M00] Götz Alefeld, Günter Mayer. *Interval analysis: theory and applications.* Journal of Computational and Applied Mathematics 121, 421 - 464, 2000. 9, 10, 13

[C&W71] Lothar Collatz, Wolfgang Wetterling. *Optimierungsaufgaben.* Springer-Verlag, ISBN 3-540-05616-5, 1971. 13

[G&L02] Nicholas I. M. Gould and Sven Leyffer. *An Introduction to Algorithms for Nonlinear Optimization.* In James Blowey, Alan Craig, Tony Shardlow, editors, Frontiers in Numerical Analysis, Springer-Verlag, ISBN 3-540-44319-3, Durham 2002. 27

[H62] R. W. Hamming. *Numerical Methods for Scientists and Engineers.* New York, McGraw-Hill, 1962. 1, 5, 15, 133, 165

[H02] Gareth I. Hargreaves. *Interval Analysis in MATLAB.* Numerical Analysis Report No. 416, Manchester Center for Computational Mathematics, 2002.
`http://www.maths.man.ac.uk/~nareports/narep416.pdf` 9, 10

[H99] Angelika C. Hailer. *Bisektion im $\mathbb{R}^n$ - Betrachtung eines Verfahrens von M. N. Vrahatis.* Master's thesis, Universität Ulm, 1999. 16

[H&W04] Eldon Hansen, G. William Walster. *Global Optimization Using Interval Analysis.* Second Edition, Marcel Dekker, Inc. , New York, ISBN 0-8247-4059-9, 2004. 9, 10, 14, 26

[HJ85] Roger A. Horn, Charles R. Johnson. *Matrix Analysis.* Cambridge University Press, ISBN 0-521-30586-1, 1985. 17, 18

[J94] Christian Jansson. *On Self-Validating Methods for Optimization Problems.* In J. Herzberger, editor, Topics in Validated Computations, Proceedings of the IMACS-GAMM International Workshop on Validated Computation, Oldenburg, 1993, pp. 381-438. Elsevier Science, ISBN 0-444-81685-2, 1994. 132

[J02] Christian Jansson. *Rigorous Lower and Upper Bounds in Linear Programming.* Technical Report 02.1, Forschungsschwerpunkt Informations- und Kommunikationstechnik, TU Hamburg-Harburg, 2002. http://www.ti3.tu-harburg.de/paper/jansson/verification.ps 29, 30, 119, 122

[J04] Christian Jansson. *Rigorous Lower and Upper Bounds in Linear Programming.* SIAM Journal on Optimization, 14, No. 3, pp. 914-935, 2004. http://www.siam.org/journals/siopt/14-3/41683.html 2, 6, 23, 29

[J04c] Christian Jansson. *A Rigorous Lower Bound for the Optimal Value of Convex Optimization Problems.* Journal of Global Optimization, 28, pp. 121-137, 2004. 29

[J99] Dieter Jungnickel. *Optimierungsmethoden.* Springer-Verlag, ISBN 3-540-66057-7, 1999. 31

[Ke96] R. Baker Kearfott. *Rigorous Global Search: Continuous Problems.* Kluver Academic Publishers, ISBN 0-7923-4238-0, 1996. 26

[K95] Olaf Knüppel. *Einschießungsmethoden zur Bestimmung der Nullstellen nichtlinearer Gleichungssysteme und ihre Implementierung.* Dissertation, Technische Universität Hamburg-Harburg, 1995. 10

[Ku96] Ulrich Kulisch. *Memorandum über Computer, Arithmetik und Numerik.* Universität Karlsruhe, Institut für Angewandte Mathematik. Berichte aus dem Forschungsschwerpunkt Computerarithmetik, Intervallrechnung und Numerische Algorithmen mit Ergebnisverifikation. Karlsruhe 1996. http://www.ubka.uni-karlsruhe.de/cgi-bin/psgunzip/1996/mathematik/6/6.pdf 121

[M62] R. E. Moore. *Interval Arithmetic and Automatic Error Analysis in Digital Computing.* Ph.d. Dissertation, Department of Mathematics, Stanford University, Stanford, California, 1962, published as Applied Mathematics and Statistics Laboratories Technical Report No. 25, 1962. http://interval.louisiana.edu/Moores_early_papers/disert.pdf 10

[N01] Arnold Neumaier. *Introduction to Numerical Analysis.* Cambridge University Press, ISBN 0-521-33610-4, 2001. 22

[N&S04] Arnold Neumaier and Oleg Shcherbina. *Safe bounds in linear and mixed-integer programming.* Math. Programming, A 99, 283-296, 2004. http://www.mat.univie.ac.at/~neum/ms/mip.pdf 2, 6, 29, 31

[N03] Arnold Neumaier. *Complete Search in Continuous Global Optimization and Constraint Satisfaction.* Acta Numerica 2004 (A. Iserles, ed.), Cambridge University Press, pp. 271-369, 2004. http://www.mat.univie.ac.at/~neum/ms/glopt03.pdf 27, 28, 114

[NW88] George L. Nemhauser, Laurence A. Wolsey. *Integer and Combinatorial Optimization.* ISBN 0-471-82819-X, John Wiley &Sons, Inc. 1988. 54

[R83] Siegfried M. Rump. *Solving Algebraic Problems with High Accuracy.* Habilitationsschrift. In Ulrich W. Kulisch, Willard L. Miranker, editors, A New Approach to Scientific Computation, pp. 51 -120. Academic Press, New York, 1983. 10, 15, 17, 18, 19, 24

[R96] Siegfried M. Rump. *Improved Iteration Schemes for Validation Algorithms for Dense and Sparse Nonlinear Systems.* Computing, 57, 77 - 84, 1996. 19, 22

[R98] Siegfried M. Rump. *A Note on Epsilon-Inflation.* Reliable Computing, 4, 371 - 375, 1998. 22, 23

[R98] Siegfried M. Rump. *INTLAB - Interval Laboratory.* Technical Report 98.4, Forschungsschwerpunkt Informations- und Kommunikationstechnik, TUHH, 1998. 10

[R99] Siegfried M. Rump. *Verified Solution of Large Linear and Nonlinear Systems.* In H. Bulgak and C. Zenger, editors, Error Control and adaptivity in Scientific Computing, pages 279-298. Kluwer Academic Publishers, 1999. 9, 10

[R01] Siegfried M. Rump. *Self-validating methods.* Linear Algebra and its Applications 324, 3 - 13, 2001. 10, 12

[R03] Siegfried M. Rump. *INTLAB - Interval Laboratory, a MATLAB toolbox for verified computations.* Version 4.1.2, 2004 and Version 5.2, 2005. `http://www.ti3.tu-harburg.de/rump/intlab/index.html` 9, 19, 26, 118

[Z86] Eberhard Zeidler. *Nonlinear Functional Analysis and its Applications. Part I: Fixed-Point Theorems.* Springer-Verlag, New York, ISBN 0-387-90914-1, 1986. 17

Water Distribution Network Design

[A&S77] E. Alperovits, U. Shamir. *Design of Optimal Water Distribution Systems.* Water Resources Research, 13(6), 885 - 900, 1977. 36, 42, 43, 47, 48, 149, 150, 167

[A&P05] Adel R. Awad, Ingo von Poser. *Genetic Algorithm Optimization of Water Supply Networks.* Water Intelligence Online, IWA Publishing, 2005. http://www.iwaponline.com/wio/2005/08/wio200508018.htm 47, 48, 154

[B03] Volker Bartsch, University of Technology Hamburg-Harburg, private communication, 2003. 80

[B&W03] Michael R. Bloomberg, Christopher O. Ward. *NEW YORK CITY 2003 Drinking Water Supply and Quality Report.* DEP New York City Department of Environmental Protection, 2003. `http://www.ci.nyc.ny.us/html/dep/pdf/wsstat03.pdf` 43

[B00] Gerhard Bollrich. *Technische Hydromechanik, Band 1, Grundlagen.* Verlag Bauwesen, Berlin, 5. Auflage 2000, ISBN 3-345-00744-4. 45, 77, 80

[B&a04] Jens Burgschweiger, Bernd Gnädig, Marc C. Steinbach. *Optimization Models for Operative Planning in Drinking Water Networks.* ZIB-Report 04-48, Konrad-Zuse Zentrum für Informationstechnik Berlin, 2004. 79

[C38] Cyril Frank Colebrook. *Turbulent Flow in Pipes, with particular reference to the Transition Region between the Smooth and Rough Pipe Laws.* Journal of the Institution of Civil Engineers London, 0368-2455, pp. 133-156, 11, 1938 / 1939. 77, 78, 87

[C36] Hardy Cross. *Analysis of flow in networks of conduits or conductors.* Bulletin of the Engineering Experiment Station, University of Illinois, No. 286, Urbana III, 1936. 48

[C&S01] M. da C. Cunha, J. Sousa. *Hydraulic Infrastructures Design Using Simulated Annealing.* Journal of Infrastructure Systems, American Society of Civil Engineerings, 7(1), 32 - 39, New York, 2001. 48, 157

[DC98] Helmut Damrath, Klaus Cord-Landwehr. *Wasserversorgung.* Teubner Stuttgart, 11. Auflage, ISBN 3-519-15249-5, 1998. 77, 78, 80

[D&a96] Graeme C. Dandy, Angus R. Simpson, Laurence J. Murphy. *An improved genetic algorithm for pipe network optimization.* Water Resources Research, 32(2), 449 - 485, 1996. 48, 154

[D95] DVGW Deutscher Verein des Gas- und Wasserfaches e.V. *Maschinelle und elektrische Anlagen in Wasserwerken.* wiss. Leitung Otto-Gerhard Ebel. Oldenbourg Verlag, ISBN 3-486-26339-0, 1995. 45

[D99] DVGW Deutscher Verein des Gas- und Wasserfaches e.V. *Wassertransport und -verteilung.* wiss. Leitung Wolfram Hirner, Hans Fleckner, Robert Sattler. Oldenbourg Verlag, ISBN 3-486-26219-X, 1999. 157

[E&a94] Gideon Eiger, Uri Shamir, Aharon Ben-Tal. *Optimal design of water distribution networks.* Water Resources Research, 30(9), 2637 - 2646, 1994. 48, 150, 157

[F&a87] O. Fujiwara, B. Jenchaimahakoon, N. C. P. Edirisinghe. *A Modified Linear Programming Gradient Method for Optimal Design of Looped Water Distribution Networks.* Water Resources Research, 23(6), 977 - 982, 1987. 47, 114

[F&D87] Okitsugu Fujiwara, Debashis Dey. *Two Adjacent Pipe Diameters at the Optimal Solution in the Water Distribution Network Models.* Water Resources Research, 23(8), 1457 - 1460, 1987. 33, 63, 68, 70, 150

[F&K90] Okitsugu Fujiwara, Do Ba Khang. *A Two-Phase Decomposition Method for Optimal Design of Looped Water Distribution Networks.* Water Resources Research, 26(4), 539 - 549, 1990. 43, 48, 156, 167

[F&K91] Okitsugu Fujiwara, Do Ba Khang. *Correction to "A Two-Phase Decomposition Method for Optimal Design of Looped Water Distribution Networks".* Water Resources Research, 27(5), 985 - 986, 1991. 48, 157

[G00] Zong Woo Geem. *Optimal Design of Water Distribution Networks using Harmony Search.* Ph. D. Thesis, Department of Civil and Environmental Engineering, Korea University, 2000.
`http://jsbach.netian.com/data/research/hs_thesis.pdf` 36, 48, 150, 153, 154, 157

[G&a00] Zong Woo Geem, Tae Gyun Kim, Joong Hoon Kim. *Optimal Layout of Pipe Networks using Harmony Search.* International Conference on Hydroscience and Engineering, Seoul, 2000.
`http://jsbach.netian.com/data/research/c_2000_iche.pdf` 48

[G&a01] Zong Woo Geem, Joong Hoon Kim, G. V. Loganathan. *A New Heuristic Optimization Algorithm: Harmony Search.* Simulation, 76(2), 60-68, 2001.
`http://jsbach.netian.com/data/research/j_2001_simulation.pdf` 48

[G92] I. C. Goulter. *Systems Analysis in Water-Distribution Network Design: From Theory to Practice.* Journal of Water Resources Planning and Management, 118(3), 1992. 36, 38

[K00] Bryan W. Karney. *Hydraulics of Pressurized Flow.* In Larry W. Mays ed., *Water Distribution Systems Handbook,* McGraw-Hill, ISBN 0-07-134213-3, 2000. 77

[K&S89] Avner Kessler, Uri Shamir. *Analysis of the Linear Programming Gradient Method for Optimal Design of Water Supply Networks.* Water Resources Research, 25(7), 1469 - 1480, 1989. 47, 150

[K02] Imad Kordab. *Simulation and Analysis of Small-Scale Water Distribution Systems: The Case of Escheburg (Germany).* Master Thesis am Arbeitsbereich Wasserwirtschaft und Wasserversorgung, University of Technology Hamburg-Harburg, 2002. 38, 42, 80, 81

[LC64] F. P. Linaweaver, C. Scott Clark. *Costs of Water Transmission.* Journal American Water Works Association, 56, 12, pp. 1549 - 1560, 1964. 44

[L98] Chyr Pyng Liou. *Limitations and Proper Use of the Hazen-Williams Equation.* Journal of Hydraulic Engineering, pp. 951 - 954, 1998. 81

[L&a95] G. V. Loganathan, J. J. Greene, T. J. Ahn. *Design Heuristic for Globally Minimum Cost Water-Distribution Systems.* Journal of Water Resources Planning and Management, ASCE, 121(2), 182-192, 1995. 34, 48, 149, 150, 154

[L&a02] G. V. Loganathan, H. D. Sherali, S. Park, S. Subramanian. *Optimal Design-Rehabilitation Strategies for Reliable Water Distribution Systems.* Virginia Water Resources Research Center, Special Report SR-20-2002.
`http://www.vwrrc.vt.edu/publications/Loganthan%20report%20special.pdf`
38, 47, 53

[M99] Larry W. Mays. *Hydraulic Design Handbook.* L. W. Mays ed., McGraw-Hill, ISBN 0-07-041152-2, 1999. 36, 77

[M44] Lewis F. Moody. *Friction Factors for Pipe Flow.* Transactions of the American Society of Mechanical Engineers. American Society of Mechanical Engineers, New York, 66, pp. 671 - 678, 1944. Water Resources Research, 21(5), 642 - 652, 1985. 78

[M&G85] D. R. Morgan, I. C. Goulter. *Optimal Urban Water Distribution Design.* Water Resources Research, 21(5), 642 - 652, 1985. 48, 154

[M&S02] Johann Mutschmann, Fritz Stimmelmayr. *Taschenbuch der Wasserversorgung.* Vieweg Verlag, 13. Auflage 2002, ISBN 2-528-22554-8. 77, 80

[Q&a79] G. E. Quindry, E. D. Brill, J. C. Liebman, A. R. Robinson. *Comment on "Design of Optimal Water Distribution Systems" by E. Alperovits and U. Shamir.* Water Resources Research, 15(6), 1651 - 1654, 1979. 47, 150

[R&B82] William F. Rowell, J. Wesley Barnes. *Obtaining Layout of Water Distribution Systems.* Journal of the Hydraulics Division, Proceeding of the American Society of Civil Engineers, ASCE, 108, 137 - 148, 1982. 48

[S&W95] D. A. Savic, G. A. Walters. *Genetic Operators and constraint handling for pipe network optimization.* Evolutionary Computing 2, Fogarty,T.C. (ed), Springer-Verlag, 154-165, 1995. `http://citeseer.nj.nec.com/savic95genetic.html` 48

[S&L69] John C. Schaake, Dennis Lai. *Linear Programming and Dynamic Programming Application to Water Distribution Network Design.* MIT Department of Civil Engineering, Hydrodynamics Laboratory, Report No. 116, 1969. 4, 39, 42, 43, 44, 47, 152, 154, 165

[S&S93] Hanif D. Sherali, Ernest P. Smith. *An Optimal Replacement-Design Model for a Reliable Water Distribution Network System.* Integrated Computer Applications in Water Supply, (1) Methods and Procedures for Systems Simulation and Control, 1993. 34, 38, 150, 151

[S&S97] Hanif D. Sherali, Ernest P. Smith. *A Global Optimization Approach to a Water Distribution Network Design Problem.* Journal of Global Optimization, 11, 107 - 132, 1997. 48, 150

[S&a98] Hanif D. Sherali, Rajiv Totlani, G. V. Loganathan. *Enhanced lower bounds for the global optimization of water distribution networks.* Water Resources Research, 34(7), 1831 - 1841, 1998. 39, 41, 44, 48, 56, 150, 151, 157

[S&a01] Hanif D. Sherali, Shivaram Subramanian, G. V. Loganathan. *Effective Relaxations and Partitioning Schemes for Solving Water Distribution Network Design Problems to Global Optimality.* Journal of Global Optimization, 19, 1 - 26, 2001. x, 34, 35, 39, 42, 43, 48, 49, 53, 54, 56, 57, 59, 60, 77, 84, 101, 105, 120, 137, 150, 152, 154, 155, 157, 167

[S01] Shivaram Subramanian. *Optimization Models and Analysis of Routing, Location, Distribution and Design Problems on Networks.* Ph. D. Thesis, Virginia Polytechnic Institute and State University, 1998. `http://scholar.lib.vt.edu/theses/available/etd-042499-225537` 39, 43, 48, 49, 52, 53, 56, 59, 114, 120, 140, 150, 151, 154, 155, 156, 157

[S&J76] Prabhata K. Swamee, Akalank K. Jain. *Explicit Equations for Pipe-Flow Problems.* Journal of the Hydraulics Division, Proceedings of the American Society of Civil Engineers, New York, 102, 5, pp. 657 - 664, 1976. 77

[W&a95] Thomas M. Walski, E. Downey Brill jr., Johannes Gessler, Ian C. Goulter, Roland M. Jeppson, Kevin Lansey, Han-Lin Lee, Jon C. Liebman, Larry Mays, David R. Morgan, Lindell Ormsbee. *Battle of the Network Models: Epilogue.* Journal of Water Resources Planning and Management, ASCE 113(2), 191 - 203, 1987. 38, 47

[W95] Thomas M. Walski. *Optimization and Pipe-Sizing Decisions.* Journal of Water Resources Planning and Management, ASCE 121(4), 340 - 343. 1995. 34, 38

[W&a03] Thomas M. Walski, Donald V. Chase, Dragan A. Savic, Walter Grayman, Stephen Beckwith, Edmundo Koelle. *Advanced Water Distribution Modeling and Management.* Haestad Press, Waterbury, CT USA, 2003, esp. Chapter 8: *Using Models for Water Distribution System Design,* and Chapter 2: *Modeling Theory.*
`http://www.haestad.com/library/books/awdm/awdm.pdf` 33, 35, 36, 77, 78, 80, 85

[WH05] Gardner S. Williams, Allen Hazen. *Hydraulic Tables. Showing the loss of head due to the friction of water flowing in pipes, aqueducts, sewers, etc. and the discharge over weirs.* New York, John Wiley & Sons, 1905. 85, 107

[W&S02] Zheng. Y. Wu, Angus R. Simpson. *A self-adaptive boundary search genetic algorithm and its application to water distribution systems.* Journal of Hydraulic Research, 40(2), 191 - 203, 2002.
`http://www.iahr.org/publications/assets/jhr40-2/wu.pdf` 47, 48

Index